Fundamentals of Ordinary Differential Equations

Uri Elias

Fundamentals of Ordinary Differential Equations

Uri Elias
Department of Mathematics
Technion – Israel Institute of Technology
Haifa, Israel

ISBN 978-3-031-86531-2   ISBN 978-3-031-86532-9   (eBook)
https://doi.org/10.1007/978-3-031-86532-9

© The Editor(s) (if applicable) and The Author(s), under exclusive license to Springer Nature Switzerland AG 2025

This work is subject to copyright. All rights are solely and exclusively licensed by the Publisher, whether the whole or part of the material is concerned, specifically the rights of translation, reprinting, reuse of illustrations, recitation, broadcasting, reproduction on microfilms or in any other physical way, and transmission or information storage and retrieval, electronic adaptation, computer software, or by similar or dissimilar methodology now known or hereafter developed.
The use of general descriptive names, registered names, trademarks, service marks, etc. in this publication does not imply, even in the absence of a specific statement, that such names are exempt from the relevant protective laws and regulations and therefore free for general use.
The publisher, the authors and the editors are safe to assume that the advice and information in this book are believed to be true and accurate at the date of publication. Neither the publisher nor the authors or the editors give a warranty, expressed or implied, with respect to the material contained herein or for any errors or omissions that may have been made. The publisher remains neutral with regard to jurisdictional claims in published maps and institutional affiliations.

This book is published under the imprint Birkhäuser, www.birkhauser-science.com by the registered company Springer Nature Switzerland AG
The registered company address is: Gewerbestrasse 11, 6330 Cham, Switzerland

If disposing of this product, please recycle the paper.

Preface

This book is based on lectures in various courses of ordinary differential equations that were taught by the author at the Technion—Israel Institute of Technology for students of engineering and science departments. It intends to be a high-level course for engineering and science students and beginning math majors that have already learned basic calculus and linear algebra courses. The students should have a working ability with derivatives and integrals. Linear algebra prerequisites include matrices and vectors, solution of systems of linear equations and determinants, linear dependence of vectors, and basis for n-dimensional vector space and the definition of eigenvalues. Diagonalization of a matrix will be helpful.

Our aim was to make the book friendly to the students. The explanations are detailed and presented in the style a lecturer speaks in front of a class. An effort was made to guess what are the subtle points that a student can come across and fail, and we tried to get ahead of it. Examples and guiding exercises placed within the text are an integral part of this book. Some of the examples are routine, while others are part of the text written in exercise form. In many examples we try to motivate why this example was chosen and what is its goal. Students are encouraged to work through the problems at the ends of chapters, which are written in the same style. Some problems are of the formulated as "false proofs" where the student is asked to find where an error is hidden.

The material described here is broader than what can be taught in a course of two weekly hours. There are hints, comments, helpful motivations, and explanations why we will do this way and not another. When alternative approaches exist, sometimes both are described and enables the instructor to choose his favorite one. Some sections are for enrichment only and should not be taught in a standard course. These subjects are nevertheless included, since a respectable ODE book cannot skip them. In general, all those extras that a lecturer would like to add if he was not limited by time. Time, as we know, is always too short.

Solving differential equations involves routine technical work. Moreover, most equations cannot be solved by explicit formulas. Even when formulas are found, they do not always help to understand the behavior of the solution. Therefore we prefer to emphasize the qualitative aspects rather than specialized methods and tricks, which solve particular differential equations of historical importance but are not very useful in real life. In this book, we tried to add the ideas from behind the scenes, something of the intuition that allows to understand the qualitative behavior of equations and their solutions and also a little bit of the beauty of understanding "How this being called a differential equation works?"

Haifa, Israel   Uri Elias
January 2025

Contents

1	**What Is an Ordinary Differential Equation?**	1
	1.1 What Is This Being?	1
	1.2 Examples and Models	2
2	**First-Order Differential Equations**	7
	2.1 First-Order Linear Equations	7
	2.2 Separable Equations	16
	2.3 Solution by Substitution	20
	2.4 Exact Equations	23
	2.5 Integrating Factors	29
	2.6 Equations from the Form $y'' = f(y)$	31
	2.7 Geometric Topics: Direction Field	33
	2.8 Families of Orthogonal Curves	37
3	**Existence and Uniqueness Theorems**	47
	3.1 Why Existence and Uniqueness?	47
	3.2 Outline of the Proof of Existence Theorem	54
	3.3 Proof of the Existence of the Solution	56
	3.4 Proof of the Uniqueness of the Solution	61
	3.5 Applications of the Uniqueness of Solution	63
	3.6 Local Solutions	68
	3.7 Qualitative Investigation of Solutions	71
	3.8 Global Existence Theorem	77
	3.9 The Concept of Stability	80
	3.10 Existence and Uniqueness for Higher-Order Equations	83
4	**Linear Equations of Higher Order**	95
	4.1 Linear Algebra Reminder	95
	4.2 Existence and Uniqueness Theorem for Linear Equations	96
	4.3 Homogeneous Linear Differential Equations	99
	4.4 Wronskian and Abel's Formula	107
	4.5 Reduction of the Order of an Equation	113

	4.6	Equations with Constant Coefficients	117
	4.7	Qualitative Behavior and Stability	126
	4.8	Euler's Equation	129
	4.9	Nonhomogeneous Linear Equations	132
	4.10	The Method of Variation of Parameters	134
	4.11	The Method of Undetermined Coefficients	141

5 Systems of Differential Equations ... 159
    5.1 Existence and Uniqueness Theorem for Systems ... 159
    5.2 Systems of Linear Equations ... 162
    5.3 Homogeneous System and the Structure of Its Solutions ... 164
    5.4 Abel's Formula for the System ... 169
    5.5 Fundamental Matrix ... 171
    5.6 Systems with Constant Coefficients ... 173
        5.6.1 Systems with a Basis of Eigenvectors ... 175
        5.6.2 Systems That Have No Basis of Eigenvectors ... 180
    5.7 Non-Homogeneous Systems ... 189

6 The Qualitative Theory and the Phase Plane ... 199
    6.1 Trajectories and Critical Points in the Phase Plane ... 199
    6.2 Classification of Linear Systems in the Phase Plane ... 205
    6.3 Small Linear Perturbations of Linear Systems ... 218
    6.4 Notes on Nonlinear Systems in the Plane ... 220

7 Solution of Differential Equations by Power Series ... 233
    7.1 Reminder About Power Series ... 233
    7.2 Solution of Differential Equations by Power Series ... 236
    7.3 Solutions Around a Regular-Singular Point ... 244
    7.4 Equal Indices ... 254
    7.5 Indices That Differ by a Positive Integer ... 259
    7.6 The Bessel Equation ... 262

8 The Laplace Transform ... 271
    8.1 Definition of the Laplace Transform ... 271
    8.2 Initial Value Problems ... 275
    8.3 Additional Features of the Laplace Transform ... 280
    8.4 Convolution ... 285
    8.5 Step Functions ... 287
    8.6 The Dirac Function ... 292
    8.7 Some Nice Examples ... 297

A The Orbits of the Planets ... 303

B Historical Notes ... 309

Index ... 313

Chapter 1
What Is an Ordinary Differential Equation?

1.1 What Is This Being?

A central idea of mathematics is the concept of an equation. In previous studies, we have already encountered algebraic equations, in which the unknowns are real or complex numbers. Here we first meet equations that deal with unknown functions.

An *ordinary differential equation* is a relation that involves an unknown function of a single variable (say, $y(x)$ or $u(t)$), its derivatives, and, of course, the independent variable $(x, t, \ldots)$:

$$F(x, y(x), y'(x), \ldots, y^{(n)}(x)) = 0 , \tag{1.1}$$

where F is a real function of $n + 1$ variables. The *order* of a differential equation is the order of the highest derivative that appears in the equation. In many cases, we will prefer to extract the highest derivative and to present the equation in a *normalized form*:

$$y^{(n)}(x) = f(x, y(x), y'(x), \ldots, y^{(n-1)}(x)) .$$

The most simple differential equation is $y' = 0$. The differential equation $x^2(y''')^7 + \arcsin(xy'' + y) = 17$ is of order three, but we do not know anything interesting or significant about it.

Contrary to an ordinary differential equation, in a *partial differential equation* the unknown is a function of several variables, such as $u(x, y, z, \ldots)$, and the equation consists of the unknown function and its partial derivatives $u, u_x, u_y, u_z, u_{xx}, u_{xy}, \ldots$. Partial differential equations are not discussed in this text.

The differential equation (1.1) is called *linear* if F is a linear function of the unknown function and of all its derivatives (but not of the independent variable). For example, a linear equation of order n with an independent variable x and an unknown function $y(x)$ is

$$a_0(x)y^{(n)} + a_1(x)y^{(n-1)} + \cdots + a_{n-1}(x)y' + a_n(x)y = b(x) .$$

If the right-hand side is 0, the equation is called *homogeneous*.

What are the possible objectives in the study of differential equations?

1. *Solving differential equations explicitly.* Unfortunately, this goal is not achievable in most cases and we cannot write the solutions of a differential equation by a finite number of familiar functions and operations, even if the very existence of the solution is known. Despite this, there exist numerous useful equations whose explicit solutions are well known. Therefore, a significant part of our work is devoted to studying solution methods. There exist encyclopedic collections that list equations whose solutions have been found through the cumulative effort of generations.[1]
2. *Qualitative understanding of the equation.* We try to understand the behavior of solutions without finding them explicitly, solely of the form of the equation.
3. *Numerical solution of equations.* This is an important approach, which is used and implemented by computers, but its place is not in this text, but rather in numerical analysis studies.

1.2 Examples and Models

A derivative describes a rate of change, so differential equations usually describe dynamic processes, physical laws, feedback systems, etc. We present here some examples to which we will return later for further discussions.

Example 1.1 (Growth of Population) We mark by $N(t)$ the size of a population (microbes, humans) at time t. Although population size is measured by integer numbers (discrete values), when it comes to a large population we ignore this restriction and refer to $N(t)$ as a function of real variable (a continuum of values). The most simplistic assumption is that the population growth (ΔN) per unit of time (Δt) is a fixed percentage of the population size at that time, i.e.,

$$\frac{\Delta N}{\Delta t} = kN(t) ,$$

---

[1] For example, Polyanin, A., Zaitsev, V., Handbook of Exact Solutions for Ordinary Differential Equations, CRC Press, 1995.

1.2 Examples and Models

when k is a positive constant. When we look at smaller and smaller intervals of time, we get at the limit $\Delta N/\Delta t \to N'(t)$. So

$$N'(t) = kN(t) \,.$$

This is a first-order linear differential equation. To follow the size of the population, it is not enough to know the rate of growth, but we also need to know the size of the population at the beginning of the measurement. We append to the differential equation information about the status at the start time $t = t_0$, an *initial value condition* $N(t_0) = N_0$.

This model is too simplistic. Usually, resources are limited and all the individuals of the population compete with each other. Therefore, it is reasonable to subtract from a rate of growth a quantity proportional to the product $N \times N$:

$$N'(t) = kN(t) - mN^2 \,, \qquad k, m > 0 \,.$$

This is also an equation of first order, but it is no longer linear due to the expression N^2. This differential equation is called the *logistic equation*. As before, we add to the equation an initial value condition $N(t_0) = N_0$.

Of course, these models only approximate reality. In the following chapters, we will return to them for further discussion.

Example 1.2 (Radioactive Decay) Let $y(t)$ be the number of particles in a sample at time t. The change of the number of particles Δy per time unit Δt is proportional to the sample size $y(t)$. The particles decompose and $\Delta y < 0$; therefore, in this case

$$\frac{\Delta y}{\Delta t} = -ky(t) \,, \quad k > 0 \,.$$

The differential equation and its initial value condition are

$$y'(t) = -ky(t) \,, \quad y(t_0) = y_0 \,.$$

Example 1.3 (Predator-Prey Model) When two species live in the same environment, we have to take into account their interaction and their growth model is more complicated. Let $x(t)$ be the population of a vegetarian species, say rabbits, and $y(t)$ denote the predators (wolves). Without predators and with nonrestricted food supply, the vegetarians would multiply, as in Example 1.1, according to $x'(t) = ax(t)$, $a > 0$. But the growth of the vegetarians is reduced by the predators proportionally to the product of their abundance, xy. So we assume that $x' = ax - bxy$, $a, b > 0$.

Now we turn to the predators. Without rabbits, the wolves would starve and their population will decline at a fixed rate, $y' = -cy$, $c > 0$. At the presence of rabbits, the encounter of predators and vegetarians is proportional to the product xy, the wolves are satisfied and multiply, and their population growth is ruled by $y' =$

$-cy + dxy$, $c, d > 0$. This system of two differential equations with two unknown functions $x(t)$, $y(t)$,

$$x' = ax - bxy ,$$
$$y' = -cy + dxy , \quad a, b, c, d > 0,$$

is called the *Volterra-Lotka model*. It is a system of nonlinear differential equations and as such it is not studied thoroughly in this book.

Example 1.4 (Spread of Infectious Diseases) The following model describes the development of infections in a population. We divide the population into three classes: Let $s(t)$ denote the relative proportion at time t of those who are susceptible to the disease and may become infected, $i(t)$ denotes the infectious, and $r(t)$ are those who were removed from the process, such as recovered and deceased patients. Of course, $s(t) + r(t) + i(t) \equiv 1$.

The infection rate is proportional to the number of encounters between infectious individuals and susceptible individuals. The infected leave class s and move to class i. Therefore, $ds/dt = -\beta si$, with the infection coefficient $\beta > 0$. The change of $i(t)$ is due to two factors: the addition of new infected patients and the exit of cured and deceased patients from this class. Hence, $di/dt = \beta si - \gamma i$, where the positive constant γ is the exit rate. Those leaving i move to class r, and thus, $dr/dt = \gamma i$. Thus, the transitions $s \longrightarrow i \longrightarrow r$ are described by the system

$$\frac{ds}{dt} = -\beta si ,$$
$$\frac{di}{dt} = \beta si - \gamma i , \qquad (1.2)$$
$$\frac{dr}{dt} = \gamma i , \qquad \beta, \gamma > 0 .$$

This model is called *the SIR model* (Why?). It ignores other changes in population size such as births and natural deaths during the process. The system (1.2) is defined everywhere, but our domain of interest is only $D = \{s \geq 0, \ i \geq 0, \ r \geq 0\}$.

Newton's second law states that the acceleration of a body is proportional to the force exerted on it. Since acceleration is the second derivative of the body's location, numerous laws of motion in mechanics are expressed by second-order differential equations. Here are some examples.

Example 1.5 (A Missile Fired Upwards from the Surface of the Earth) Let us choose the vertical axis y starting at the center of the earth, which is modeled as a sphere of radius R. Gravity is inversely proportional to the square of the distance from the center of the earth and points downward, so the acceleration is

$$y'' = -\frac{k}{y^2} .$$

1.2 Examples and Models

It is known that on the surface of the earth, i.e., at $y = R$, the acceleration downward is g, so $g = k/R^2$ and

$$y'' = -g\frac{R^2}{y^2}\ .$$

At the starting time, $t = 0$, we have $y(0) = R$ and let the initial speed be $v(0) = y'(0) = v_0$. Therefore, our model consists of a second-order nonlinear differential equation and two initial value conditions.

Example 1.6 (The Motion of a Body Attached to a Spring) By the Hooke law, the force exerted by a spring on a body attached to it is proportional to its deviation from the equilibrium point and acts in the opposite direction. If the mass of the body is m and $y(t)$ denotes its motion, then

$$my'' = -ky\ .$$

This is a second-order linear differential equation and it describes *harmonic motion*.

Example 1.7 (Nonlinear Friction) A body is dropped from a height h along the vertical axis y. The resistance that the air exerts on it (drag) is proportional to the square of the speed, $(y')^2$. The equation of motion is therefore

$$my''(t) = -mg + k(y')^2\ .$$

If the body is dropped from height h with no initial speed, the two initial value conditions are $y(0) = h$ and $y'(0) = 0$. The differential equation is of second order, but the substitution $v(t) = y'(t)$ transforms it into a first-order equation:

$$mv'(t) = -mg + kv^2,\qquad v(0) = 0\ .$$

Both equations are, of course, nonlinear. In Example 3.10 we shall see how the speed $v(t)$ behaves after a "long time", i.e., when $t \to \infty$.

Example 1.8 (A Pendulum) A body of mass m hangs at the end of a massless rod of length L. The angle of its deviation from the vertical is indicated by $\theta(t)$ (measured by radians) and is a function of the time t. The length of the arc that the body passes is $L\theta$. We are interested in the component of the gravity which points in the direction tangent to the motion. See Fig. 1.1.

Newton's second law leads to the equation

$$m(L\theta)'' = -mg\sin\theta\ ,\quad \text{i.e.,}\quad \theta'' = -\frac{g}{L}\sin\theta\ .$$

This is a nonlinear equation of second order. For small values of θ we may take approximately $\sin\theta \approx \theta$ and the equation is replaced by a linear equation, $\theta'' =$

Fig. 1.1 A pendulum

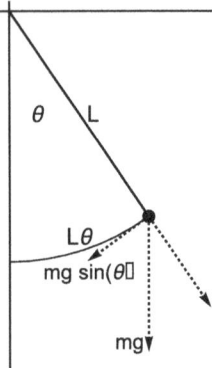

$-\frac{g}{L}\theta$. For large oscillations or even complete rotations, the original equation should be used. We shall return to the nonlinear pendulum equation in Example 6.9.

Examples 1.5, 1.6, and 1.8 are of the form $y''(t) = f(y)$ with different functions $f(y)$, in each of them the independent variable t and the first derivative do not appear explicitly. We shall learn about equations of this type in Sect. 2.6.

Example 1.9 (A Simple Electric Circuit) A current I in a serial resistor-inductor-capacitor electric circuit (RLC) with a voltage source $E(t)$ is modeled by

$$LI''(t) + RI'(t) + C^{-1}I(t) = E'(t),$$

where L is inductance, R is resistance, and C is capacitance. It is a second-order linear equation with constant coefficients L, R, C^{-1}.

Example 1.10 (The Schrödinger Equation) The Schrödinger equation with a single spatial dimension and independent of time is

$$-\frac{\hbar^2}{2m}\frac{d^2\psi}{dx^2} + V(x)\psi = E\psi.$$

Here $V(x)$ is a potential function, E indicates energy, and $\hbar$ is the reduced Planck constant. This is a linear second-order differential equation for the wave function $\psi(x)$.

Chapter 2
First-Order Differential Equations

The general form of a first-order differential equation is

$$F(x, y, y') = 0,$$

when F is a real-valued function that depends on three variables. When possible, we try to extract the derivative $y'(x)$ and write the equation in a *normalized form*:

$$y'(x) = f(x, y). \tag{2.1}$$

We have already mentioned in the introduction that most differential equations cannot be solved explicitly by "closed formulas," that is, expressions consisting of a finite number of known functions and their integrals. Despite this, we will start the study of differential equations by presenting some useful solution methods that are suitable for several well-known equations.

2.1 First-Order Linear Equations

In this section we take advantage of the opportunity to introduce some of the basic concepts of differential equations, as general solution, initial value problem, integrating factor, etc.

Recall that the models of population growth and radioactive decomposition are

$$N'(t) - kN(t) = 0, \tag{2.2}$$

$$y'(t) + ky(t) = 0, \tag{2.3}$$

where $k > 0$ is constant. We will first discuss linear, homogeneous equations

$$\text{(H)} \qquad y' + p(x)y = 0 \qquad (2.4)$$

and later linear, nonhomogeneous equations

$$\text{(NH)} \qquad y' + p(x)y = q(x) \,. \qquad (2.5)$$

The origin of the terms "homogeneous" and "nonhomogeneous" is in linear algebra. Note that both equations are already written in a normalized form, where the coefficient of $y'(x)$ is 1.

Homogeneous Linear Equation

Suppose that $y(x)$ is a solution of $y' + p(x)y = 0$. Assuming that $y(x)$ is nowhere zero, we divide by it and get

$$\frac{y'(x)}{y(x)} = -p(x).$$

We integrate both sides (a non-definite integral) according to the variable x:

$$\int \frac{y'(x)}{y(x)} \, dx = \int -p(x) \, dx \,.$$

This results in

$$\ln |y(x)| = -\int p(x) \, dx + \ln C \,,$$

where $\ln C$ is a constant of integration. This is possible since any real number can be written in the form $\ln C$ with some suitable positive C. Therefore,

$$|y(x)| = C e^{-\int p(x) \, dx} \,.$$

According to our assumption, the solution $y(x)$ may be either positive or negative, and accordingly, $|y(x)| = \pm y(x)$. Let us mark $\pm C$ by K and conclude that

$$y(x) = K e^{-\int p(x) \, dx} \,, \qquad (2.6)$$

where K is some constant of an arbitrary sign. For example, the solutions of (2.2), (2.3) are $N(t) = K e^{kt}$, $y(t) = K e^{-kt}$, respectively.

Our declared goal was to find a solution of equation (H). However, (2.6) presents an infinite number of different solutions that correspond to different values of

2.1 First-Order Linear Equations

Fig. 2.1 The general solution of Eq. (2.2)

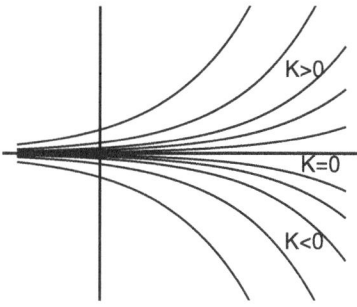

K. (2.6) is an one-parameter family of solutions and is called a *general solution* of (H). Such a family of solutions is illustrated graphically in Fig. 2.1.

Behind the innocent calculation above, there are hidden several ideas.

Remark 2.1 (Separability of Differential Equation) If in the integral $\int y'(x)/y(x)\,dx$ we make a change of variable $y = y(x)$, $dy = y'(x)\,dx$, then the integration can be written as

$$\int \frac{dy}{y} = \int \frac{y'(x)\,dx}{y(x)} = \int -p(x)\,dx \ . \tag{2.7}$$

This justifies a formal treatment: We "multiply the equation $\dfrac{dy}{dx} = -p(x)y$ by dx" and "divide it by y" and so get, symbolically,

$$\frac{dy}{y} = -p(x)\,dx \ .$$

Now both sides are integrated, each according to the appropriate variable, and we reach the solution that was found above.

The key to the success of this solution method is the possibility to separate among the variables x and y in Eq. (H) and move each of them to another side of the equation. This method will be called later the *separation of variables*.

Remark 2.2 (Singular Solutions) To rewrite Eq. (H) as $y'(x)/y(x) = -p(x)$ it is necessary to assume that $y(x) \neq 0$, since it is impossible to divide by 0. Also, once we write the integration constant in the form $\ln C$, it is assumed (without explicitly stating) that $C > 0$ and in particular that $C \neq 0$. However, a glance at (H) reveals that $y(x) \equiv 0$ is a solution and it is also included in the formula (2.6) for the choice $C = K = 0$! Thus, although on the way it had been assumed that $y(x) \neq 0$, in retrospect the solution $y(x) \equiv 0$ is nevertheless included in the family of solutions (2.6). So, in (2.6) it is possible to take any $-\infty < K < \infty$. Later on, we will meet this phenomenon and the corresponding "singular solutions" again.

Remark 2.3 (The Trivial Solution) In Remark 2.2 we mentioned the identically zero solution $y(x) \equiv 0$. Since it is a solution of every homogeneous linear equation, it will be called the *trivial solution*. On the other hand, we saw solutions that are nowhere zero. And what about a solution of the equation (H) that vanishes at one point but differs from zero at other points? This question is answered in Sect 3.5, in the context of the existence and uniqueness theorem.

Remark 2.4 (Permitted Coefficients) In the formal calculations that led us to solution (2.6) there hides the assumption that the integral $\int p(x)\,dx$ exists. For this to be the case it may be assumed, for example, that the function $p(x)$ is continuous.

Initial Value Problems

A real-life problem consists of a differential equation and the initial value of the required solution at the starting point. In our case, let us look at

$$y' + p(x)y = 0, \qquad y(x_0) = y_0.$$

Such a problem is called an *initial value problem*. This simple problem is easily solved: In the general solution (2.6) take $x = x_0$, $y = y_0$ and find the single suitable value of K. In this way, we identify among the infinitely many different solutions included in the general solution, one which obeys the given initial value condition. Note that in our case the equation is first order, the general solution contains one free parameter, and we require a single initial value condition. This is not a coincidence but an intrinsic feature that will be studied later.

Geometrically, the initial value condition $y(x_0) = y_0$ means that the graph of the solution passes through the point $(x, y) = (x_0, y_0)$. Therefore, among the infinitely many trajectories of the general solution shown in Fig. 2.2, the one which passes through (x_0, y_0) represents the solution of the initial value problem.

In the above example, we selected the solution of the initial value problem from the general solution (2.6), so that $y = y_0$ will correspond to $x = x_0$. Alternatively,

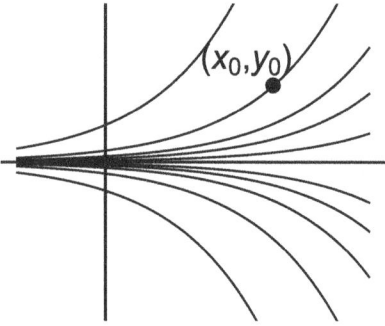

Fig. 2.2 An initial value problem for Eq. (2.2) and its solution

2.1 First-Order Linear Equations

the problem can be solved directly and elegantly by the use of definite integrals. For $y_0 \neq 0$ we write

$$\int_{y_0}^{y} \frac{dy}{y} = -\int_{x_0}^{x} p(x)\, dx \ .$$

This choice of the limits of the integrals guarantees automatically the initial value condition $y(x_0) = y_0$. Later on, we get

$$\ln|y| - \ln|y_0| = \ln|y|\Big|_{y_0}^{y} = -\int_{x_0}^{x} p(x)\, dx \ ,$$

i.e.,

$$|y/y_0| = e^{-\int_{x_0}^{x} p(x)\, dx} \ .$$

For $x = x_0$ the initial value is $y = y_0$ (and not $y = -y_0$!); therefore, we can get rid of the absolute value and finally get

$$y(x) = y_0\, e^{-\int_{x_0}^{x} p(x)\, dx} \ . \tag{2.8}$$

Note that although during the calculations we assumed that $y_0 \neq 0$, in retrospect (2.8) yields even for $y_0 = 0$ the trivial solution $y(x) \equiv 0$.

Exercise 2.1 Solve the initial value problem of Example 1.1, $N' = kN$, $N(t_0) = N_0$.

Nonhomogeneous Linear Equation

The homogeneous equation $y' + p(x)y = 0$ had been solved by moving all x-s to one side of the equation and all y-s to the other one. For the nonhomogeneous equation

$$(NH) \qquad y' + p(x)y = q(x),$$

such separation is impossible. (Try and see!) Therefore, a different method is needed.

We approach Eq. (NH) according to the following considerations: On the left-hand side of (NH) there appear y', y which also occur in the derivative of a product:

$$(uy)' = uy' + u'y \ .$$

Is it possible to multiply the two sides of (NH) by some suitable function of x that will turn its left-hand side into a derivative of a product? If the answer is affirmative, $y(x)$ will be obtained by integrating both sides.

Let us try to multiply both sides of (NH) by an unknown function, traditionally denoted by $\mu(x)$:[1]

$$\mu(x)y' + \mu(x)p(x)y = \mu(x)q(x) . \tag{2.9}$$

If we are lucky and the left-hand side of (2.9) happens to be equal to

$$\mu(x)y' + \mu'(x)y ,$$

then Eq. (2.9) becomes

$$(\mu(x)y)' = \mu(x)y' + \mu'(x)y = \mu(x)y' + \mu(x)p(x)y = \mu(x)q(x) ,$$

which can be straightforward integrated, as we hope. Consequently, we try to determine $\mu(x)$ so that $\mu(x)y' + \mu(x)p(x)y \equiv \mu(x)y' + \mu'(x)y$, that is to say

$$\mu'(x) = \mu(x)p(x) .$$

But this is just a homogeneous linear differential equation of the type that had been solved in the previous section. We get

$$\int \frac{\mu'(x)}{\mu(x)} \, dx = \int p(x) \, dx + C ,$$

or $\ln|\mu(x)| = \int p(x)\,dx + C$. Since we only need a single function $\mu(x)$, we take $C = 0$ (or any other number that we like) and get

$$\mu(x) = e^{\int p(x)\,dx} .$$

The function $\mu(x)$ is called an *integrating factor* of (NH). Note that the integrating factor which we found corresponds only to a normalized equation, where the coefficient of y' is 1. We will encounter the idea of multiplying an equation by a suitable integrating factor later again.

Once the factor $\mu(x)$ is found, we multiply both sides of (NH) by it:

$$e^{\int p\,dx} y' + e^{\int p\,dx} p(x)y = e^{\int p\,dx} q(x) .$$

As expected, the left-hand side is exactly a derivative of the product $e^{\int p\,dx} y$, so

$$\left(e^{\int p\,dx} y\right)' = e^{\int p\,dx} q(x) .$$

---

[1] μ—the Greek letter μ, pronounced as "mu"—is equivalent to Latin m.

2.1 First-Order Linear Equations

After integration

$$e^{\int p\,dx} y = \int e^{\int p\,dx} q(x)\,dx + C.$$

Finally,

$$y(x) = e^{-\int p\,dx} \int e^{\int p\,dx} q(x)\,dx + Ce^{-\int p\,dx}. \tag{2.10}$$

The existence of these integrals is guaranteed if we assume, for example, that $p(x), q(x)$ are continuous functions.

It is possible to achieve the same solution formula (2.10) by a completely different method. The way of thought is as follows: The general solution of the homogeneous equation is of the form $C\exp\left(-\int p(x)\,dx\right)$, where C is a constant parameter. Accordingly, we suggest to look for a solution of the nonhomogeneous equation of the form $c(x)\exp\left(-\int p(x)\,dx\right)$, where $c(x)$ is an unknown function. Since the parameter C is replaced by a function $c(x)$, this technique is called the *variation of parameter method*. We will meet this method, applied to other equations and systems, in the following chapters.

Before we utilize the method, we pose two questions: Why is it possible to present a solution of the nonhomogeneous equation as a product of the form $c(x)\exp\left(-\int p\,dx\right)$? And why should one look for a solution of this specific form?

The answer to the first question is immediate: Any function $f(x)$ can be written in the form $f(x) = c(x)\exp\left(-\int p\,dx\right)$ for a suitable $c(x)$, since we can divide by $\exp\left(-\int p\,dx\right) \neq 0$ and choose $c(x) = f(x)\exp\left(\int p(x)\,dx\right)$. In particular, this is true for our solution. The second question is answered if we show the effectiveness of the method. Try to substitute

$$y(x) = c(x)e^{-\int p\,dx}, \quad y'(x) = c'(x)e^{-\int p\,dx} + c(x)(-p(x))e^{-\int p\,dx}$$

into the equation $y' + p(x)y = q(x)$. We get

$$\left(c'(x)e^{-\int p\,dx} - c(x)p(x)e^{-\int p\,dx}\right) + p(x)c(x)e^{-\int p\,dx} = q(x).$$

After cancellation, there remains

$$c'(x) = e^{\int p\,dx} q(x),$$

so

$$c(x) = \int e^{\int p\,dx} q(x)\,dx + C.$$

Therefore, the so-obtained solution $y(x) = c(x) \exp\left(-\int p \, dx\right)$ is exactly the one that was found in (2.10).

The general solution (2.10) of (NH) that was found has a typical structure: The last term, $Ce^{-\int p}$, is exactly the general solution of (H). The first term is a particular solution of (NH), the one that corresponds to the choice $C = 0$. Therefore, any solution of (NH) consists of one particular solution of (NH) plus all solutions of (H):

$$y_{\text{NH}} = y_{\text{p}} + y_{\text{H}} \, .$$

A similar principle exists for linear algebraic equation systems and we will also encounter it in solving linear differential equations of any order.

When we attach to Eq. (NH) the initial value condition $y(x_0) = y_0$, then putting $x = x_0$, $y = y_0$ in the general solution (2.10) determines a unique value of C (since $\exp\left(-\int p \, dx\right) \neq 0$). So, a single solution is obtained for the initial value problem.

We summarize these arguments as a theorem:

Theorem 2.1 *If the coefficients $p(x), q(x)$ are continuous in an interval $[a, b]$, and x_0 belongs to $[a, b]$, then the initial value problem which consists of the differential equation (NH) and the initial value condition $y(x_0) = y_0$ has a single solution and it is defined in the whole interval $[a, b]$.*

In this theorem we have emphasized the existence and the uniqueness of a solution of an initial value problem for the equation (NH). For this equation, the conclusions follow from the explicit formulas for the general solution, but they are far from being granted for other equations. The question of existence and uniqueness for general initial value problems will be discussed in the next chapter. In the meantime, this topic is illustrated by some problematic equations and initial value conditions.

Example 2.1 The normalized equation

$$y' + \frac{2}{x} y = x^5$$

has the coefficient $p(x) = 2/x$, which is continuous both in the interval $(0, \infty)$ and in $(-\infty, 0)$, but not at the point $x = 0$. In fact, the equation is not defined for $x = 0$. The appropriate integrating factor is

$$\mu(x) = e^{\int p \, dx} = e^{\int 2/x \, dx} = e^{2 \ln x} = x^2 \, .$$

Multiplying $\mu(x) = x^2$, we have

$$(x^2 y)' = x^2 y' + 2xy = x^7 \, ,$$

2.1 First-Order Linear Equations

and after integration we get $x^2 y = x^8/8 + C$. Therefore, the general solution is

$$y(x) = x^6/8 + Cx^{-2} .$$

If an initial value condition $y(2) = 7$ is chosen, then to $x = 2$, $y = 7$ corresponds to $C = -4$ and it determines a single solution:

$$y(x) = x^6/8 - 4x^{-2}, \quad 0 < x < \infty .$$

On the other hand, setting initial value conditions at a point $x = 0$ where the coefficient $p(x)$ is noncontinuous, e.g., $y(0) = 7$, rises difficulties. If $C \neq 0$, it is impossible to put $x = 0$ in the general solution $y = x^6/8 + C/x^2$, and if $C = 0$, the resulting solution $y = x^6/8$ does not satisfy $y(0) = 7$. Therefore, no solution of the equation can satisfy the initial value condition $y(0) = 7$.

Finally, the initial value problem $y(0) = 0$ is satisfied by a single solution $y = x^6/8$. As the equation is not defined at $x = 0$, it is better to replace the initial value condition $y(0) = 0$ by $\lim_{x \to 0} y(x) = 0$.

Example 2.2 The equation $y' - \dfrac{2}{x} y = x^5$ has the integrating factor

$$\mu(x) = e^{\int (-2/x) \, dx} = x^{-2} .$$

We multiply the equation by it and get

$$(x^{-2} y)' = x^{-2} y' - 2x^{-3} y = x^3 ,$$

i.e., $x^{-2} y = x^4/4 + C$, so the general solution is

$$y = x^6/4 + Cx^2 .$$

For the initial value condition $y(2) = 8$, there corresponds $C = -2$ and a single solution $y = x^6/4 - 2x^2$. On the other hand, for an initial value condition at the problematic point $x = 0$, say $y(0) = 8$, there is no solution, since for every choice of C, $y(0) = 0/4 + C \cdot 0 \neq 8$.

Finally, the initial value condition $y(0) = 0$ leads to $y(0) = 0/4 + C \cdot 0 = 0$, which holds for every C. So this initial value problem has infinitely many solutions, all of them from the form $y = x^6/4 + Cx^2$, for any constant C.

The purpose of these two examples was not to show how to solve nonhomogeneous linear equations but rather to point out that the existence of a solution of an initial value problem and the uniqueness of such a solution are not self-evident. This topic will be discussed in Chap. 3, which is dedicated to *existence and uniqueness theorems* for initial value problems.

2.2 Separable Equations

The linear, homogeneous equation $\dfrac{dy}{dx} + p(x)y = 0$ was solved in (2.7) by bringing all occurrences of the variable x to one side and all occurrences of the unknown y to the other side of the equation. The resulting expression was easily integrated. This is an example of a general method that is called *separation of variables*. It is applicable for equations of the form

$$y' = \frac{dy}{dx} = g(x)h(y) , \qquad (2.11)$$

which are naturally called *separable equations*. We write (2.11) as

$$\int \frac{dy}{h(y)} = \int g(x)\, dx ,$$

integrate it, and get

$$H(y) = G(x) + C . \qquad (2.12)$$

Example 2.3

$$\frac{dy}{dx} = (x^2+1)(y^2+1)$$

Here $\int dy/(y^2+1) = \int (x^2+1)\, dx$, that is, $\arctan(y) = x^3/3 + x + C$.

A solution that is given by an implicit function $H(y) - G(x) = C$ poses no theoretical difficulty. This is a single-parametric family of functions, a general solution containing an infinite number of solutions, so that different solutions match different values C. Geometrically, these are the level curves of the function $z = H(y) - G(x)$.

On the other hand, an implicit solution given by $H(y) = G(x) + C$ has a disadvantage: Sometimes it is easy to extract the solution $y(x)$; in other cases it is impractical. For instance, in Example 2.3, we have $y = \tan(x^3/3 + x + C)$. But if the function $H(y)$ is complicated, it may prove impractical to extract an explicit y.

The treatment of initial value conditions $y(x_0) = y_0$ is convenient: Substituting $x = x_0$, $y = y_0$ in Eq. (2.12) determines a single value of C. Another way to handle the initial value condition is to guarantee it by the lower and upper limits of a definite integral. If we take the limits of the integrals as

$$\int_{y_0}^{y} \frac{dy}{h(y)} = \int_{x_0}^{x} g(x)\, dx ,$$

2.2 Separable Equations

it automatically ensures that the initial value conditions are met. The dependence of the solution on the initial value point (x_0, y_0) is also emphasized. This idea is utilized in the following example.

Example 2.4 Given the initial value problem $y' = (y^4 - 1)x^2$, $y(2) = 5$. At which point x will the solution $y(x)$ have the value $y = 7$?

By separating variables we get that the solution which satisfies $y(2) = 5$ by

$$\int_5^y \frac{dy}{y^4 - 1} = \int_2^x x^2 \, dx \, .$$

At the point x_1 where y gets the value 7, we have

$$\int_5^7 \frac{dy}{y^4 - 1} = \int_2^{x_1} x^2 \, dx = \frac{1}{3}\left(x_1^3 - 2^3\right) \, ,$$

i.e.,

$$x_1 = \left(3 \int_5^7 \frac{dy}{y^4 - 1} + 8\right)^{1/3} \, .$$

The above integral can be calculated explicitly, but we will not deal with it here. It is important to note that the question was answered with no need to find the solution $y(x)$ explicitly.

When the method of separation of variables is used, one has to pay attention to an additional consideration. We divided both sides of $y' = \frac{dy}{dx} = g(x)h(y)$ by $h(y)$ to get

$$\int \frac{dy}{h(y)} = \int g(x) \, dx \, ,$$

provided that $h(y) \neq 0$ and we do not divide by zero. What happens if $h(y_1) = 0$ for a certain value y_1? For $y = y_1$ the expression $\int dy/h(y)$ is questionable but there is no need at all to write it. When we return to the original differential equation $y' = g(x)h(y)$, it turns out that the constant function $y(x) \equiv y_1$ satisfies this equation, since $h(y_1) = 0$ on its right-hand side and the derivative of the constant y_1 is zero on its left-hand side. Therefore, in addition to the general solution,

$$H(y) - G(x) = C \, ,$$

there is another solution of the form $y(x) \equiv y_1$. Since this solution was obtained in a particular way, different from the general solution, it is referred to as a *singular solution*. If $h(y)$ vanishes for several values of y, we get for each such value a different singular solution.

Example 2.5

$$y' = \sin y \cos x.$$

When $\sin y \neq 0$, $y \neq n\pi$, then

$$\int \frac{dy}{\sin y} = \int \cos x \, dx + C$$

and explicitly

$$\ln\left|\frac{1 - \cos y}{\sin y}\right| = \sin x + C.$$

In addition to this general solution, for each value of y such that $\sin y = 0$, there corresponds a singular solution $y(x) \equiv n\pi$, $n = 0, \pm 1, \pm 2, \ldots$.

May a singular solution be included in the general solution or not? The answer depends on "how the general solution is written." We will demonstrate this in the next example.

Example 2.6 The equation $y' = 2xy^2$ is rewritten as $\int dy/y^2 = \int 2x \, dx$, whenever $y \neq 0$. This leads to a general solution $-1/y = x^2 + C$ which depends on one parameter C:

$$y(x) = -\frac{1}{x^2 + C}. \qquad (2.13)$$

In addition to this general solution, there exists also a singular solution $y(x) \equiv 0$ that is not included in the general solution (2.13) for any choice of C.

On the other hand, a solution that satisfies the condition $y(0) = y_0$ requires that $y_0 = -1/C$, so it can be written as

$$y = \frac{y_0}{1 - y_0 x^2}. \qquad (2.14)$$

Disregarding the way that (2.14) was obtained, it can be considered a general solution that depends on the parameter y_0. (2.14) is well defined even if $y_0 = 0$, for which we get the solution $y(x) = 0$. From this point of view the solution $y(x) = 0$ is included in the general solution (2.14). Behind the process above stands, of course, the replacement of the parameter C by $(-1/y_0)$.

The conclusion is that a singular solution may be included in a general solution and may also be excluded. It all depends on the choice of parametrization, the form of writing the general solution, and the method that we found it.

2.2 Separable Equations

Example 2.7 In Example 1.1 we mentioned the logistic equation for growth of population:

$$N'(t) = kN - mN^2, \qquad k, m > 0,$$

combined with initial value conditions $N(t_0) = N_0$. We return to this problem and find its explicit solution. For the sake of simplicity we let $L = k/m$ and formulate the problem as

$$\frac{dN}{dt} = kN\left(1 - \frac{N}{L}\right), \qquad N(t_0) = N_0. \tag{2.15}$$

This equation fits naturally into the scheme of separation of variables, since the independent variable t does not appear explicitly anywhere. Assuming that $N(t) \neq 0, L$,

$$\int \frac{dN}{N(1 - N/L)} = \int k \, dt.$$

We separate the left-hand side to partial fractions and integrate between limits that guarantee the initial value conditions:

$$\int_{N_0}^{N} \left(\frac{1/L}{1 - N/L} + \frac{1}{N}\right) dN = \int_{t_0}^{t} k \, dt.$$

This yields

$$\left(-\ln|1 - N/L| + \ln|N|\right)\Big|_{N_0}^{N} = k(t - t_0)$$

and

$$\left(\ln\left|\frac{1}{N} - \frac{1}{L}\right|\right)\Big|_{N_0}^{N} = -k(t - t_0).$$

In the exponential form we eliminate the absolute values while ensuring that $N = N_0$ when $t = t_0$:

$$\frac{1}{N} - \frac{1}{L} = \left(\frac{1}{N_0} - \frac{1}{L}\right) e^{-k(t - t_0)}.$$

Finally,

$$N(t) = \frac{N_0}{N_0/L + (1 - N_0/L) e^{-k(t - t_0)}}. \tag{2.16}$$

In addition to (2.16), there exist also two singular solutions:

$$N(t) \equiv 0, \qquad N(t) \equiv L.$$

(2.16) may be interpreted also as a general solution that depends on the parameter N_0. It is defined also for negative values of N_0, although they have no meaning for a growth of population model.

The graphic sketch of the solutions is postponed to Sect. 2.7, where geometric considerations will be used. In the meantime, we ask some guiding questions:

Exercise 2.2

(i) What is $\lim_{t \to +\infty} N(t)$ when $0 < N_0 < L$? And when $N_0 = 0$ or $N_0 = L$?

(ii) Let $N_0 > L$. What happens to $N(t)$ when t increases from the initial point t_0 to $+\infty$? Show that when $N_0 > L$, the denominator of Eq. (2.16) vanishes at $t = t_0 + \ln(1 - L/N_0) < t_0$ and $N(t) \to +\infty$ as t decreases from t_0 towards this point.

(iii) Let $N_0 < 0$. What happens to $N(t)$ when t increases from t_0 towards the point $t_0 + \ln(1 - L/N_0) > t_0$?

2.3 Solution by Substitution

In this section we discuss a few types of differential equations which can be solved by some suitable substitutions.

Under the title *equation of a homogeneous type* we mean first-order equations $y' = f(x, y)$ where the right-hand side depends only on the ratio y/x:

$$\frac{dy}{dx} = F\left(\frac{y}{x}\right). \qquad (2.17)$$

The term "homogeneous" expresses the feature that if x and y are replaced by kx and ky (rescaling of the axes), the equation does not change. This "homogeneous" has nothing to do with "homogeneous linear equations" which is used in linear algebra, but rather to the concept of "homogeneous function" in analysis. A function $f(x, y)$ is called *homogeneous of degree h* if $f(tx, ty) = t^h f(x, y)$ for all $t \neq 0$. Thus, $F(y/x)$ on the right-hand side of Eq. (2.17) is a homogeneous function of degree 0. Examples of differential equations of a homogeneous type are as follows:

$$y' = \frac{y^2 + 2xy}{x^2},$$

$$y' = \ln x - \ln y = -\ln\left(\frac{y}{x}\right).$$

2.3 Solution by Substitution

For equations of the form (2.17) we introduce a new unknown function $z(x) = y(x)/x$. Put $y(x) = xz(x)$ and $\dfrac{dy}{dx} = x\dfrac{dz}{dx} + z(x)$ into the original equation (2.17):

$$x\frac{dz}{dx} + z = F(z).$$

The equation

$$x\frac{dz}{dx} = F(z) - z \qquad (2.18)$$

is suitable for the separation of variables as

$$\int \frac{dz}{F(z) - z} = \int \frac{dx}{x}, \qquad (2.19)$$

i.e., $G(z) = \ln|x| + C$. This implicit function depends on a single parameter C and defines the general solution $z(x)$. Finally, we return from $z(x)$ to the original unknown $y(x)$.

As in any process of separation of variables, we must check whether the expression $F(z) - z$ which appears in the denominator of (2.18) is nonzero, or perhaps there is a value z_0 such that $F(z_0) - z_0 = 0$. If this is the case, then Eq. (2.18) has a singular solution $z(x_0) \equiv z_0$, because its right-hand side is 0 for $z = z_0$ and the left-hand side is zero being a derivative of a constant. In terms of the original problem, this singular solution is equivalent to $y(x)/x \equiv z_0$, i.e., $y(x) = z_0 x$, a straight line through the origin.

Example 2.8

$$y' = \frac{y^2 + 2xy}{x^2}.$$

The substitution $y = xz$, $y' = xz' + z$ leads to $xz' + z = z^2 + 2z$, that is, $x\dfrac{dz}{dx} = z^2 + z$. As long as $z \neq 0, -1$, its solution is

$$\int \frac{dx}{x} = \int \frac{dz}{z^2 + z} = \int \left(\frac{1}{z} - \frac{1}{z+1}\right) dz.$$

Therefore,

$$\ln|x| = \ln|z| - \ln|z+1| + \ln C,$$

$$x = \frac{z}{z+1} C = \frac{Cy/x}{y/x + 1} = \frac{Cy}{y+x},$$

and a general, one-parametric solution is

$$y = \frac{Cx^2}{1 - Cx} . \tag{2.20}$$

To the exceptional values $z = 0, -1$ there correspond two singular solutions $z_1(x) \equiv 0$ and $z_2(x) \equiv -1$. The corresponding singular y-solutions are $y_1(x) \equiv 0$, $y_2(x) = -x$. Check this directly!

Note that the singular solution $y_1(x) \equiv 0$ is included in the general solution (2.20) for $C = 0$. The other one, $y_2(x) = -x$, is not included in the general solution for any finite value of C, but it is obtained as a limit when $C \to \infty$.

The following example can also be solved by an appropriate substitution.

Example 2.9 Consider

$$y' = f(ax + by + c)$$

where a, b, c are constants and $f(\ )$ is a function of a single variable, say

$$y' = \sin(2x - 3y + 4) .$$

This time we replace the unknown y by z through the substitution $z = ax + by + c$, or more precisely, $z(x) = ax + by(x) + c$, $z'(x) = a + by'(x) = a + bf(z)$. The so-obtained equation

$$\frac{dz}{dx} = a + bf(z)$$

is solved by separation of variables. The integral

$$\int \frac{dz}{a + bf(z)} = \int dx + C$$

is an implicit function that defines the general solution $z(x)$. From $z(x)$ we return to the original unknown $y(x)$.

Problem 2.20 at the end of this chapter introduces another example, the Bernoulli equation, that is also solved by an appropriate substitution.

We end this section with another trick related to the change of the unknown function:

Example 2.10 The equation $\dfrac{dy}{dx} = \dfrac{y}{x + y^3}$ is nonlinear in y. Now we change our point of view, and instead of considering y as a function of x, we look for the inverse function $x(y)$. According to the inverse function theorem,

$$\frac{dy}{dx} = 1 \bigg/ \frac{dx}{dy} ,$$

so our equation becomes

$$\frac{dx}{dy} - \frac{1}{y}x = y^2, \qquad y \neq 0,$$

which is a nonhomogeneous linear equation for $x(y)$. Multiplying by the integrating factor $\mu(y) = \exp\left(\int(-1/y)\,dy\right) = y^{-1}$, it becomes

$$\frac{d}{dy}\left(\frac{1}{y}x(y)\right) = \frac{1}{y}\frac{dx}{dy} - \frac{1}{y^2}x = y.$$

Integration according to y and simplification yields the general solution

$$x = \frac{1}{2}y^3 + Cy.$$

It is inconvenient to extract y as a function of x and the parameter C from this implicit function. Note that the original equation has also the singular solution $y(x) \equiv 0$.

2.4 Exact Equations

In this section, we deal with a family of differential equations that are related to Green's formula and the concept of a conservative vector field. Let us start with a short reminder of some definitions and ideas from calculus.

If $F(x, y)$ is a differentiable function, then moving from a point (x, y) to a nearby point $(x + \Delta x, y + \Delta y)$, the value of F varies by

$$\Delta F = F(x + \Delta x, y + \Delta y) - F(x, y) = \frac{\partial F}{\partial x}\Delta x + \frac{\partial F}{\partial y}\Delta y + R,$$

where R denotes a remainder whose order of magnitude is less than $\sqrt{\Delta x^2 + \Delta y^2}$. Therefore,

$$\Delta F \approx \frac{\partial F}{\partial x}\Delta x + \frac{\partial F}{\partial y}\Delta y$$

and we define symbolically the *complete differential* of F:

$$dF = \frac{\partial F}{\partial x}dx + \frac{\partial F}{\partial y}dy. \qquad (2.21)$$

Since the symbols dx, dy have various meanings in different contexts, this equality must be given an interpretation. One way to do this is to refer to x, y as dependent on some parameter t. So $F(x, y) = F(x(t), y(t))$ depends on t and its rate of change along the path $(x(t), y(t))$ is given by the chain rule

$$\frac{dF}{dt} = \frac{\partial F}{\partial x}\frac{dx}{dt} + \frac{\partial F}{\partial y}\frac{dy}{dt} . \qquad (2.22)$$

This holds for any smooth parametrization, so (2.21) may be considered as a symbolic presentation of (2.22) along any path.

When a point moves along a level curve of F, $F(x, y) = $ const, the value of F does not change at all, $dF = 0$. Therefore, symbolically

$$\frac{\partial F}{\partial x} dx + \frac{\partial F}{\partial y} dy = 0 . \qquad (2.23)$$

This fits well with the implicit function theorem, which states for the extracted $y(x)$ that $\dfrac{dy}{dx} = -\dfrac{\partial F}{\partial x} \Big/ \dfrac{\partial F}{\partial y}$.

We also need the concept of *line integral*. Let $C = \{(x(t), y(t)) | a \le t \le b\}$ be a smooth curve. The line integral $\int_C P\,dx + Q\,dy$ is a shorthand for

$$\int_a^b \left(P(x(t), y(t))\frac{dx}{dt} + Q(x(t), y(t))\frac{dy}{dt} \right) dt .$$

The *vector field* $(P(x, y), Q(x, y))$ is *conservative* if the line integral $\int_C P\,dx + Q\,dy$ depends only on the endpoints of the path but not on the choice of the connecting path C.

After these preparations, we study a family of differential equations:

$$P(x, y)dx + Q(x, y)dy = 0 \qquad (2.24)$$

where $P(x, y)$, $Q(x, y)$ are two given functions. This is a symbolic presentation of the first-order differential equation:

$$\frac{dy}{dx} = -\frac{P(x, y)}{Q(x, y)} . \qquad (2.25)$$

Our goal is to take advantage of the similarity between the differential equation (2.24) and the level curve equation (2.23) to try to solve the equation. This is not always possible, but in some cases we can use the similarity for our benefit. If there exists a function $F(x, y)$ so that

$$\frac{\partial F(x, y)}{\partial x} = P(x, y) , \qquad \frac{\partial F(x, y)}{\partial y} = Q(x, y) , \qquad (2.26)$$

2.4 Exact Equations

then Eq. (2.24) is identical with the level curve equation (2.23) and

$$P\,dx + Q\,dy = F_x dx + F_y dy = 0.$$

In this case (2.24) is called an *exact differential equation*. Since (2.23) holds along a level curve $F(x, y) = $ const, the implicit function

$$F(x, y) = C$$

defines a one-parametric general solution of the exact differential equation (2.24).

Example 2.11 Given the equation

$$(3x^2 + y^2)\,dx + 2xy\,dy = 0.$$

Since

$$3x^2 + y^2 = \frac{\partial}{\partial x}\left(x^3 + xy^2\right), \qquad 2xy = \frac{\partial}{\partial y}\left(x^3 + xy^2\right),$$

the equation is of the form $F_x dx + F_y dy = 0$ with $F(x, y) = x^3 + xy^2$. Thus, $x^3 + xy^2 = C$ is a general solution of our equation. By the way, our equation is equivalent to

$$\frac{dy}{dx} = -\frac{3x^2 + y^2}{2xy}$$

which can be solved also as an equation of a homogeneous type.

Under which conditions there exists a function $F(x, y)$ so that (2.26) holds? We recall the following theorem of calculus which is closely related to Green's formula:

Theorem 2.2 *Suppose that two functions $P(x, y)$, $Q(x, y)$ have continuous first-order partial derivatives in a simply connected domain of the (x, y) plane. The following three conditions are equivalent:*

1. $\dfrac{\partial P(x, y)}{\partial y} \equiv \dfrac{\partial Q(x, y)}{\partial x}.$
2. *The vector field $(P(x, y), Q(x, y))$ is a conservative field.*
3. *There exists a scalar function $F(x, y)$, called the potential function of the field $(P(x, y), Q(x, y))$ such that*

$$\frac{\partial F(x, y)}{\partial x} = P(x, y), \qquad \frac{\partial F(x, y)}{\partial y} = Q(x, y).$$

In other words, the vector field $\big(P(x, y), Q(x, y)\big)$ is the gradient of $F(x, y)$:

$$\nabla F(x, y) = (F_x, F_y) = \big(P(x, y), Q(x, y)\big).$$

If these three equivalent conditions hold, a corresponding potential function $F(x, y)$ is given up to an additive constant by

$$F(x, y) = \int_{(x_0, y_0)}^{(x, y)} P\,\mathrm{d}x + Q\,\mathrm{d}y, \tag{2.27}$$

where the integration is performed along any piecewise smooth curve from the point (x_0, y_0) to the point (x, y) within the simple connected domain.

Recall that a *domain* is an open and connected set. A planar domain is called *simply connected* if it "has no holes within." The simply connectedness is of critical importance for the validity of Green's formula and for the study of two-dimensional conservative vector fields.

Our main consequence is

Theorem 2.3 *Under the assumptions of Theorem 2.2, the equation*

$$P(x, y)\,\mathrm{d}x + Q(x, y)\,\mathrm{d}y = 0$$

is exact if and only if the three equivalent properties in Theorem 2.2 hold. If that is the case, a general solution of the equation is given implicitly by $F(x, y) = C$, where $F(x, y)$ is the corresponding potential function.

Remark (Implicit Solution) Note that $F(x, y)$ is a function of two variables and it *is not* a solution of our differential equation. On the other hand, the implicit function $F(x, y) = C$ usually defines $y(x)$ as a function of the variable x and the parameter C, even if it is difficult to find it explicitly, and the resulting $y = y(x, C)$ is a general solution of the differential equation.

The most direct way to check that Eq. (2.24) is exact is to examine whether

$$\frac{\partial P(x, y)}{\partial y} \equiv \frac{\partial Q(x, y)}{\partial x} \tag{2.28}$$

holds in the simply connected domain. If that is the case, then it is practical to calculate the potential function $F(x, y)$ by integrating the pair of equations

$$\begin{aligned}\frac{\partial F(x, y)}{\partial x} &= P(x, y), \\ \frac{\partial F(x, y)}{\partial y} &= Q(x, y),\end{aligned} \tag{2.29}$$

2.4 Exact Equations

(and not by the line integral (2.27)). Integrate the first equation of (2.29) according to x. The constant of integration is a constant in terms of x but not necessarily with respect to y. Therefore,

$$F(x, y) = \int P(x, y)\, dx + h(y),$$

where $h(y)$ is a yet unknown function that we shall find later. We differentiate this expression with respect to y and exchange the order of differentiation (with respect to y) and integration (according to x):

$$\frac{\partial F}{\partial y} = \int \frac{\partial P(x, y)}{\partial y}\, dx + h'(y).$$

We compare this with the second equation of (2.29), $F_y = Q(x, y)$, and get

$$Q(x, y) = \int \frac{\partial P(x, y)}{\partial y}\, dx + h'(y)$$

i.e.,

$$h'(y) = Q(x, y) - \int \frac{\partial P(x, y)}{\partial y}\, dx. \tag{2.30}$$

On the left-hand side of (2.30), $h'(y)$ is independent of x, while the right-hand side of (2.30) seems to be dependent of x, which is impossible. However, this "contradiction" is not a real one. The derivative of the right-hand side according to x is, by assumption (2.28),

$$\frac{\partial}{\partial x}\left(Q(x, y) - \int \frac{\partial P(x, y)}{\partial y}\, dx\right) = \frac{\partial Q(x, y)}{\partial x} - \frac{\partial P(x, y)}{\partial y} \equiv 0.$$

Consequently, the right-hand side is independent of x and depends only on y. Thus, $h(y)$ is calculated by a straightforward integration, and the potential function $F(x, y)$ is found.

The solution of an initial value problem $y(x_0) = y_0$ is obtained by putting $(x, y) = (x_0, y_0)$ into $F(x, y) = C$. This leads to $C = F(x_0, y_0)$ and the solution of the initial value problem is

$$F(x, y) = F(x_0, y_0).$$

Example 2.12 Consider the equation

$$(3x^2 y + 8xy^2)\, dx + (x^3 + 8x^2 y + \sin y)\, dy = 0$$

which is equivalent to

$$\frac{dy}{dx} = -\frac{3x^2y + 8xy^2}{x^3 + 8x^2y + \sin y}.$$

In this form, it is not of any type that we discussed previously (linear, separable, homogeneous type). But in its original form, it is an exact equation since

$$P(x, y) = 3x^2y + 8xy^2, \quad Q(x, y) = x^3 + 8x^2y + \sin y$$

satisfy

$$\frac{\partial P}{\partial y} = 3x^2 + 16xy = \frac{\partial Q}{\partial x}.$$

Therefore, look for a potential function $F(x, y)$ so that

$$\frac{\partial F}{\partial x} = P(x, y) = 3x^2y + 8xy^2,$$

$$\frac{\partial F}{\partial y} = Q(x, y) = x^3 + 8x^2y + \sin y.$$

By integration of the first equation according to x,

$$F(x, y) = \int (3x^2y + 8xy^2)\, dx = x^3y + 4x^2y^2 + h(y),$$

whose derivative with respect to y is

$$F_y(x, y) = x^3 + 8x^2y + h'(y).$$

Comparison to $Q(x, y) = x^3 + 8x^2y + \sin y$ gives $h'(y) = \sin y$, $h(y) = -\cos y$. Thus, the potential function is

$$F(x, y) = x^3y + 4x^2y^2 - \cos y$$

and a general solution of the differential equation is

$$x^3y + 4x^2y^2 - \cos y = C.$$

There is no practical way to extract y implicitly from here as a function of x and C. To solve an initial value problem $y(x_0) = y_0$, put $(x, y) = (x_0, y_0)$ to get the value of C.

2.5 Integrating Factors

Given a differential equation of the form (2.24),

$$P(x, y)\,dx + Q(x, y)\,dy = 0$$

which is not exact, i.e., the condition $P_y = Q_x$ does not hold. Does there exist some function $\mu(x, y)$ so that after multiplying (2.24) by it, the so-obtained

$$\mu(x, y)P(x, y)\,dx + \mu(x, y)Q(x, y)\,dy = 0$$

will be exact? If such $\mu(x, y)$ exists, it is called an *integrating factor*. The condition for this is of course

$$\frac{\partial}{\partial x}\Big(\mu(x, y)Q(x, y)\Big) = \frac{\partial}{\partial y}\Big(\mu(x, y)P(x, y)\Big),$$

i.e.,

$$\frac{\partial \mu}{\partial x}Q(x, y) - \frac{\partial \mu}{\partial y}P(x, y) + \mu(x, y)(Q_x - P_y) = 0. \qquad (2.31)$$

This is a first-order, linear, and homogeneous partial differential equation. Unfortunately, it is not more simple to solve this partial equation than solving the original ordinary equation. In fact, during the study of linear partial differential equations of the first order, it turns out that solving (2.31) is equivalent to solving the original equation (2.24). What is then the benefit of seeking integrating factors? We look for them because in some cases it is possible to find a simple integrating factor and to take advantage of it to solve the given equation.

Example 2.13 When does there exist an integrating factor that depends only on the variable x? If $\mu = \mu(x)$ and it does not depend on y, the partial equation (2.31) becomes the ordinary equation

$$\frac{d\mu}{dx}Q(x, y) + \mu(x)(Q_x - P_y) = 0,$$

i.e.,

$$\frac{\mu'(x)}{\mu(x)} = \frac{P_y(x, y) - Q_x(x, y)}{Q(x, y)}.$$

The left-hand side is independent of y. This is possible only if the right-hand side is also independent of y. Thus, it is possible to find an integrating factor $\mu = \mu(x)$ if and only if the expression $(P_y - Q_x)/Q$ is independent of y.

Similarly, one can find an integrating factor $\mu = \mu(y)$ that depends only on y by solving

$$\frac{\mu'(y)}{\mu(y)} = \frac{Q_x(x,y) - P_y(x,y)}{P(x,y)},$$

provided that the right-hand side is independent of x.

Example 2.14

$$(3y^4 - 7x^4y^2)\,dx + (4xy^3 - 2x^5y)\,dy = 0, \quad y(0) = 0.$$

Here

$$\frac{\mu'(x)}{\mu(x)} = \frac{P_y - Q_x}{Q} = \frac{(12y^3 - 14x^4y) - (4y^3 - 10x^4y)}{4xy^3 - 2x^5y} = \frac{2}{x}$$

depends only on x. Therefore, $\ln \mu(x) = 2\ln x$, $\mu(x) = x^2$. We do not add a constant of integration because only a single integrating factor is needed. Multiplying the equation by $\mu(x) = x^2$ gives the exact equation

$$(3x^2y^4 - 7x^6y^2)\,dx + (4x^3y^3 - 2x^7y)\,dy = 0,$$

which has a general solution $x^3y^4 - x^7y^2 = C$. Check by differentiation!

The initial value condition $y(0) = 0$ determines $C = 0$. In the implicit solution $x^3y^4 - x^7y^2 = 0$ there are three hidden solutions of the initial value problem: $y \equiv 0$, $y = x^2$, $y = -x^2$. (To verify that $y = x^2$ is indeed a solution, put in the original equation $y = x^2$, $dy = 2x\,dx$.)

An equation of the form (2.24) may have many integrating factors. Let us look, for example, at the equation

$$y\,dx - x\,dy = 0. \tag{2.32}$$

(This is equivalent to the equation $dy/dx = y/x$ which is separable and easy to solve.) Here $P(x,y) = y$, $Q(x,y) = -x$, so $P_y \neq Q_x$. We list some suitable integrating factors:

1. $\mu_1(y) = y^{-2}$ is an integrating factor of (2.32). Indeed, multiplication by it yields $\frac{1}{y}\,dx - \frac{x}{y^2}\,dy = 0$, which is an exact equation in any simply connected domain where $y \neq 0$, since

$$\frac{\partial}{\partial y}\left(\frac{1}{y}\right) = -\frac{1}{y^2} = \frac{\partial}{\partial x}\left(-\frac{x}{y^2}\right).$$

2.6 Equations from the Form $y'' = f(y)$

The corresponding potential function is $F_1(x, y) = x/y$, since $(F_1)_x = 1/y$, $(F_1)_y = -x/y^2$. A general solution of (2.32) is given by the explicit function

$$F_1(x, y) = x/y = \text{const}.$$

2. Multiplication of (2.32) by $\mu_2(x, y) = \dfrac{1}{xy}$, $x, y \neq 0$, leads to $\dfrac{1}{x} dx - \dfrac{1}{y} dy = 0$.
   This is also an exact equation and has general solution

$$F_2(x, y) = \ln |x| - \ln |y| = \text{const},$$

because $(F_2)_x = 1/x$, $(F_2)_y = -1/y$.

3. Multiplying (2.32) by the factor $\mu_3(x, y) = \dfrac{1}{x^2 + y^2}$ yields

$$\frac{y}{x^2 + y^2} dx + \frac{-x}{x^2 + y^2} dy = 0,$$

which is exact in simply connected domains that do not contain $(x, y) = (0, 0)$. The suitable general solution is now $F_3(x, y) = -\arctan\left(\dfrac{y}{x}\right) = \text{const}$, because

$$\left(-\arctan \frac{y}{x}\right)_x = \frac{y}{x^2 + y^2}, \quad \left(-\arctan \frac{y}{x}\right)_y = \frac{-x}{x^2 + y^2}.$$

Note that all general solutions that were found differ only by their appearance but lead to the same solutions.

4. Check that $\mu_4(x, y) = \dfrac{1}{x^2 - y^2}$ is also an integrating factor.

Exercise 2.3 Find an integrating factor for the equation

$$(y^2 + yx^2) dx + (x^3 - 3xy^2) dy = 0$$

of the form $\mu(x, y) = x^\alpha y^\beta$, α, β unknown constants, and solve the equation.

2.6 Equations from the Form $y'' = f(y)$

The nonlinear pendulum equation $\theta'' = -\dfrac{g}{L} \sin \theta$ in Example 1.8, the equation of motion $y'' = -\dfrac{k}{y^2}$ in Example 1.5, and the linear spring equation $y'' = -\dfrac{k}{m} y$ in Example 1.6 are all second-order equations of the form

$$y'' = f(y), \tag{2.33}$$

where the independent variable and the first derivative y' do not appear explicitly. We discuss the second-order Eq. (2.33) in this chapter, since it is easily transformed into a first-order equation which is then solved by separation of variables.

Suppose, for example, that the independent variable in Eq. (2.33) is t and there are given two initial value conditions $y(t_0) = \alpha$, $y'(t_0) = \beta$. Assuming that the solution $y(t)$ is not identically constant and its derivative $y'(t)$ is not identically zero, we multiply the two sides of (2.33) by $y'(t)$ and integrate according to t from t_0 to t:

$$\int_{t_0}^{t} y''(t) y'(t)\, dt = \int_{t_0}^{t} f(y) y'(t)\, dt . \tag{2.34}$$

As $\left(\frac{1}{2} y'^2\right)' = y'' y'$, the left-hand side of (2.34) equals $\frac{1}{2} y'^2(t) - \frac{1}{2} y'^2(t_0)$. The integral on the right-hand side of (2.34) is reduced, by the change of integration variable $y = y(t)$, $dy = y'(t) dt$, to $\int_{y(t_0)}^{y(t)} f(y)\, dy$. With the help of these comments and thanks to the initial value conditions, we get

$$\frac{1}{2} y'^2 - \frac{1}{2} \beta^2 = \int_{\alpha}^{y} f(y)\, dy .$$

For the sake of convenience we denote $F(y) = \int_{\alpha}^{y} f(y)\, dy$. Of course $F(y(t_0)) = F(\alpha) = 0$. With this notation $y'^2 = 2F(y) + \beta^2$,

$$\frac{dy}{dt} = \sqrt{2F(y) + \beta^2} .$$

The sign of the square root must be selected according to the sign of β so that $y'(t_0) = \beta$. The so-obtained first-order equation can be solved by separation of variables:

$$\int_{t_0}^{t} dt = \int_{\alpha}^{y} \frac{dy}{\sqrt{2F(y) + \beta^2}} ,$$

$$t = \int_{\alpha}^{y} \frac{dy}{\sqrt{2F(y) + \beta^2}} + t_0 .$$

Although we have not been able to find explicitly $y(t)$, we have obtained its inverse function, t expressed in terms of y. The solution of the second-order equation that we found depends, of course, on the two parameters α, β.

To justify multiplication by $y'(t)$, we assumed that the solution $y(t)$ is not identically constant. A constant function $y(t) \equiv y_0$ is a solution of Eq. (2.33) provided that $f(y_0) = 0$. If such y_0 exists, it is a singular solution that is not included in the two-parametric solution above.

Whereas the Eq. (2.33) is theoretically solved, the calculation of integrals maybe difficult. Multiply, for example, the pendulum equation

$$\theta'' = -\frac{g}{L} \sin \theta$$

by $\theta'(t)$ and integrate

$$\frac{1}{2}\theta'^2(t) = \int \theta''(t)\theta'(t) \, dt = \int -\frac{g}{L} \sin \theta(t) \theta'(t) \, dt$$

$$= -\frac{g}{L} \int \sin \theta \, d\theta = \frac{g}{L}(\cos \theta + c_1) \ .$$

Consequently,

$$\frac{d\theta}{dt} = \left(\frac{2g}{L}\right)^{1/2} \sqrt{\cos \theta + c_1} \ , \tag{2.35}$$

$$t = \left(\frac{L}{2g}\right)^{1/2} \int \frac{d\theta}{\sqrt{\cos \theta + c_1}} + c_2 \ .$$

The last integral is an "elliptic integral," and for $c_1 \neq \pm 1$, it cannot be expressed in terms of elementary functions. We will return to the pendulum equation from another point of view in Example 6.9.

2.7 Geometric Topics: Direction Field

So far, we have referred to the differential equation (2.1),

$$y' = f(x, y),$$

analytically and tried to find explicit solutions $y(x)$. This equation can also be given a geometric interpretation. $y'(x)$ is the slope of the graph of the solution $y(x)$ and the differential equation means that when a solution passes through a point (x, y), its slope at this point is $f(x, y)$. We draw therefore at every point (x, y) an arrow of slope $f(x, y)$ (the length of an arrow is insignificant). This covers the domain of the definition of the equation with a continuum of "directional arrows" called the *direction field* of the equation. See Fig. 2.3.

From a geometric point of view, finding a general solution to a differential equation means to find a family of curves covering the domain of definition so that at any point (x, y) the slope of the curve that passes there equals the required slope

Fig. 2.3 A direction field and a solution of an initial value problem

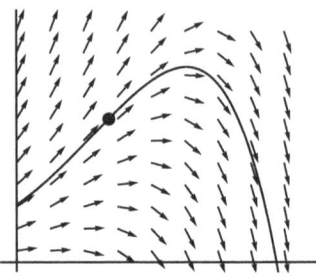

$f(x, y)$. Solving the initial value problem means finding a curve that departs from the given initial point, and during its progress, it follows the given direction field.

The main benefit of a direction field is the possibility to visualize the behavior of the solutions. This is of great importance because, as we have already mentioned, it is rarely possible to find explicitly the solutions of a differential equation. Even in cases where the solutions are explicitly known, it is convenient to look at their graphic description because, as we know, a single image is worth a thousand formulas. Moreover, it is possible to draw the direction field, even for a complicated $f(x, y)$. A plenty of free direction field plot programs are available on the Internet.

Example 2.15 Consider

$$\frac{dy}{dx} = \frac{y}{x}$$

(which is equivalent to (2.32)). At any point (x, y) the slope of the requested solution is y/x, which is also the slope of a straight line that connects (x, y) to the origin. Therefore, the general solution is the family of straight lines passing through the origin, $y = Cx$. Verify this by a direct solution.

Exercise 2.4

$$\frac{dy}{dx} = -\frac{x}{y}. \tag{2.36}$$

The slope at point (x, y) is $-x/y$ and this direction is perpendicular to the straight line that connects the point to the origin. See Fig. 2.4. What are the curves which are perpendicular at any point to the radial direction? Solve the given equation directly as well.

Example 2.16 Let us draw the direction field that corresponds to the logistic equation (2.15):

$$y'(t) = ky\left(1 - \frac{y}{L}\right), \quad k, L > 0.$$

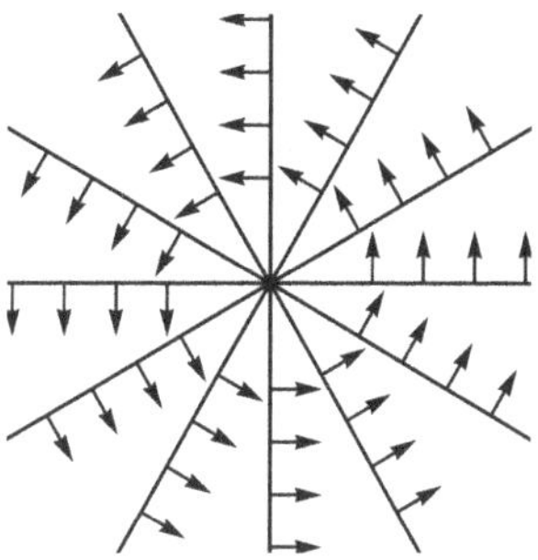

2.7 Geometric Topics: Direction Field

Fig. 2.4 Direction field for Eq. (2.36)

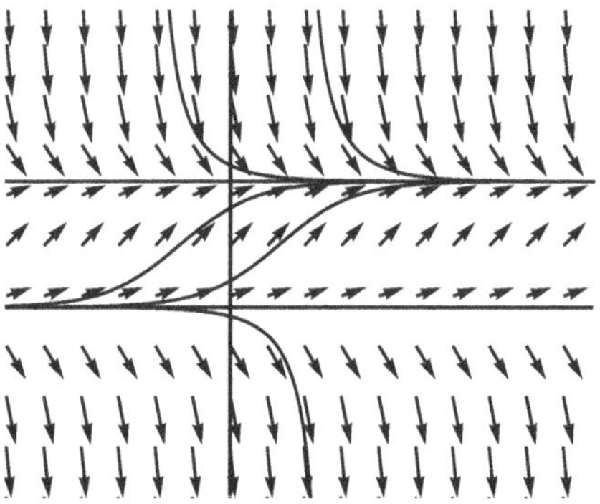

Fig. 2.5 Direction field and some solutions of the logistic equation

We are interested also in negative y values which are irrelevant to a population growth model.

For $0 < y < L$, the derivative y' is positive. Therefore, within this horizontal strip, each solution increases. When in this strip y is close to L or to 0, then y' is positive but small, and therefore, each solution increases there slowly. Conversely, for $y > L$ and for $y < 0$, the derivative y' is negative and in these domains all solutions decrease. In addition, the slope y' at a point (t, y) is independent of t and depends only on y because t does not appear on the right-hand side of the differential equation. Therefore, shifting a solution to the right or the left (replacing t with $t + c$) creates additional solutions. See Fig. 2.5.

The geometric considerations are helpful to guess the form of the solution family without solving the equation. However, they are not sufficient to decide, for example, whether a solution that increases towards the straight line $y = L$ meets it at a certain point or maybe it just gets closer to the line but never touches it. We cannot rely our own eyes as they may be deceiving. The question will be answered in the next chapter, with the help of the theorem of existence and uniqueness.

Another convenient way is to draw arrows of the direction field of Eq. (2.1) only at points (x, y) where $f(x, y) = k$, the level curves of the function $f(x, y)$. At these points, the direction field has a fixed slope $y' = k$ and the curves are called *isoclines* (from Greek: of equal slope). For example, in Fig. 2.4, the straight lines through the

Fig. 2.6 The isoclines $y - x = -4, -3, \ldots, 3, 4$ of the equation $y' = y - x$, the direction field and some graphic solutions

Fig. 2.7 Isoclines of Eq. (2.37) for slopes $k = 4, 2, 0, -2, -4, -6$

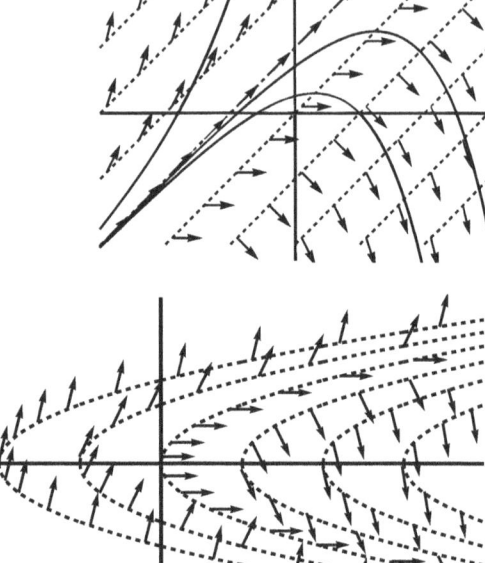

origin are the isoclines of Eq. (2.36). In Fig. 2.5 the isoclines are the horizontal lines $y = $ const.

Example 2.17 Let us look at the linear equation $y' = y - x$ whose general solution is $y = Ce^x + x + 1$. The isoclines of this equation are the straight lines $y - x = k$, and along it the slope of the direction field is k. Figure 2.6 depicts several isoclines, the direction that corresponds to each isocline, and some solutions.

Example 2.18 Consider the equation

$$y' = y^2 - x \,, \tag{2.37}$$

the solutions of which we do not know to write explicitly. The isoclines of (2.37) are the parabolas $y^2 - x = k$. In Fig. 2.7 we show several isoclines and on each of them we mark direction arrows of suitable slope k.

Is it possible to guess from these direction arrows the behavior of the solutions of the differential equation? Try to sketch them.

In Fig. 2.8 we also add the graphs of some numerically calculated solutions.

2.8 Families of Orthogonal Curves

Fig. 2.8 Isoclines and solutions of Eq. (2.37)

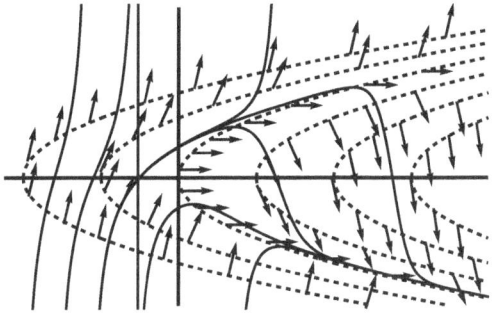

2.8 Families of Orthogonal Curves

We end Chap. 2 with another geometric application. Given a single-parametric family of curves, we would like to find another family of curves so that at each intersection point, curves from the two families intersect at a right angle. Two such families are called *orthogonal* to each other.

We have seen in examples that the general solution of a first-order differential equation is a single-parametric family of curves. Now we will try to do the opposite and attach to a single-parametric family of curves a first-order differential equation. We demonstrate this for the family of parabolas:

$$y = cx^2, \quad -\infty < c < \infty. \tag{2.38}$$

The derivative according to x is $y' = 2cx$. From the two equations $y = cx^2$, $y' = 2cx$, we eliminate the parameter c by extracting it from one equation and substituting it into the other one and get the relation

$$y' = \frac{2y}{x}.$$

This is a first-order differential equation which is independent of the parameter c, so it is satisfied by every curve from the given family (2.38), for every value of c.

The derivative $y'(x)$ of a solution is also the slope k_1 of the tangent to the curve $y = y(x)$ at (x, y). Now, two lines with slopes k_1, k_2, respectively, are perpendicular to each other provided that $k_2 = -1/k_1$. Therefore, a curve from the requested orthogonal family must have at the point (x, y) a slope $k_2 = -1/k_1 = -x/2y$ and the differential equation which corresponds to the orthogonal family is

$$y' = -\frac{x}{2y}.$$

Fig. 2.9 Orthogonal families of curves

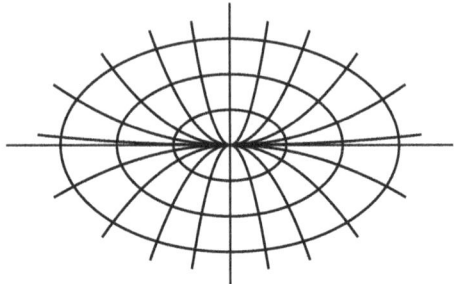

Separation of variables leads to $\int 2y\,dy = \int -x\,dx$ and to the family of ellipses:

$$\frac{x^2}{2} + y^2 = k \,.$$

The constant of integration k in the second family is unrelated to the constant c in the first family. The two families of orthogonal curves are shown in Fig. 2.9.

Let us summarize the steps of the process:

1. Given a family of curves that depends on one parameter c, say

$$F(x, y, c) = 0 \,. \tag{2.39}$$

2. Differentiate it with respect to x using the chain rule to get the equation

$$F_x(x, y, c) + F_y(x, y, c)\frac{dy}{dx} = 0 \,. \tag{2.40}$$

3. Eliminate the parameter c from the two Eqs. (2.39), (2.40). For example, extract c from one of the equations and substitute it into the other one. The result is a relationship between the quantities x, y, y', which is independent of c, say $G(x, y, y') = 0$. We extract y' and get

$$y' = f(x, y).$$

In this way, we build for the given single-parametric family of curves (2.39) a first-order differential equation which is independent of the parameter c and fits to all the curves in the given family. The steps above are theoretically possible according to the theorem of the implicit functions, but may be practically impossible.

4. The orthogonality condition for the trajectories of the requested equation is

$$y' = -1/f(x, y) \,.$$

2.8 Families of Orthogonal Curves

The solutions of this differential equation are the family of curves which are perpendicular to the given family.

Exercise 2.5 Prove that the curves which are orthogonal to the family $y = ce^x$, $-\infty < c < \infty$, are the parabolas $y^2 = -2x + k$.

Problems

2.1 Solve the following equations:

(a) $y' - \dfrac{2}{x^3} y = 0$.

(b) $y' + \dfrac{2}{x^3} y = 0$, $y(2) = 3$.

(c) $y' + \dfrac{2}{x^3} y = 0$, $y(0) = 3$.

Explain the difference between problems (b) and (c).

2.2 Solve the initial value problem $y' + \dfrac{2}{x} y = \dfrac{\cos x}{x^2}$, $y(\pi) = 3$.

2.3 Solve the equation $y' + \tan(x) y = x \sin(2x)$.

2.4 Given a piecewise continuous function

$$f(x) = \begin{cases} x, & x \leq 2, \\ x + 2, & x > 2. \end{cases}$$

Find a solution of the initial value problem $(x^2 - 1)y' + 2xy = f(x)$, $y(2) = 1$, which is continuous at the point $x = 2$. Is this solution differentiable at $x = 2$? Explain your answer.

2.5 Solve the initial value problem $y'' = e^{-|x|}$, $y(0) = y'(0) = 0$. Note that the solution and its first derivative must be continuous at each point. Is the second derivative also continuous everywhere?

2.6 Given the initial value problem $y' + \dfrac{c}{x} y = x^\alpha$, $y(0) = 0$.

For which values of the constants α, c the problem has a single solution, no solution, or infinitely many solutions defined for $x \geq 0$?

2.7 The same question as above for the initial value condition $y(0) = 1$.

2.8 Find a solution of the initial value problem $y' + \dfrac{\ln^2 x}{\sin^2 x} y = 0$, $y(5) = 0$.

2.9

(a) Solve $\dfrac{dy}{dx} = \dfrac{y}{\sin y - 2x}$. Hint: A linear equation is hidden here.

(b) Solve $(2xy - e^{-2y})\dfrac{dy}{dx} + y = 0$.

2.10

(a) Find all the solutions of the equation $y' - \tan(x)y = 1$. Does the initial value problem $y(\pi/2) = 3$ have a solution? And of the problem $y(\pi/3) = 2$?
(b) The same question for the equation $y' + \tan(x)y = \cos^2(x)$.

Note: If $\mu(x) = |m(x)|$ is an integrating factor, so are $+m(x)$ and $-m(x)$.

2.11

(a) Find a solution of the equation $y' + 2y = x - 4$ which is tangent to the x axis at some point.
(b) Find a solution of the same equation that is tangent to the straight line $y = 5x$ at some point.

2.12 Given the initial value problem $y' - 2y = 2 + 5\sin x$, $y(0) = y_0$. For which value of y_0 the solution is bounded when $x \to +\infty$? (A function $y(x)$ is called bounded if there exists a constant M so that $|y(x)| \leq M$.)

2.13

(a) Given any three solutions $y_1(x)$, $y_2(x)$, $y_3(x)$ of the nonhomogeneous differential equation $y' + p(x)y = q(x)$. Prove that the quotient $\dfrac{y_3 - y_1}{y_2 - y_1}$ is constant.
(b) Show that if $y_1(x)$, $y_2(x)$ are two different solutions of $y' + p(x)y = q(x)$, then any solution of this equation can be written in the form

$$y(x) = y_1(x) + C(y_2(x) - y_1(x)).$$

2.14 Recall that a general solution of a linear, nonhomogeneous equation $y' + p(x)y = q(x)$ can be written as the sum of one particular solution plus all the solutions of the corresponding homogeneous equation, $y_{NH}(x) = y_p(x) + y_H(x)$.

(a) Solve the equation $y' - 2y = 3e^{4x}$ by carefully "guessing" one particular solution and adding to it all the solutions of the corresponding homogeneous equation.
(b) The same question for the equation $y' - 2y = 3\sin x$. Try to guess a particular solution of the shape $A \sin x + B \cos x$.
(c) The same question for the equation $y' - 2y = 3x^2 + 4x + 5$. What will be the form of a particular solution this time?

2.15 In the previous question, we had to guess a particular solution of equations from the form $y' - 2y = q(x)$ when $q(x)$ is an exponential, trigonometric, or

2.8 Families of Orthogonal Curves

polynomial function. Explain why the method works precisely in these cases. Can you suggest additional equations for which a particular solution can be obtained based on the same considerations?

2.16 Solve $yy' \sin x = (\sin x - y^2) \cos x$. Hint: Try to use a substitution.

2.17 Solve $y' = \dfrac{1}{x^3 y - xy}$.

2.18 Solve the equations

(a) $\dfrac{dy}{dx} = x(y^2 + y)$.

(b) $x \dfrac{dy}{dx} = y^2 + y$.

2.19 Solve the equation $y' - 2y = y^2 - 3$.

2.20 Let $\alpha \neq 0, 1$. The Bernoulli equation[2] is an equation of the form

$$y' + p(x)y = q(x)y^\alpha .$$

Divide the two sides of the equation by y^α and show that the substitution $z(x) = \bigl(y(x)\bigr)^{1-\alpha}$ transforms Bernoulli's equation into the linear equation

$$z'(x) + (1 - \alpha)p(x)z = (1 - \alpha)q(x) .$$

Why was it assumed that $\alpha \neq 0, 1$? In which cases the function $y(x) \equiv 0$ is a solution of the Bernoulli equation?

2.21 Solve $t^2 \dfrac{dy}{dt} + 2ty - y^3 = 0$.

2.22 Solve the equations

(a) $y' = \dfrac{2}{x}y + \dfrac{x}{y^2}$.

(b) $y' = \dfrac{y}{x} + \sqrt{y}$.

2.23 Solve the equation $y' - y = y^3$ by two different methods.

2.24 We are looking for a curve $y = y(x)$ such that its tangent lines in the first quadrant cut from the x-axis a segment of length $a > 0$ and from the y axis a segment of length $b > 0$ so that $a + b = 1$.
Find a differential equation for the unknown function $y(x)$. There is no need to solve the differential equation you found.

---

[2] Jacob (James) Bernoulli, 1654–1705.

Check and verify that $y(x) = x + 1 - 2\sqrt{x}$ is one of the solutions of the equation you found.

2.25 Find the curves $y = y(x)$ in the first quadrant for which a point (x, y) divides the segment of the tangent line between the x-axis and the y-axis into two parts of equal length.

2.26 A child walks to the right along the x-axis on the (x, y)-plane and drags behind him a toy tied to a thread of a fixed length k.

(a) Find the differential equation that describes the trajectory along which the toy moves.
(b) Assume that the child begins to move to the right from the point $(0, 0)$ and the toy is initially placed at the point $(0, k)$. What is the equation of the trajectory of movement? The equation may be of the form $y = y(x)$ or $x = x(y)$.
(c) An equivalent mathematical problem is formulated as: Find a curve so that the length of the tangent line from the tangency point to its intersection with the x-axis is a constant k. This curve is called *tractrix*. The problem was first proposed by Leibniz.

2.27

(a) Solve $y' = \dfrac{y + 8x}{x + 2y}$, $y(1) = 1$.
(b) Solve the equation with the initial value condition $y(1) = 2$.

2.28

(a) Solve $y' = \dfrac{2y - x}{2x - y}$, $y(2) = 3$.
(b) Explore and solve this differential equation with the initial value condition $y(0) = 0$ and explain what is going on there.

2.29 Find all solutions of the initial value problem $y' = \dfrac{y + x}{y - x}$, $y(0) = 0$.

2.30 Solve $y' = \dfrac{x^2 + xy + y^2}{x^2}$.

2.31 Prove that if $y(x)$ is a solution of an equation of a homogeneous type $y' = g\left(\dfrac{y}{x}\right)$, then for every $\alpha \neq 0$ also $z(x) = \dfrac{1}{\alpha} y(\alpha x)$ is a solution. What is the geometric significance of this feature?

2.32 For which pairs of numbers α, β the equation $y' = x^\alpha + y^\beta$ can be transformed into an equation of a homogeneous type by a change of variable of the form $y = u^m$?

2.33 Find a number α so that the substitution $y = z^\alpha$ transforms the equation $y' = \dfrac{x^2 + y^6}{6x^2 y^2}$ into an equation of a homogeneous type. Solve the given equation.

2.8 Families of Orthogonal Curves

2.34 Find a substitution $y = z^n$ with an appropriate value of n, which transforms the equation $y' = \dfrac{-8x^3 y}{x^4 + y^2}$ to an equation of a simpler shape. Solve the equation.

2.35

(a) Find a method to solve a differential equation of the form $y' = \dfrac{y}{x} f\left(\dfrac{y}{x^2}\right)$ by a suitable substitution.

(b) Solve the equation $y' = \dfrac{y}{x} \ln\left(\dfrac{y}{x^2}\right)$.

2.36

(a) Solve the equation $y' = \cos(y - x + 5)$, $y(2) = -3$ by substituting a new unknown $v = y - x + 5$.

(b) The same question as above for $y' = (2x + 3y + 4)^2$, $y(0) = 0$.

2.37 Verify that the following equations are exact and solve them:

(a) $(-cx + ay)\,dx + (ax + by)\,dy = 0$, a, b, c constants.
(b) $(2xy^3 - y^2)\,dx + (3x^2 y^2 - 2xy)\,dy = 0$.
(c) $(3z - 7y - 3)\,dy + (3y - 7z + 7)\,dz = 0$.

2.38 Find integrating factors for the following equations and solve them:

(a) $(x^2 + y^2 + 2x)\,dx + 2y\,dy = 0$.
(b) $(x\cos t - t\sin t)\,dt + (x\sin t + t\cos t)\,dx = 0$.
(c) $(x^3 y^2 - y)\,dx + (x^2 y^4 - x)\,dy = 0$, with an integrating factor of the from $x^m y^n$.

2.39 Find an integrating factor for the equation $y\sin 2x\,dx - (\sin^2 x - y^2)\,dy = 0$ and find all its solutions. Find explicitly the solution $y = y(x)$ which passes through the point $(x, y) = (0, 1)$ and the solution that passes through the point $(x, y) = (1, 0)$.

2.40 For the differential equation $y\,dx + (x + x^2 y^4)\,dy = 0$ find an integrating factor of the form $\mu = x^\alpha y^\beta$. Find all the solutions of the equation and find one that satisfies the initial value condition $y(3) = 0$.

2.41 Show that if a first-order differential equation can be solved by separation of variables, then it can be written also as an exact equation.

2.42

(a) Prove: If the equation $P(x, y)\,dx + Q(x, y)\,dy = 0$ (not necessarily exact) has a general solution from the form $F(x, y) = \text{const}$, then

$$\frac{F_x}{F_y} = \frac{P}{Q}.$$

(b) Show that if the equation $P\,dx + Q\,dy = 0$ above has an integrating factor $\mu(x, y)$, then the expression $\mu(x, y)F(x, y)$ is an integrating factor as well.

2.43 Solve the equation $2xyy' + x^2 - y^2 = 0$ using several methods. How many different methods did you manage to find?

2.44 Solve the equation $(3y^2 - 2xy)\,dx + (-x^2 + 6xy)\,dy = 0$ by as many various methods as possible.

2.45 The same question for the equation $y' = -\dfrac{xy^3}{1 + 2x^2y^2}$.

2.46 Solve the equation $x^2 y\,dx - (x^3 + y^3)\,dy = 0$ by three different methods.

2.47 Solve the initial value problem $y''(t) = \dfrac{2}{y^3} - \dfrac{1}{y}$, $y(0) = 2$, $y'(0) = 1$ by the method described in Sect. 2.6.

2.48

(a) Draw the direction field of the equation $y' = (y-1)(y-2)(y-3)$.
(b) Prove that if $y(x)$ is a solution, then $y(x+c)$, c constant, is a solution as well. What is the geometric significance of this property?
(c) Show that the isoclines corresponding to this equation are the horizontal straight lines.
(d) How does the solution of the initial value problem $y(0) = y_0$, $1 < y_0 < 2$, behave when $x \to +\infty$? And how if $2 < y_0 < 3$?

2.49 The same question for the equation $y' = (y-1)(y-2)^2$.

2.50 The same question for the equation $y' = (y-1)(y-2)^3$. What is the difference between the graphs of the solutions of this equation and the graphs of the previous one?

2.51 Draw the direction field of the equation $y' = \dfrac{\cos(y)}{x\sin(y)}$ and find all its solutions.

2.52

(a) Draw the direction field of the equation $y' = |y-x|$ and an approximate sketch of some typical solutions.
(b) Calculate the solution $y(x)$ of the initial value problem $y(0) = 1/2$ for all $-\infty < x < \infty$.

2.53 Draw the direction field of the equation $y' = |x| - y$. Calculate the solution of the initial value problem $y(0) = 1$ for all $-\infty < x < \infty$ and draw it on the direction field. Describe in detail the behavior of this solution when $x \to \pm\infty$.

2.54 Find a family of curves that are orthogonal to the family of circles:

$$x^2 + (y+c)^2 = c^2 - 1, \qquad |c| > 1.$$

2.55 Find a family of curves that are orthogonal to the family $y^2 - 2x^2 = kx^3$.

2.8 Families of Orthogonal Curves

Find the curve of the new family that passes through (3, 1).

2.56 The same question for the family $2x^2 + y^2 = cx$ and the initial point (1, 1).

2.57 The same question for the family of parabolas $y^2 = 2(x - c)$.

2.58

(a) Draw the family of curves $\sqrt{x^2 + y^2} - y = c$, $c > 0$. What are these curves?
(b) Find a first-order differential equation that corresponds to this family of curves.
(c) Find the orthogonal family of curves.

2.59

(a) Find a first-order differential equation that corresponds to the family of parabolas $y^2 = 4c(x + c)$, $-\infty < c < \infty$.
(b) In the differential equation that you found, replace each appearance of y' by $(-1/y')$ and check that the differential equation remains unchanged. It seems that the family of our parabolas is orthogonal to itself! Explain how is this possible.
Hint: Draw several parabolas from the family that correspond to various values of c.

2.60 Find the family of curves that intersect the family of circles $x^2 + y^2 = c$ at an angle $45°$.
Hint: The angle α between the curves $y = y_1(x)$, $y = y_2(x)$ is given by the formula
$$\tan \alpha = \frac{y_1'(x) - y_2'(x)}{1 + y_1'(x) y_2'(x)}.$$
Note that the angle can be measured in two opposite directions.

Chapter 3
Existence and Uniqueness Theorems

3.1 Why Existence and Uniqueness?

In this chapter, we will discuss initial value problems of the form

$$y' = f(x, y), \quad y(x_0) = y_0, \tag{3.1}$$

and try to answer a few questions:

1. Does there exist a solution of the initial value problem?
   It should be emphasized that the question is not whether we can find the solution explicitly but its mere existence or nonexistence.
2. Is it possible that the initial value problem has more than one solution?
3. If there exists a solution of the problem, what is its domain of definition?
   That is, for which values of x the solution $y(x)$ is defined and meets all requirements.

We emphasize that all questions relate exclusively to initial value problems (which consist of a differential equation plus an initial value condition). For a differential equation alone, without an initial value condition, there is no use to ask about uniqueness, since usually there are infinitely many solutions that depend on a parameter.

Let us start with a few examples that aim to illustrate that the above questions are essential and the answers are not at all self-evident. The first two examples are about the question of the existence of solutions and uniqueness of solution.

Example 3.1 In Example 2.1 we discussed the initial value problem

$$y' + \frac{2}{x} y = x^5, \quad y(0) = 7,$$

which is of the form (3.1) with $f(x, y) = -2y/x + x^5$ and saw that it has no solution.

Example 3.2 The initial value problem

$$y' - \frac{2}{x} y = x^5, \quad y(0) = 0,$$

is also of the form (3.1), this time with $f(x, y) = 2y/x + x^5$. It was considered in Example 2.2 and we saw that it has an infinite number of different solutions.

The following two examples discuss the question of the scope of the definition of solutions.

Example 3.3

$$y' = -\frac{x}{y}, \quad y(1) = 2. \tag{3.2}$$

We have already encountered this equation in (2.36) in the context of its direction field. In this equation, the right-hand side $f(x, y) = -x/y$ is defined for every $-\infty < x < \infty$ and for $y > 0$, that is, in the upper half-plane, and $f(x, y)$ is continuous and even differentiable there. $f(x, y)$ is defined also in the lower half-plane $y < 0$, but it is impossible to connect the initial point $(x_0, y_0) = (1, 2)$ to the lower half-plane without crossing the line $y = 0$, where the equation is not defined. Therefore, we will deal with the initial value problem only in the upper half-plane:

$$D = \{-\infty < x < \infty, \quad y > 0\}.$$

The equation is immediately solved by separation of variables: $\int y \, dy = -\int x \, dx$, i.e., $x^2 + y^2 = \text{const}$. The initial value condition $y(1) = 2$ implies $x^2 + y^2 = 5$ and the only solution $y(x)$ of the initial value problem is

$$y(x) = +\sqrt{5 - x^2}.$$

It is a solution of the differential equation only for $-\sqrt{5} < x < \sqrt{5}$. At the endpoints $x = \pm\sqrt{5}$ the function $y(x)$ has no derivative, and therefore, it cannot be considered there as a solution of a differential equation.

We witness a surprising phenomenon: Although the differential equation (3.2) is defined for all x and all positive y and the right-hand side is continuous and differentiable for these values, the solution of the initial value problem is defined only for $-\sqrt{5} < x < \sqrt{5}$ and ceases to exist at $x = -\sqrt{5}, \sqrt{5}$. This is even though in Eq. (3.2) there is no visible sign that something unusual may happen precisely for $x = \sqrt{5}$ and not for other x. If we append to the same differential equation other initial value conditions, we get solutions that are defined in other intervals, each in a certain neighborhood of its initial point.

3.1 Why Existence and Uniqueness?

Example 3.3 demonstrates that the *domain of definition* of a solution cannot be taken for granted and it is not simply derived from the *domain of definition of the equation*. This is a new phenomenon: We obtained a *local solution* which is defined only in some interval containing the initial point x_0. These features must be explained.

Example 3.4 Consider the initial value problem

$$y' = y^2, \qquad y(2) = 1.$$

By separating variables we get $\int dy/y^2 = \int dx$, i.e., $-1/y = x + C$. The initial value conditions $y(2) = 1$ require that $C = -3$ and we get the function $y = \dfrac{1}{3-x}$.

Where is this function a solution to the problem?

For each $-\infty < x < 3$ it is a differentiable function that satisfies the differential equation and the initial value conditions at $x = 2$, so it is certainly a solution in the interval $(-\infty, 3)$. As one approaches $x = 3$ from the left, $\lim_{x \to 3^-} y(x) = +\infty$, that is, this function has a vertical asymptote.

We again encounter the same surprising phenomenon: The differential equation has its right-hand side $f(x, y) = y^2$ defined, continuous, and differentiable of any order for every x, y (x does not appear explicitly in the equation!). Nevertheless, a solution ceases to exist at $x = 3$ which is an unremarkable value in context of the equation.

In $(3, \infty)$ the function $y = 1/(3 - x)$ continues to satisfy the same differential equation. However, checking the initial value conditions $y(2) = 1$ is meaningless, because the point $x = 2$ is not in this interval at all. Therefore, $y = 1/(3 - x)$ is clearly not a solution of the initial value problem in $(3, \infty)$. See Fig. 3.1.

And what about the set $(-\infty, 3) \cup (3, \infty)$, that is, all the x-axis except $x = 3$? Here the differential equation holds, and the initial value condition is met, but how does the solution pass from $(-\infty, 3)$ into $(3, \infty)$? Here we have to make a decision what we consider as a solution. Since a differential equation usually describes a dynamic process that starts from a certain point and progresses from it as much as possible, it is unlikely that a solution that "explodes" on the left of $x = 3$ and $\lim_{x \to 3^-} y(x) = +\infty$ reappears on the right of $x = 3$ with $\lim_{x \to 3^+} y(x) = -\infty$.

Fig. 3.1 The solution of $y' = y^2$, $y(2) = 1$

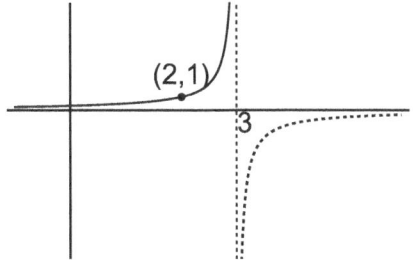

Therefore, we limit the domain of definition of a solution to be a connected set, i.e., an interval of x-axis (open or closed, finite or infinite), that contains the initial point. In our case, we take the domain of definition of the solution as the interval $(-\infty, 3)$.

The idea is summarized in the following definition.

Definition 3.1 The *domain of definition* of a solution $y(x)$ of the initial value problem (3.1) is the maximal interval of x values that contains the initial point x_0 and in which $y(x)$ is differentiable and satisfies the differential equation and the initial value condition.

Exercise 3.1 Each solution has its own unique domain of definition. Find the domain of definition of the initial value problem $y' = y^2$, $y(3) = 2$ and compare it to that of Example 3.4.

After this introduction we are ready to formulate a theorem about the existence and uniqueness of initial value problems.

Theorem 3.1 *Given the initial value problem (3.1),*

$$y' = f(x, y), \quad y(x_0) = y_0,$$

where the function $f(x, y)$ and its partial derivative according to y, $\dfrac{\partial f(x, y)}{\partial y}$, are continuous in a two-dimensional domain D in the (x, y) plane and the point (x_0, y_0) is in the interior of D. Then there is a neighborhood $[x_0 - h, x_0 + h]$ of the point x_0 on which the initial value problem has a solution. This solution is unique.

Note that while in the theorem it is assumed that the partial derivative $\partial f / \partial y$ is continuous, there is no assumption about the partial derivative $\partial f / \partial x$. This difference should not surprise us since the roles of the unknown y and the independent variable x are completely different.

This is a version of the existence and uniqueness theorem of Picard-Lindelöf.[1] It deals with the question of existence and the question of uniqueness at the same time (Fig. 3.2).

Fig. 3.2 An initial point and a local solution in its neighborhood

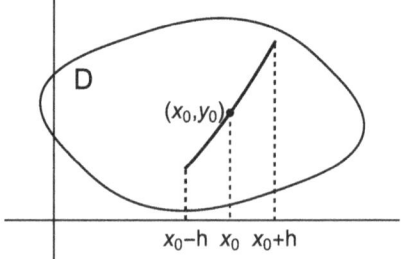

---

[1] Émile Picard, 1856–1941; Ernst Lindelöf, 1870–1946.

3.1 Why Existence and Uniqueness?

We highlight two essential features of the theorem:

1. **(Local existence)** The results of Theorem 3.1 are local and they ensure the existence of a solution only in a certain neighborhood of the initial point. We do not claim that the neighborhood guaranteed by the theorem is the maximal domain of definition of the solution. It is just an estimate of the domain of definition, which holds for many equations. When we discuss one specific equation and a specific solution, the domain of definition of the solution will likely be greater than what is promised by the theorem. For instance, for the initial value problem $y' = y^2$, $y(2) = 1$ in Example 3.4, the theorem ensures the existence of a solution in some finite interval around $x = 2$ while in practice the solution is defined in $(-\infty, 3)$.

   Note that it is important to distinguish between the domain of definition of the equation (which is a two-dimensional domain D in the (x, y) plane) and the domain of definition of a solution (which is an interval of the x-axis).

2. **(Sufficient conditions)** The assumptions of Theorem 3.1 are sufficient conditions for the existence and uniqueness of a solution, but they are not always necessary. That is to say, it is possible that the conditions of the theorem are not met, and yet a certain problem has a solution and it is unique. Therefore, when the assumptions of the theorem are not met, all we can say is that we do not know whether there exists a solution neither we know if there is one solution or many.

Theorem 3.1 (Picard-Lindelöf's existence and uniqueness theorem) is but one of several results in the literature that discuss the existence of solutions and their uniqueness. Other theorems prove, with certain assumptions, only the existence of a solution, and still other theorems deal only with the question of uniqueness for initial value problems. Among other theorems, we mention the existence theorem of Cauchy: *"If the function $f(x, y)$ is continuous in the domain D and (x_0, y_0) is an internal point of D, then initial value problem $y' = f(x, y)$, $y(x_0) = y_0$ has a solution $y(x)$ defined in a certain neighborhood of the point x_0."* The assumptions of Cauchy's theorem are more simple than those of Theorem 3.1, but on the other hand, it does not discuss at all the question of solution uniqueness. In this context, we will not elaborate on other existence theorems and uniqueness theorems.

A complete proof of Theorem 3.1 appears in Sects. 3.3 and 3.4. Here we only bring some notes and examples that explain the concepts of the theorem and its conclusions.

In Example 3.1 we have seen that the initial value problem

$$y' + \frac{2}{x} y = x^5, \quad y(0) = 7,$$

has no solution. In this equation $f(x, y) = -2y/x + x^5$ and the function $f(x, y)$ is discontinuous at the initial point $(x_0, y_0) = (0, 7)$. Here the assumptions of the existence and uniqueness theorem do not hold in any neighborhood of the given initial point.

In Example 3.2, we saw that the initial value problem

$$y' - \frac{2}{x} y = x^5, \quad y(0) = 0,$$

has infinitely many different solutions. Again, the function $f(x, y) = 2y/x + x^5$ is discontinuous at the initial point $(x_0, y_0) = (0, 0)$, and the assumptions of the theorem do not hold in any neighborhood of this point.

Exercise 3.2 What does the existence and uniqueness theorem say about the initial value problem

$$\frac{dy}{dx} = \frac{y}{x}, \quad y(0) = 0 \;?$$

Solve the problem also directly and compare it to Example 2.15.

Remark (Discontinuity) It is not a coincidence that in the recent three counterexamples the discontinuous $f(x, y)$ behaves "wildly" at the initial point. In Problem 2.4, for example, where the discontinuity of $f(x, y)$ is a jump, there exists a solution that is differentiable everywhere except at the jump point.

Example 3.5

$$y' = y^{1/3}, \quad y(0) = 0.$$

In this problem the function $f(x, y) = y^{1/3}$ is continuous for all x and for all y in the plane R^2 (x is not mentioned at all in f). But its derivative with respect to y,

$$\frac{\partial f(x, y)}{\partial y} = \frac{1}{3y^{2/3}},$$

is not continuous at $y = 0$. That is why the assumptions of the existence and uniqueness theorem do not hold in any neighborhood of the point $(0, 0)$ nor in any neighborhood of points of the form $(x_0, 0)$. Therefore, we cannot deduce anything about this problem from Theorem 3.1.

We will solve the problem explicitly. If $y \neq 0$, we separate the variables as $\int y^{-1/3} dy = \int dx$, that is, $\frac{3}{2} y^{2/3} = x - c$. A general solution is

$$y = \pm \left[\frac{2}{3}(x - c) \right]^{3/2}.$$

To get $y(0) = 0$ we select $c = 0$ and get two solutions

$$y(x) = \pm \left(\frac{2}{3} x \right)^{3/2}, \quad x \geq 0.$$

In addition to these solutions there exists a singular solution $y(x) \equiv 0$.

3.1 Why Existence and Uniqueness?

Fig. 3.3 Several solutions of the initial value problem

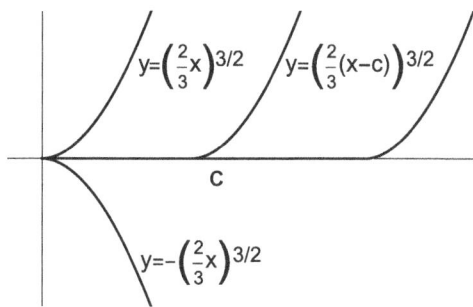

In fact, there are many more solutions. For any $c \geq 0$ we define

$$y(x) = \begin{cases} 0, & -\infty < x \leq c, \\ \left(\dfrac{2}{3}(x-c)\right)^{3/2}, & c \leq x < \infty. \end{cases}$$

See Fig. 3.3. Any such function is differentiable at any point, including at the point $x = c$, since the one-sided right and left derivatives at c are both zero. All functions satisfy the differential equation (check separately in each interval!) and the initial value condition. Thus, we have found infinitely many different solutions of this problem.

Example 3.6 Example 3.5 shows that when $\partial f / \partial y$ is discontinuous, the solution may be nonunique. But is it true that whenever $\partial f / \partial y$ is discontinuous, nonuniqueness must occur? It is not necessarily so.

Let us look, for example, at the problem:

$$y' = y^{1/3} + 1, \qquad y(0) = 0.$$

In this equation the function $f(x, y) = y^{1/3} + 1$ is continuous and $\partial f / \partial y$ is discontinuous at $y = 0$. By separation of variables

$$\int_0^y \frac{dy}{y^{1/3} + 1} = \int_0^x dx.$$

The denominator is not zero close to $y = 0$; therefore, no singular solution passes through $(x, y) = (0, 0)$. The integral on the left-hand side exists (put $y = z^3$ and calculate it!) and we get an implicit function of the form $G(y) = x$. Now we proceed according to the inverse function theorem. $G'(y) = 1/\left(y^{1/3} + 1\right)$, $G'(0) = 1 > 0$, so $G(y)$ is strictly increasing on some neighborhood of $(0, 0)$. According to the inverse function theorem, there exists a unique solution $y = G^{-1}(x)$ and it has a continuous derivative. All this happens even though the conditions of the existence and uniqueness theorem do not hold. Thus, the assumptions of Theorem 3.1 are sufficient but not necessary.

3.2 Outline of the Proof of Existence Theorem

In this section, we do not prove the existence and uniqueness theorem quoted above but only hint at the main ideas of the proof. A full proof is given in the following two sections.

Suppose that $y(x)$ is a solution of the initial value problem (3.1) and integrate $y'(x) = f(x, y(x))$ on the interval $[x_0, x]$. x marks the upper limit of the integral and we do not like to use it also as the variable of the integration.[2] So we choose another letter s to mark the variable of integration :

$$\int_{x_0}^{x} y'(s)\, ds = \int_{x_0}^{x} f(s, y(s))\, ds \ .$$

But $\int_{x_0}^{x} y'(s)\, ds = y(x) - y(x_0) = y(x) - y_0$; therefore,

$$y(x) = y_0 + \int_{x_0}^{x} f(s, y(s))\, ds \ . \tag{3.3}$$

Thus, the differential equation and the given initial value conditions are replaced by an *integral equation*. This integral equation automatically takes care of the initial value conditions, since the choice of $x = x_0$ as the upper limit of integration results in $y(x_0) = y_0 + 0$.

The integral equation is solved by the method of *successive iterations*. As a first step, we take some continuous function $y_0(x)$ and place it into the right-hand side (only!) of the integral equation (3.3). The resulting expression is a new function that will be denoted as $y_1(x)$:

$$y_1(x) = y_0 + \int_{x_0}^{x} f(s, y_0(s))\, ds \ .$$

Since the functions $f(x, y)$ and $y_0(x)$ are continuous, the integral on the right-hand side is differentiable, and therefore, the new function $y_1(x)$ is continuous and differentiable. $y_1(x)$ even satisfies the initial value condition $y_1(x_0) = y_0$.

Let us repeat this process: We put the $y_1(x)$ into the right-hand side of the integral equation (3.3) and name the resulting expression $y_2(x)$:

$$y_2(x) = y_0 + \int_{x_0}^{x} f(s, y_1(s))\, ds \ .$$

---

[2] Recall that in integral calculus, the value of a definite integral does not depend on the choice of the variable of integration. $\int_a^b F(x)\, dx$ and $\int_a^b F(t)\, dt$ are identical and their value does not depend on either "dummy variable" x or t but only on the lower and upper limits.

3.2 Outline of the Proof of Existence Theorem

The function $y_2(x)$ is also differentiable because $f(x, y)$ and $f(s, y_1(s))$ are continuous. $y_2(x)$ also satisfies the initial value condition. We repeat this process and define more iterations, step by step. At the nth step we define

$$y_{n+1}(x) = y_0 + \int_{x_0}^{x} f(s, y_n(s))\, ds \,. \tag{3.4}$$

Thus, we get a sequence of functions $y_0(x), y_1(x), y_2(x), \ldots$. The main part of the proof of the theorem is to show that when $n \to \infty$, this sequence of functions converges uniformly in a certain interval around x_0 to some limit, $\lim_{n\to\infty} y_n(x) = y(x)$ and the two sides of (3.4) converge to

$$y(x) = y_0 + \int_{x_0}^{x} f(s, y(s))\, ds \,.$$

But this is exactly the integral equation (3.3) that is equivalent to the differential equation and the given initial value conditions. The proof of the convergence is the complicated part of the proof (for example, why is there convergence only in a certain interval $[x_0 - h, x_0 + h]$ and not for all x?). Complete proofs of the existence and of the uniqueness appear in Sects. 3.3 and 3.4, respectively.

Example 3.7 Let us calculate the successive iterations for the initial value problem

$$y' = y + x, \quad y(0) = 0\,.$$

With the choice of $y_0(x) \equiv 0$, the first iteration is

$$y_1(x) = 0 + \int_0^x (y_0(s) + s)\, ds = \int_0^x s\, ds = x^2/2\,.$$

The following iterations are

$$y_2(x) = 0 + \int_0^x (y_1(s) + s)\, ds = \int_0^x \left(\frac{s^2}{2} + s\right) ds = \frac{x^3}{3 \cdot 2} + \frac{x^2}{2},$$

$$y_3(x) = 0 + \int_0^x (y_2(s) + s)\, ds = \int_0^x \left(\frac{s^3}{3 \cdot 2} + \frac{s^2}{2} + s\right) ds$$

$$= \frac{x^4}{4 \cdot 3 \cdot 2} + \frac{x^3}{3 \cdot 2} + \frac{x^2}{2},$$

and so, the nth iteration is $y_n(x) = \sum_{i=2}^{n+1} x^i/i!$ and $y(x) = \lim_{n\to\infty} y_n(x) = \sum_{i=2}^{\infty} x^i/i!$.
But we know that $e^x = \sum_{i=0}^{\infty} x^i/i!$ for every x; therefore, $y(x) = e^x - (x + 1)$.

Compare this limit to the solution that is obtained by solving the linear nonhomogeneous equation directly.

Why do we iterate an integral equation and not a differential equation? Applying iteration to an integral equation is a reasonable strategy because integration smooths out functions while differentiation reduces their smoothness. If, for example, $f(x, y)$, $y(x)$ are continuous functions, then the integral $\int_{x_0}^{x} f(s, y(s))\,ds$ is already a differentiable function of x. Integration also improves convergence while differentiation reduces it. For example, the sequence of functions $f_n(x) = \sin(n^2 x)/n$ converges uniformly to $f(x) \equiv 0$ as $n \to \infty$ and the integrals $F_n(x) = \int_0^x \sin(n^2 s)/n\,ds = (1 - \cos(n^2 x))/n^3$ converge to $F(x) \equiv 0$ even faster. On the other hand, the sequence of derivatives $f_n'(x) = n\cos(n^2 x)$ does not converge at all. Another advantage of the integral equation is that it automatically takes care of the initial value condition.

3.3 Proof of the Existence of the Solution

The proof of the existence and uniqueness theorem is often not included in basic differential equations courses because of its length. In this section, we will prove the existence of a solution of the initial value problem (3.1) under the assumptions of Theorem 3.1, and in the next section we will prove its uniqueness.

We saw that the initial value problem (3.1) is equivalent to the integral equation (3.3). Our goal is to construct a solution to this integral equation as a limit of a sequence of iterations which converge in a certain neighborhood of the initial point.

Around the given initial point (x_0, y_0) we select a rectangle

$$B = \{(x, y) \mid x_0 - a \leq x \leq x_0 + a,\ y_0 - b \leq y \leq y_0 + b\}$$

whose side lengths are $2a$, $2b$, respectively, and which is contained in the domain D. According to the assumption of Theorem 3.1, the function $|f(x, y)|$ is continuous in the rectangle B; hence, it attains its maximal value there. Let

$$M = \max_{B} |f(x, y)|.$$

Since

$$|y'| = |f(x, y)| \leq M$$

in B, the slope of the solution graph in rectangle B is bounded between M and $(-M)$. In particular, the slope of a solution graph passing through the initial point (x_0, y_0) lies between the two straight lines:

$$y - y_0 = \pm M(x - x_0).$$

3.3 Proof of the Existence of the Solution 57

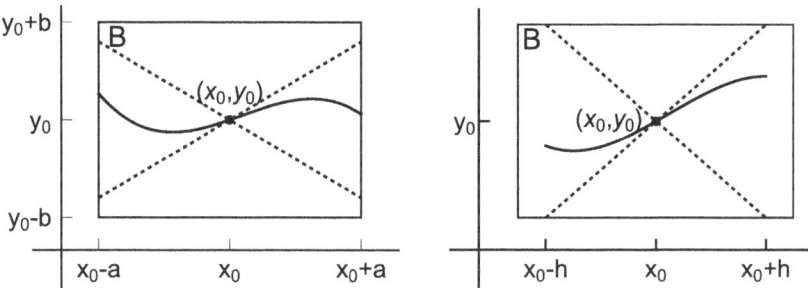

Fig. 3.4 The rectangle B, the restricting lines, and a graph of a solution

We draw these two restricting straight lines on the rectangle B. If $M \geq b/a$, the restricting lines meet the upper side and the lower side of B at the points $x = x_0 \pm b/M$ (See the right-hand side drawing in Fig. 3.4 with $h = b/M$.) When $M \leq b/a$, the restricting lines meet the right and left sides of the rectangle B at $x = x_0 \pm a$ (the left drawing in Fig. 3.4).

Since we do not know in advance which of the two cases will occur, we mark the smaller of the two horizontal distances by h:

$$h = \min\left\{a, \frac{b}{M}\right\}.$$

Our goal is to prove the existence of a solution of the integral equation in the interval $x_0 - h \leq x \leq x_0 + h$.

Step 1 As hinted in the previous section, we define a sequence of iterations. The first term may be some arbitrary continuous function $y_0(x)$ defined in the interval $[x_0 - h, x_0 + h]$ and whose graph is contained in the rectangle B, namely, $y_0 - b \leq y_0(x) \leq y_0 + b$. The consecutive iterations are defined by (3.4):

$$y_{n+1}(x) = y_0 + \int_{x_0}^{x} f(s, y_n(s))\,ds, \qquad n = 0, 1, 2, \ldots .$$

First, we show that for $x_0 - h \leq x \leq x_0 + h$, the graphs of all iterations $y_n(x)$ are located in the rectangle B, that is, $|y_n(x) - y_0| \leq b$. For convenience, we will deal first with the right half of the interval, $x_0 \leq x \leq x_0 + h$. The left half may be discussed later similarly.

$y_0(x)$ satisfies the claim according to its definition. The next approximation $y_1(x)$ is defined by

$$y_1(x) = y_0 + \int_{x_0}^{x} f(s, y_0(s))\,ds .$$

Therefore, thanks to $x_0 \leq s \leq x$,

$$|y_1(x) - y_0| = \left| \int_{x_0}^{x} f(s, y_0(s)) \, ds \right| \leq \int_{x_0}^{x} |f(s, y_0(s))| \, ds \, .$$

Since for $x \leq x_0 + h$ the graph of the function $y_0(x)$ is located in the rectangle B, $|f(s, y_0(s))| \leq \max_B |f(x, y)| \equiv M$, and in the interval of integration $x_0 \leq s \leq x \leq x_0 + h$,

$$|y_1(x) - y_0| \leq \int_{x_0}^{x} M \, ds = M(x - x_0) \leq Mh \leq b \, . \tag{3.5}$$

Hence, for $x_0 \leq x \leq x_0 + h$ we have $|y_1(x) - y_0| \leq b$ and the graph of $y_1(x)$ stays in the rectangle B.

The treatment of subsequent iterations is similar. From

$$y_2(x) = y_0 + \int_{x_0}^{x} f(s, y_1(s)) \, ds \, ,$$

it follows that

$$|y_2(x) - y_0| \leq \int_{x_0}^{x} |f(s, y_1(s))| \, ds \, .$$

But we have just proved that the graph of $y_1(x)$ lies in the rectangle B; therefore, $|f(s, y_1(s))| \leq M$ for $x_0 \leq s \leq x \leq x_0 + h$. Now the proof continues as above and the same reasoning applies to all subsequent iterations.

Step 2 We will estimate the sequence $\{y_n(x)\}$ more precisely. The first two iterations are

$$y_1(x) = y_0 + \int_{x_0}^{x} f(s, y_0(s)) \, ds \, ,$$

$$y_2(x) = y_0 + \int_{x_0}^{x} f(s, y_1(s)) \, ds,$$

and for $x_0 \leq x$, their difference is

$$|y_2(x) - y_1(x)| \leq \left| \int_{x_0}^{x} \left(f(s, y_1(s)) - f(s, y_0(s)) \right) ds \right|$$

$$\leq \int_{x_0}^{x} |f(s, y_1(s)) - f(s, y_0(s))| \, ds \, . \tag{3.6}$$

3.3 Proof of the Existence of the Solution

According to the intermediate value theorem applied to the variable y (while x is hold fixed),

$$f(x, y_1) - f(x, y_0) = \frac{\partial f(x, \tilde{y})}{\partial y} (y_1 - y_0) , \quad (3.7)$$

where $\tilde{y}$ is a certain intermediate value between y_0 and y_1. The point $(x, \tilde{y})$ is in the rectangle B. By the assumptions of the theorem, the function $\frac{\partial f}{\partial y}$ is continuous in the closed rectangle B and therefore it is bounded there:

$$\left| \frac{\partial f(x, y)}{\partial y} \right| \leq K \quad \text{for all } (x, y) \text{ in } B . \quad (3.8)$$

In particular $|\partial f(x, \tilde{y})/\partial y| \leq K$ and it follows from (3.7) that

$$|f(x, y_1) - f(x, y_0)| \leq K|y_1 - y_0| .$$

With this estimate inequality (3.6) becomes

$$|y_2(x) - y_1(x)| \leq \int_{x_0}^{x} K|y_1(s) - y_0(s)| \, ds . \quad (3.9)$$

As the function $y_0(x)$ is quite arbitrary, there is no good estimate for the difference $y_1(x) - y_0(x)$. We only know from Step 1 that the graphs of both are in the rectangle B whose height is $2b$, so $|y_1(x) - y_0(x)| \leq 2b$. Therefore, inequality (3.9) becomes

$$|y_2(x) - y_1(x)| \leq \int_{x_0}^{x} K \cdot 2b \, ds = 2bK(x - x_0) . \quad (3.10)$$

We repeat similar estimates for the following successive pair of iterations. From the subtraction of

$$y_2(x) = y_0 + \int_{x_0}^{x} f(s, y_1(s)) \, ds ,$$

$$y_3(x) = y_0 + \int_{x_0}^{x} f(s, y_2(s)) \, ds ,$$

we get as above that

$$|y_3(x) - y_2(x)| \leq \int_{x_0}^{x} |f(s, y_2(s)) - f(s, y_1(s))| \, ds \leq \int_{x_0}^{x} K|y_2(s) - y_1(s)| \, ds .$$

According to (3.10), $|y_2(x) - y_1(x)| \leq 2bK(x - x_0)$; therefore, it follows that

$$|y_3(x) - y_2(x)| \leq \int_{x_0}^{x} K \cdot 2bK(s - x_0)\, ds = 2bK^2 \frac{(s-x_0)^2}{2}\bigg|_{s=x_0}^{x} = 2bK^2 \frac{(x-x_0)^2}{2}.$$

Repeating the same method, we obtain after n integrations that

$$|y_{n+1}(x) - y_n(x)| \leq 2bK^n \frac{(x - x_0)^n}{n!}. \tag{3.11}$$

These inequalities depend on the fact that the graphs of the iterates do not leave the rectangle B. As proved in Step 1, this is indeed the case as long as $x_0 \leq x \leq x_0 + h$, $h = \min\{a, b/M\}$.

The left-hand side of the interval, $x_0 - h \leq x \leq x_0$, is treated similarly. But this time, because of the reversed order of the points, we have to change the limits of integration and write

$$\left|\int_{x_0}^{x} f\, ds\right| \leq \int_{x}^{x_0} |f|\, ds.$$

Step 3 At this stage, we will prove the uniform convergence of the sequence of functions $\{y_n(x)\}_{n=0}^{\infty}$ on the interval $|x - x_0| \leq h$.

Due to the identity

$$y_n(x) = y_0(x) + (y_1(x) - y_0(x)) + \cdots + (y_n(x) - y_{n-1}(x)), \tag{3.12}$$

a question about the sequence of functions $\{y_n(x)\}_{n=0}^{\infty}$ is translated into a question about the series of functions $\sum [y_n(x) - y_{n-1}(x)]$. The advantage of replacing a sequence by a series is that for a series of functions, there is at our disposal the Weierstrass[3] test for uniform convergence:

"*If $|g_n(x)| \leq c_n$ in the interval $[a, b]$ and the series of nonnegative constants $\sum_{n=0}^{\infty} c_n$ converges, then the series of functions $\sum_{n=0}^{\infty} g_n(x)$ converges uniformly in $[a, b]$.*"

It was proved in (3.11) that

$$|y_{n+1}(x) - y_n(x)| \leq 2b\, K^n \frac{|x - x_0|^n}{n!} \leq 2b \frac{(Kh)^n}{n!}.$$

The series of constants $\sum_{n=0}^{\infty} (Kh)^n/n!$ converges for all K, h (being the exponential series or with the help of the quotient test); therefore, the series of functions $\sum [y_n(x) - y_{n-1}(x)]$ converges absolutely and uniformly in the interval $|x - x_0| \leq$

---

[3] Karl Weierstrass, 1815–1897.

h. Hence, thanks to the identity (3.12), also the sequence of functions $\{y_n(x)\}_{n=0}^{\infty}$ converges uniformly.

Step 4 In Step 3 we proved that the sequence of iterations $\{y_n(x)\}$ converges uniformly for $|x - x_0| \leq h$. Let us denote the limit function by $y(x)$,

$$\lim_{n \to \infty} y_n(x) = y(x) , \quad |x - x_0| \leq h .$$

The functions $\{y_n(x)\}$ are continuous and, as well known from the differential calculus, the limit function $y(x)$ is continuous as well in the interval $|x - x_0| \leq h$. Now we pass to the limit in the definition of the iterations, Eq. (3.4):

$$y_{n+1}(x) = y_0 + \int_{x_0}^{x} f(s, y_n(s)) \, ds .$$

According to integral calculus, in the case of uniform convergence, "the limit of integrals is equal to the integral of the limit"; thus, the limit function $y(x)$ satisfies the integral equation (3.3):

$$y(x) = y_0 + \int_{x_0}^{x} f(s, y(s)) \, ds .$$

$y(x)$, $f(x, y)$, and $f(x, y(x))$ are continuous functions; hence, the integral on the right-hand side is a function with a continuous derivative. So also $y(x)$ on the left-hand side has a continuous derivative. We can therefore differentiate the integral equation (3.3) and deduce that $y(x)$ satisfies the original differential equation $y' = f(x, y)$. By placing the initial point $x = x_0$ in (3.3) it turns out that $y(x)$ also fulfills the initial value condition $y(x_0) = y_0$. This completes the proof of the existence of a solution of the initial value problem (3.1) in the interval $|x - x_0| \leq h$.

3.4 Proof of the Uniqueness of the Solution

The proof of the uniqueness of a solution is shorter than the proof of its existence. Suppose, on the contrary, that the initial value problem $y' = f(x, y)$, $y(x_0) = y_0$ has two solutions $u(x)$, $v(x)$ that are defined in a certain interval $x_0 \leq x \leq x_0 + a$. They both satisfy the same integral equation:

$$u(x) = y_0 + \int_{x_0}^{x} f(s, u(s)) \, ds ,$$

$$v(x) = y_0 + \int_{x_0}^{x} f(s, v(s)) \, ds .$$

Our goal is to prove that $u(x) \equiv v(x)$. Subtract the equations from each other:

$$u(x) - v(x) = \int_{x_0}^{x} \left(f(s, u(s)) - f(s, v(s)) \right) ds .$$

We repeat some of the ideas that were used in Eqs. (3.6), (3.7), and (3.8). According to the intermediate value theorem there is a value $\tilde{y}$ between $u(x)$ and $v(x)$ so that $f(x, u(x)) - f(x, v(x)) = \dfrac{\partial f(x, \tilde{y})}{\partial y}(u(x) - v(x))$ and let, as in Eq. (3.8), $K = \max\limits_{(x,y)\in B} \left| \dfrac{\partial f(x, y)}{\partial y} \right|$. Thus,

$$0 \leq |u(x) - v(x)| \leq \int_{x_0}^{x} \left| \frac{\partial f(s, \tilde{y})}{\partial y} \right| |u(s) - v(s)| \, ds$$
$$\leq \int_{x_0}^{x} K |u(s) - v(s)| \, ds , \quad x_0 \leq x \leq x_0 + a . \quad (3.13)$$

We denote $w(x) = \int_{x_0}^{x} |u(s) - v(s)| \, ds$. So $w'(x) = |u(x) - v(x)|$ and, of course, $w(x_0) = 0$. Inequality (3.13) becomes

$$0 \leq w'(x) \leq K w(x) .$$

At first sight, it was expected here to divide the two sides by $w(x)$, as we did in the separation of variables, but this cannot be done since we may divide by 0. (Do not forget that our goal is to prove that $w(x) = u(x) - v(x) \equiv 0$.) Instead, we write $w'(x) - K w(x) \leq 0$ and multiply this differential inequality by an integrating factor $e^{-Kx} > 0$:

$$\left(e^{-Kx} w(x) \right)' = e^{-Kx} w'(x) - K e^{-Kx} w(x) \leq 0 .$$

The two sides are now integrated from x_0 to x, $x_0 \leq x$:

$$e^{-Kx} w(x) - e^{-Kx_0} w(x_0) = \int_{x_0}^{x} \left(e^{-Ks} w(s) \right)' ds \leq 0 .$$

Since $w(x_0) = 0$, there remains $e^{-Kx} w(x) \leq 0$. On the other hand, by its definition, $w(x) \geq 0$. Therefore, necessarily $w(x) \equiv 0$ and hence also $|u(x) - v(x)| = w'(x) \equiv 0$. For $x_0 - a \leq x \leq x_0$ the proof is similar. Thus, we proved that $u(x) \equiv v(x)$ and the initial value problem cannot have two different solutions. In the literature this reasoning is called "Gronwall's lemma."[4]

---

[4] Thomas Gronwall, 1877–1932.

Exercise 3.3 Seemingly, a "more simple" substitution could have been chosen. Let us mark the left-hand side of inequality (3.13) by $z(x) = |u(x) - v(x)|$. So (3.13) becomes

$$0 \le z(x) \le K \int_{x_0}^{x} z(s)\, ds \ .$$

We differentiate and get $z'(x) \le Kz(x)$ and continue as before. Why is this "proof" completely wrong?

3.5 Applications of the Uniqueness of Solution

The uniqueness of the solution of an initial value problem is of great importance. Let us summarize it:

Theorem 3.2 (Nonintersection Principle) *If the first-order differential equation $y' = f(x, y)$ satisfies the conditions of the existence and uniqueness theorem in a domain D in the (x, y) plane, then any two different solutions of this equation do not intersect in D.*

Saying that "solutions intersect," we are taking creative freedom of speech. To be precise, we should say "the graphs of the solutions pass through the same point." Indeed, if two different solutions pass through some point (x_1, y_1) of D, then the initial value problem $y' = f(x, y)$, $y(x_1) = y_1$ would have two different solutions, contradicting the existence and uniqueness theorem.

Figures 2.2, 2.3, and 2.6 show families of nonintersecting solutions. For nonuniqueness cases where solutions meet, see Figs. 3.3 and 3.5.

Example 3.8 Let us return to the logistic equation (2.15) (Example 2.7):

$$y'(t) = ky\left(1 - \frac{y}{L}\right), \quad k, L > 0 \ .$$

$f(t, y) = ky(1 - y/L)$ on the right-hand side undoubtedly fulfills the assumptions of the existence and uniqueness theorem for all t, y, being a polynomial in y and independent of t.

Solving Eq. (2.15) by separation of variables, the two singular solutions $y(t) \equiv 0$ and $y(t) \equiv L$ have a special role. In Example 2.16 we drew the direction field of the equation (Fig. 2.5) and emphasized that any solution whose values are between 0 and L, $0 < y(t) < L$, inevitably increases. We asked then whether such a solution meets the horizontal line $y = L$ at a certain point or does the solution just approach it without ever touching it? Now the answer becomes clear. Since the existence and uniqueness theorem holds everywhere, no two solutions meet. In particular, the solution $y(t) \equiv L$ and any other solution $y(t)$ which is determined by an initial value condition

$$y(t_0) = y_0, \qquad 0 < y_0 < L,$$

never meet. Therefore, $0 < y(t) < L$, $-\infty < t < \infty$.

We have seen that any solution starting between 0 and L increases and is bounded from above by $y = L$; hence, it must tend to some limit. Does $y(t)$ converge to L or perhaps $y(t)$ tends to some other limit, smaller than L? If $y(t)$ increases towards some A, $A < L$, there would be

$$y'(t) \to kA(1 - A/L) > 0, \qquad t \to \infty.$$

But it is impossible that while $y(t)$ tends to a horizontal line, its derivative (slope) would be bounded from below by some strictly positive constant. Therefore, $\lim_{t \to +\infty} y(t) = L$. Similarly, we conclude that $\lim_{t \to -\infty} y(t) = 0$. Compare these conclusions with the results of Exercise 2.2 for the explicit solutions.

Exercise 3.4 Explain why the solution of the logistic equation and an initial value condition $y(t_0) = y_0$ with $y_0 > L$ also satisfies $\lim_{t \to +\infty} y(t) = L$.

We must act carefully before we decide to what limit a solution tends. The following two examples demonstrate the difficulty involved.

Example 3.9 Given the initial value problem

$$y' = -\frac{x^2 y}{x^2 + y^2}, \qquad y(1) = y_0 > 0.$$

This equation satisfies the assumptions of the existence and uniqueness theorem in the entire plane, except for the point $(0, 0)$. It has the solution $y(x) \equiv 0$ and no other solution meets it. Therefore, if $y_0 > 0$, the solution of our initial value problem is positive and $y' < 0$. So it is monotonically decreasing and bounded from below and tends to some limit as $x \to \infty$. This limit must be 0. Otherwise, if $y(x)$ decreases to some limit $L > 0$, there would be $0 < L < y(x) < y_0$ for every x and we would get that $\lim_{x \to \infty} y'(x) = \lim_{x \to \infty} \left(-\frac{x^2 y}{x^2 + y^2} \right) = -L$. But a function cannot be both bounded from below and have a slope approaching a negative constant.

Next, we look at a seemingly similar initial value problem:

$$y' = -\frac{y}{x^2 + y^2}, \qquad y(1) = y_0 > 0.$$

3.5 Applications of the Uniqueness of Solution

The solution of this initial value problem is also bounded from below by the solution $y(x) \equiv 0$; it decreases and tends to some limit when $x \to \infty$. But in this case, the limit is not 0! Indeed, for every $y > 0$,

$$\frac{dy}{dx} = -\frac{y}{x^2+y^2} > -\frac{y}{x^2}$$

and hence

$$\int_{y_0}^{y} \frac{dy}{y} \geq \int_{1}^{x} -\frac{dx}{x^2}.$$

We get that $\ln(y/y_0) \geq \frac{1}{x} - 1$, that is, $y(x) \geq y_0 e^{1/x - 1} > y_0/e$. So $y(x)$ cannot decrease to 0 when $x \to \infty$.

Example 3.10 In Chap. 1, Example 1.7, we saw that if a body falls and is slowed down by a friction proportional to the square of its speed, then its speed $v(t)$ satisfies

$$mv'(t) = -mg + kv^2, \quad v(0) = 0, \quad m, k > 0.$$

We will write it as

$$v' = \frac{k}{m}(v - \sqrt{mg/k})(v + \sqrt{mg/k}), \quad v(0) = 0.$$

This equation has a structure similar to the logistic equation and has two constant solutions:

$$v_1(t) = \sqrt{mg/k}, \quad v_2(t) = -\sqrt{mg/k}.$$

When the values of $v(t)$ are between these two constants, then its derivative $v'(t)$ is negative and $v(t)$ tends to the lower of the two fixed solutions:

$$\lim_{t \to +\infty} v(t) = -\sqrt{mg/k}.$$

This happens in particular for the solution that starts from $v(0) = 0$, and after a long time, its speed approaches a negative constant (falling downwards!). For example, the impact speed of a parachutist does not grow without limit even when he jumps from a great height.

Interesting phenomena may occur in the absence of uniqueness.

Example 3.11 Let us look at the initial value problem:

$$y' = +\sqrt{1 - y^2}, \quad y(0) = 0.$$

Solving by separation of variables leads us to

$$\int \frac{dy}{\sqrt{1-y^2}} = \int dx, \qquad y \neq 1, -1,$$

that is, $\arcsin(y) = x + c$, that is, $y = \sin(x + c)$, and to two singular solutions $y(x) \equiv 1$, $y(x) \equiv -1$. The initial value condition $y(0) = 0$ determines $c = 0$ (or $c = 2n\pi$). The singular solutions do not satisfy the initial value condition so we stay with the solution $y = \sin(x)$ to all $-\infty < x < \infty$.

Is that indeed so? The solution $y = \sin(x)$ increases and decreases periodically while the differential equation clearly implies that $y' = +\sqrt{1-y^2} \geq 0$ and its solutions never decrease!

The source of the problem is at the point where the graph of $y = \sin(x)$ reaches $y = 1$. For $y = 1$ (and for $y = -1$) our equation does not satisfy the existence and uniqueness theorem, since the function $f(x, y) = \sqrt{1-y^2}$ is not differentiable there with respect to y. Indeed, the two solutions $y = \sin(x)$ and $y(x) \equiv 1$ meet at the point $(x, y) = (\pi/2, 1)$. Since the differential equation requires $y' \geq 0$, the solution continues from here on along $y(x) \equiv 1$. The full solution is given by

$$y(x) = \begin{cases} -1, & -\infty < x \leq -\pi/2, \\ \sin(x), & -\pi/2 \leq x \leq \pi/2, \\ 1, & \pi/2 \leq x < \infty. \end{cases}$$

See Fig. 3.5. This function has a continuous derivative everywhere. Note how this solution fits the direction field of the differential equation. Explain how to determine solutions that correspond to other initial value conditions.

As mentioned above, when the assumptions of the existence and uniqueness theorem hold, two different solutions cannot meet. If it turns out that two solutions meet at a point and fulfill the same initial value condition there, the inevitable conclusion is that these are not two different solutions but the same one. It happens, for example, that these are two different forms of writing the same solution.

Fig. 3.5 Nonuniqueness and intersection of solutions

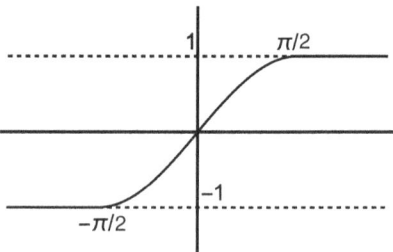

3.5 Applications of the Uniqueness of Solution

Example 3.12 Let $y_1(x)$ be any solution of the differential equation:

$$\frac{dy}{dx} + \cos^5(3\pi x)\, y = \cos^7(3\pi x).$$

Prove that the function $y_2(x) = y_1(1-x)$ is also a solution of the same differential equation. Then prove that $y_1(x) \equiv y_1(1-x)$.

We introduce a new variable $t = 1 - x$ and define

$$y_2(x) = y_1(1-x) = y_1(t).$$

According to the chain rule

$$\frac{dy_2}{dx} = \frac{dy_1}{dt}\frac{dt}{dx} = \frac{dy_1}{dt}(-1).$$

In addition also $\cos(3\pi x) = \cos(3\pi(1-t)) = \cos(3\pi - 3\pi t) = -\cos(3\pi t)$.

We move all the terms of the differential equation to one side and place the new variable in it:

$$\frac{dy_2}{dx} + \cos^5(3\pi x)\, y_2(x) - \cos^7(3\pi x) = -\frac{dy_1}{dt} - \cos^5(3\pi t)\, y_1(t) + \cos^7(3\pi t) \equiv 0.$$

It turns out that $y_2(x)$ satisfies the same differential equation as $y_1(x)$. At the point $x = 1/2$ also $t = 1 - x = 1/2$ and so $y_2(1/2) = y_1(1/2)$, i.e., the solutions y_1, y_2 satisfy the same initial value condition. It is clear that the given linear equation fulfills the requirements of the existence and uniqueness theorem in the whole (x, y) plane. Therefore, according to the existence and uniqueness theorem, these two solutions y_1, y_2 are identical, that is, $y_1(1-x) \equiv y_1(x)$ for all x.

Example 3.13 Read carefully the following statement:

"If $y(x)$ is a solution of the equation $x\dfrac{dy}{dx} = y^2 + y$, then $y(-x)$ also satisfies the same differential equation. Indeed, with $t = -x$, $u(x) = y(-x) = y(t)$,

$$\frac{du}{dx} = \frac{dy}{dt}\frac{dt}{dx} = y'(t)(-1),$$

so

$$x\frac{du}{dx} = (-t)\left(-y'(t)\right) = ty'(t) = y^2(t) + y(t) = u^2(x) + u(x)$$

and $u(x)$ satisfies the same differential equation as $y(x)$. In addition, we see that $u(0) = y(-0) = y(0)$. Therefore, $u(x)$ and $y(x)$ satisfy the same equation and the same initial value conditions and according to the existence and uniqueness theorem are the same. Consequently $y(-x) \equiv y(x)$, and every solution of the differential equation is an even function."

The above claim is false. By separation of variables $\int \frac{dy}{y^2+y} = \int \frac{dx}{x}$, and according to Problem 2.18(b), the general solution is $y(x) = x/(c-x)$, which is not even. There are also two singular solutions, $y(x) \equiv 0$ and $y(x) \equiv -1$.

Where is an error hidden in the above argument? If the equation is written in normalized form as $y' = \frac{y^2+y}{x}$, then $f(x,y) = \frac{y^2+y}{x}$ is not continuous at $x = 0$ and the differential equation does not satisfy the assumptions of the existence and uniqueness theorem!

On the other hand, it is true that if $y_c(x) = x/(c-x)$ is a solution, then also

$$y_c(-x) = \frac{-x}{c-(-x)} = \frac{x}{(-c)-x} = y_{-c}(x)$$

is solution, but a different one. All these solutions intersect at the point $(0,0)$. Try to draw several solutions, corresponding to positive and negative c values, and the two singular solutions.

Remark (Separation of Variables) The linear equation $y' + p(x)y = 0$ was solved in Sect. 2.1 by separation of variables, writing it as

$$\int \frac{y'(x)}{y(x)} dx = -\int p(x) dx,$$

provided $y(x) \neq 0$ for all x. It was also noticed that there is a singular solution $y(x) \equiv 0$ for all x. In Remark 3 of Sect. 2.1 we asked whether it is possible that a solution of this equation equals zero at some points but is nonzero at other points. Now this question is answered in the negative.

Indeed, let $y(x)$ be a solution of the homogeneous equation $y' + p(x)y = 0$ such that $y(x_0) = 0$. The identically zero function obviously satisfies the same equation and the same initial value condition. Therefore, according to the existence and uniqueness theorem, these two solutions of the initial value problem are the same, and $y(x) \equiv 0$. This verifies that equation (H) has only two types of solutions: One type is such that $y(x) \neq 0$ for all x in its domain of definition and the other type is the trivial solution $y(x) \equiv 0$.

3.6 Local Solutions

Theorem 3.1 ensures the existence of a solution of an initial value problem only in a certain neighborhood of the initial point. Why and where does a solution cease to exist?

Theorem 3.3 (Continuation of Solutions) *Suppose a solution $y(x)$ of a differential equation $y' = f(x,y)$ is defined in the interval (x_0, x_1) and $\lim y(x) = y_1$*

3.6 Local Solutions

exists when x tends to x_1 from the left-hand side. If $f(x, y)$, $f_y(x, y)$ are continuous near the point (x_1, y_1), then the solution $y(x)$ can be continued to the right of $x = x_1$. In other words, the solution of a differential equation cannot "come to an end" at a point inside the domain where the existence and uniqueness theorem holds.

Proof Assume that the graph of a solution $y(x)$ "ends" at the point (x_1, y_1), i.e., $\lim y(x) = y_1$ when x tends to x_1 from the left-hand side, and the solution $y(x)$ is not defined on the right-hand side of x_1. According to the assumption, the existence and uniqueness theorem holds in some neighborhood of (x_1, y_1); therefore, there is a solution $\tilde{y}(x)$ defined in a certain interval surrounding x_1. We "glue" $\tilde{y}(x)$ to $y(x)$ and get a continuation which is defined "a little more" on the right-hand side of x_1. Therefore, (x_1, y_1) is not the endpoint of a solution.

In this reasoning lies a tricky question. We argued that from any point (x, y) to which the solution has come, it can be continued a little longer as long as we are in the domain in which the existence theorem holds. If, say, the existence theorem holds in the whole plane, does this imply that by an infinite sequence of such continuations, we will reach any value of x? This is not necessarily the case.

Suppose, for example, that the existence theorem ensures that the solution exists on $[x_0, x_0 + h_1]$. In the next step, we extend it from $x_1 = x_0 + h_1$ until $x_2 = x_1 + h_2 = x_0 + h_1 + h_2$. Repeating this process, we arrive after n continuations from x_0 to $x_n = x_0 + h_1 + h_2 + \ldots + h_n$. Continuing the process infinitely many times, we may advance to the right by $\sum_{i=1}^{\infty} h_i$. But this sum may converge to some finite number, so the existence theorem may take us only up to a certain point, even if the theorem of existence holds in the whole plane. This happens in Example 3.4.

Well, where and how does "the solution end?" Theorem 3.3 permits several options.

(i) The solution is defined for every x and it does not "end" at all. For example, the solutions of $y' - y$ are $y = ce^x$ and they are defined for all values x.
(ii) The solution of $y' = f(x, y)$, $y(x_0) = y_0$, is defined to the right of x_0 up to a certain x_1 and there it "explodes," say $\lim y(x) = \infty$ as $x \to x_1^-$. This situation is demonstrated in Example 3.4 for the initial value problem $y' = y^2$, $y(2) = 1$. It fulfills the assumptions of the existence theorem in every rectangle in the plane but the interval of definition of its solution ends as $x \to 3^-$, since $\lim_{x \to 3^-} y(x) = +\infty$. This complies with Theorem 3.3 since the graph of the solution does not end at any finite point of the plane.
(iii) The solution ceases to exist at a point where $f(x, y)$ ceases to satisfy the conditions of the existence theorem, i.e., at the point where the solution leaves the domain D where the existence theorem holds. This situation is described in Fig. 3.6. Recall that in Fig. 3.2 we plotted a solution in some neighborhood of the initial point, that is guaranteed by the existence and uniqueness theorem. On the other hand, Fig. 3.6 describes the solution's maximal domain of definition, when it is continued as much as possible, until the boundary of the domain where the existence theorem holds.

Fig. 3.6 Domain of definition of an equation, a solution on its maximal domain of definition

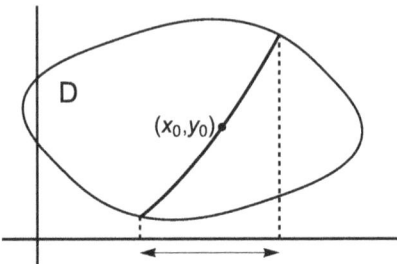

Here are two examples of solutions whose domain of definition is restricted.

The initial value problem $y' = y^2$, $y(2) = 1$ that is mentioned above is a prototype for problems of the form

$$y' = f(y), \quad y(x_0) = y_0, \tag{3.14}$$

where $f(y)$ on the right-hand side does not depend explicitly on the variable x. Such an equation is called *autonomous*. Many of our examples are of this type. An autonomous, first-order equation can be solved by separation of variables and the initial value conditions are taken care of by the limits of the integrals

$$\int_{x_0}^{x} dx = \int_{y_0}^{y} \frac{dy}{f(y)}.$$

This gives the solution $y(x)$ implicitly as

$$x = x_0 + \int_{y_0}^{y} \frac{dy}{f(y)}. \tag{3.15}$$

For the sake of simplicity, let us say that $f(y)$ is positive, $y' > 0$, and the solution increases. If the integral $\int^{\infty} dy/f(y)$ exists and is finite, then the left-hand side of (3.15) converges to some finite limit as $y \to \infty$:

$$x_1 = x_0 + \int_{y_0}^{\infty} dy/f(y).$$

This means that the solution $y(x)$ "escapes to infinity" at the finite point x_1. The "explosion point" x_1 is also the right-hand side end of the domain of definition of $y(x)$.

In Example 3.3 we saw that the initial value problem $y' = -x/y$, $y(1) = 2$ satisfies the existence and uniqueness theorem in the open upper half-plane $D = \{(x, y) \mid y > 0\}$. The solution of the initial value problem is $y(x) = +\sqrt{5 - x^2}$ and it is defined in the interval $-\sqrt{5} < x < \sqrt{5}$ which contains the initial point $x_0 = 1$.

3.7 Qualitative Investigation of Solutions

The solution extends until the points $(\sqrt{5}, 0)$ and $(-\sqrt{5}, 0)$. Indeed, these points are on the boundary of D and do not belong to D, as described in option (iii).

3.7 Qualitative Investigation of Solutions

The following example takes advantage of all our knowledge and describes the behavior of solutions even when we do not know to solve the differential equation explicitly.

Example 3.14 How does the solution of the initial value problem

$$y' = x^4 - y^4, \quad y(3) = 1, \tag{3.16}$$

look like? As we do not know to solve this equation explicitly, we will try to describe the solution with the help of geometrical considerations.

Our analysis is based on the schematic form of the direction field corresponding to the equation. See Fig. 3.7. Let us draw the two straight lines $y = -x$, $y = x$ as reference lines. These are two isoclines on which the slope of the direction field is $y' = x^4 - (\pm x)^4 = 0$. Within the sector $|y| < |x|$, we have $y' > 0$, and every solution increases. In particular, the solution starting from the initial point $(x_0, y_0) = (3, 1)$ is increasing in a neighborhood of $x_0 = 3$.

When x grows, the graph of the solution cannot cross the line $y = x$ from the bottom up. Because at a crossing point, the slope of the solution must be not less than the slope of $y = x$, i.e., at least 1, while its slope there is $y' = x^4 - y^4 = 0$. Consequently, the solution starting at the point $(3, 1)$ continues to increase as x grows but remains below the line $y = x$, that is, $1 < y(x) < x$ for $3 < x < \infty$. In particular, $y(x)$ cannot escape to $y = +\infty$ for any finite x.

If the solution is not defined for all $x > 3$, let x_1, $3 < x_1 < \infty$ be the supremum of the x-values for which it is defined. This means that the solution is defined in $[3, x_1)$ or in $[3, x_1]$ but not further to the right. Our solution increases and is bounded

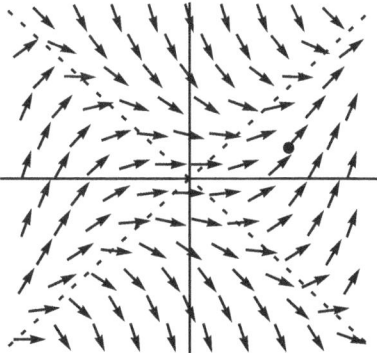

Fig. 3.7 The direction field of Eq. (3.16)

from above by the straight line $y = x$. So, when $x \to x_1$ from the left-hand side, $y(x)$ must converge to some finite limit y_1, and the graph of the solution tends to the point (x_1, y_1). Equation (3.16) satisfies the assumptions of Theorem 3.3 at every point. Thus, we conclude that the solution can be continued to the right of x_1, in contradiction with the maximality of x_1. This contradiction shows that the solution is defined for all x, $3 < x < \infty$.

The treatment in $-\infty < x < 3$ is not so simple. By looking at the direction field alone we cannot prove how does the solution through the point $(3, 1)$ advances to the left.

Example 3.14 is but a particular case of a more general principle.

Theorem 3.4 (The Funnel Theorem) *Given a differential equation $y' = f(x, y)$ and two differentiable functions $\alpha(x) < \beta(x)$ such that*

$$\alpha'(x) < f(x, \alpha(x)), \qquad (3.17)$$

$$\beta'(x) > f(x, \beta(x)), \qquad (3.18)$$

for all $x \geq a$. Suppose that the equation satisfies the conditions of the existence and uniqueness theorem in $D = \{(x, y) \mid a \leq x < \infty, \alpha(x) \leq y \leq \beta(x)\}$. Then for each (x_0, y_0) in D, the solution of the initial value problem $y(x_0) = y_0$ satisfies

$$\alpha(x) < y(x) < \beta(x) \quad \text{for all} \quad x_0 \leq x < \infty.$$

Proof From a geometric point of view, inequality (3.17) means that at every point $(x, \alpha(x))$ of the lower boundary curve $y = \alpha(x)$ of D, the slope of the direction field, $f(x, \alpha(x))$, is strictly bigger than the slope of the curve $\alpha(x)$ and the direction field crosses into D through its lower boundary. See Fig. 3.8.

A solution $y(x)$ that starts at a point (x_0, y_0) of D cannot meet $y = \alpha(x)$ for any $x > x_0$. Otherwise, if they meet, let x_1 be the first meeting point, namely, $y(x) > \alpha(x)$ for $x_0 \leq x < x_1$ and $y(x_1) = \alpha(x_1)$. By (3.17),

$$y'(x_1) = f(x_1, y(x_1)) = f(x_1, \alpha(x_1)) > \alpha'(x_1),$$

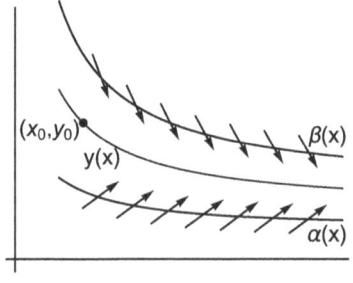

Fig. 3.8 A funnel

3.7 Qualitative Investigation of Solutions

so $y(x) - \alpha(x)$ strictly increases in the neighborhood of x_1. But this contradicts the fact that $y(x) - \alpha(x)$ is positive for $x_0 \leq x < x_1$ and is zero at x_1.

Similarly, a solution $y(x)$ does not meet $y = \beta(x)$, the upper boundary curve of D. So $\alpha(x) < y(x) < \beta(x)$ for all $x > x_0$. Because of the geometric picture, the D is called a *funnel*.[5]

Example 3.15 Let us return to the equation $y' = y^2 - x$ that was discussed in Example 2.18. The isoclines $y^2 - x = 0$ and $y^2 - x = -1$ form a funnel $\alpha(x) = -\sqrt{x}$, $\beta(x) = -\sqrt{x-1}$, because

$$\alpha'(x) = -\frac{1}{2}x^{-1/2} < f(x, \alpha(x)) = \alpha^2(x) - x = 0$$

for all $x > 0$, and

$$\beta'(x) = -\frac{1}{2}(x-1)^{-1/2} > f(x, \beta(x)) = \beta^2(x) - x = -1$$

for $x > 5/4$. Therefore, any solution $y(x)$ that starts in this funnel is defined for all $5/4 < x < \infty$ and

$$-\sqrt{x} < y(x) < -\sqrt{x-1}.$$

Since $\lim_{x \to \infty}(\sqrt{x} - \sqrt{x-1}) = 0$, this funnel gets narrower and each solution eventually behaves like $-\sqrt{x}$. See Fig. 3.10.

Exercise 3.5 Find additional funnels for the equation $y' = y^2 - x$.

The next theorem presents a more subtle and sophisticated situation.

Theorem 3.5 (Inverse Funnel) *Given a differential equation $y' = f(x, y)$ and two differentiable functions, $\alpha(x) < \beta(x)$, such that*

$$\alpha'(x) > f(x, \alpha(x)), \tag{3.19}$$

$$\beta'(x) < f(x, \beta(x)), \tag{3.20}$$

for all $x \geq a$. Suppose that the equation satisfies the conditions of the existence and uniqueness theorem in $D = \{(x, y) \mid a \leq x < \infty, \alpha(x) \leq y \leq \beta(x)\}$. Then there exists at least one solution $y(x)$ so that

$$\alpha(x) < y(x) < \beta(x), \quad a \leq x < \infty.$$

---

[5] J. H. Hubbard, B. H. West, Differential equations, A dynamical system approach, Springer Verlag, NY, 1991.

Fig. 3.9 Inverse funnel

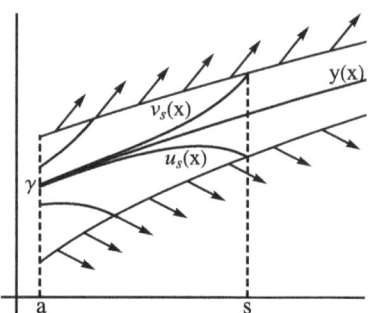

Proof Inequalities (3.19), (3.20) are the converse to (3.17), (3.18), respectively. Figure 3.9 reflects these inequalities. The geometric meaning of the inequality is that any solution that reaches a point on one of the curves $\alpha(x)$, $\beta(x)$, constituting the boundary of D, must leave D through this point, as x increases. Such D is named an *inverse funnel*.

Our goal is to prove that among all trajectories of solutions escaping from D upwards or downwards, there is at least one solution that is defined for all $a \leq x < \infty$ and remains within D.

For every s, $a < s < \infty$, let us take the solution that starts at the initial point $(s, \beta(s))$ on the curve $y = \beta(x)$, mark it as $v_s(x)$, and continue it to the left. Because of the inequalities (3.19), (3.20) it is clear that this solution cannot escape from the inverse funnel D to the left of $x = s$. It reaches $x = a$ and its value there is $v_s(a)$. See Fig. 3.9. Similarly, the solution $u_s(x)$ which starts at the point $(s, \alpha(s))$ extends to the left up to $x = a$ and its value there is $u_s(a)$.

For $a < s_1 < s_2 < \infty$ the solutions $v_{s_1}(x)$, $u_{s_1}(x)$, $v_{s_2}(x)$, $u_{s_2}(x)$ do not intersect and satisfy $u_{s_1}(a) < u_{s_2}(a) < v_{s_2}(a) < v_{s_1}(a)$, namely,

$$[u_{s_2}(a), v_{s_2}(a)] \subset [u_{s_1}(a) < v_{s_1}(a)].$$

By the Cantor intersection theorem, a sequence of closed and bounded intervals, each contained in the previous one, has an intersection that contains at least one point, say γ. The solution $y(x)$, which is defined by the initial value condition $y(a) = \gamma$, satisfies

$$\alpha(x) < u_s(x) < y(x) < v_s(x) < \beta(x)$$

for all $a < x < s$. See Fig. 3.9. Since this is true for all s, the solution $y(x)$ is defined for all $a < x < \infty$ and is trapped between $\alpha(x)$ and $\beta(x)$.

3.7 Qualitative Investigation of Solutions

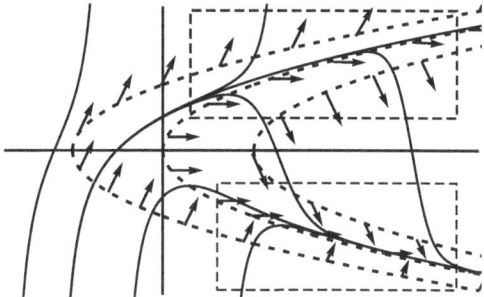

Fig. 3.10 In the lower rectangle, a funnel; in the upper rectangle, an inverse funnel

Example 3.16 Equation $y' = y^2 - x$ which is discussed in Example 2.18 has an inverse funnel bounded by the functions $\alpha(x) = +\sqrt{x}$, $\beta(x) = +\sqrt{x+1}$ (which are the isoclines $y^2 - x = 0$ and $y^2 - x = 1$). Indeed,

$$\alpha'(x) = (1/2)x^{-1/2} > f(x, \alpha(x)) = \alpha^2(x) - x \equiv 0,$$
$$\beta'(x) = (1/2)(x+1)^{-1/2} < f(x, \beta(x)) = \beta^2(x) - x \equiv 1$$

for all $x > 0$. According to Theorem 3.5 there exists at least one solution $y(x)$ which lies in this inverse funnel and satisfies $\sqrt{x} < y(x) < \sqrt{x+1}$ for all $0 < x < \infty$. See Fig. 3.10.

In the following example, we show another approach to prove the existence of a solution and at the same time also find a bound for it.

Example 3.17 Find an upper bound to the solution of the problem

$$y' = \frac{1}{x^2 + y^2}, \qquad y(1) = 1, \tag{3.21}$$

for all $1 \leq x < \infty$.

Let us compare the initial value problem (3.21) to another, more simple, initial value problem:

$$z' = \frac{1}{x^2}, \qquad z(1) = 1. \tag{3.22}$$

The solutions $y(x)$, $z(x)$ of both problems start at the same initial point $(1, 1)$. Due to $\frac{1}{x^2 + y^2} < \frac{1}{x^2}$, the slope of $y(x)$ is less than the slope of $z(x)$ and therefore the graph of $y(x)$ is below the graph of $z(x)$. The solution of (3.22) is found by direct integration to be $z(x) = 2 - 1/x$ for all $1 \leq x < \infty$. Therefore, also $y(x) \leq 2 - 1/x$ for $1 \leq x < \infty$. In particular, $y(x) < 2$ there.

This bound can be improved by a more delicate estimate of the equation. $y' > 0$ implies that $y(x)$ increases, so $y(x) \geq 1$ for all $x \geq 1$, and $\dfrac{1}{x^2 + y^2} < \dfrac{1}{x^2 + 1}$. Now we can compare initial value problem (3.21) with another initial value problem:

$$z' = \frac{1}{x^2 + 1}, \qquad z(1) = 1 . \tag{3.23}$$

According to the explanation above it follows also here that $y(x) \leq z(x)$, $x \geq 1$. The general solution for z is $z = \arctan(x) + C$ and the initial condition fixes $C = 1 - \pi/4$. In conclusion,

$$y(x) \leq z(x) = \arctan(x) + 1 - \pi/4 .$$

Since $\arctan(x) < \pi/2$, it follows that $y(x) < \pi/2 + 1 - \pi/4 = 1 + \pi/4$ which is a better upper bound than the earlier bound $y(x) < 2$.

Example 3.17 is a special case of a more general theorem.

Theorem 3.6 *Given two functions $f(x, y)$, $g(x, y)$ such that*

$$f(x, y) > g(x, y) \tag{3.24}$$

in their common domain of definition. If $y(x)$ is a solution of the initial value problem

$$y' = f(x, y), \qquad y(a) = \gamma$$

and $z(x)$ is a solution of the initial value problem

$$z' = g(x, z), \qquad z(a) = \gamma ,$$

then $y(x) > z(x)$ in each interval $a \leq x < b$ on the right-hand side of $x = a$ in which both solutions are defined.

Proof The graphs of the two solutions $z = z(x)$, $y = y(x)$ start at the point (a, γ) and at this point

$$y'(a) = f(a, \gamma) > g(a, \gamma) = z'(a) .$$

Therefore, there exists a certain neighborhood to the right of $x = a$ in which $y(x) > z(x)$.

If the statement of the theorem is false, then there is a point x_1 to the right of $x = a$ where the inequality $y(x) > z(x)$ is violated for the first time and where the graphs of $y(x)$ and $z(x)$ meet: $y(x_1) = z(x_1)$. When the graphs reach their meeting point $(x_1, y(x_1)) = (x_1, z(x_1))$ from the left-hand side, the graph of $z(x)$

approaches the graph of $y(x)$ from below, so the slope of $z(x)$ must be not smaller than that of $y(x)$:

$$z'(x_1) \geq y'(x_1) .$$

But this is impossible, because according to assumption (3.24)

$$y'(x_1) = f(x_1, y(x_1)) > g(x_1, z(x_1)) = z'(x_1) .$$

Remark 3.1 It is not necessary that the two solutions $y(x), z(x)$ will be defined exactly for the same values of x. It is possible that one of them will cease to exist before the other.

Remark 3.2 It is possible to generalize Theorem 3.6 even for cases where the strict inequality (3.24) is replaced by the weaker one, $f(x, y) \geq g(x, y)$, but in this case the proof needs finer arguments.

3.8 Global Existence Theorem

As we have already mentioned several times, Theorem 3.1 ensures the existence of a solution to the initial value problem $y' = f(x, y)$, $y(x_0) = y_0$, only in a certain neighborhood of x_0 and not necessarily for all values of x for which the functions $f(x, y)$, $f_y(x, y)$ are continuous. For this reason Theorem 3.1 is said to be a *local existence theorem*. Better results can be achieved at the cost of additional assumptions.

Theorem 3.7 *Consider the initial value problem $y' = f(x, y)$, $y(x_0) = y_0$, (Eq. (3.1)). Suppose that the functions $f(x, y)$, $\dfrac{\partial f(x, y)}{\partial y}$ are continuous in the infinite strip*

$$S = \{(x, y) \mid a \leq x \leq b, \ -\infty < y < \infty \} ; \tag{3.25}$$

the function $\dfrac{\partial f}{\partial y}$ is bounded in the entire strip S by some constant K,

$$\left| \frac{\partial f(x, y)}{\partial y} \right| \leq K , \quad a \leq x \leq b, \ -\infty < y < \infty , \tag{3.26}$$

and $a \leq x_0 \leq b$. Then the initial value problem (3.1) has a solution $y(x)$ that is defined for all $a \leq x \leq b$. This solution is unique.

Theorem 3.7 is a *global existence theorem* since the existence of the solution $y(x)$ is guaranteed to all values of x for which the assumptions of the theorem hold and not only in some small neighborhood of the initial point x_0. The assumptions (3.26) are stronger than those of Theorem 3.1. For instance, for the equation $y' = y^2$ in Example 3.4, the function $f(x, y) = y^2$ and its derivative $f_y(x, y) = 2y$ are continuous for all y, and $f_y(x, y) = 2y$ is bounded in every finite rectangle B in the (x, y)-plane, as required by Theorem 3.1. However, $f_y(x, y) = 2y$ is unbounded in any infinite strip S and does not fulfill the assumptions of Theorem 3.7. Indeed, the solution of any initial value problem in Example 3.4 exists only in some neighborhood of the initial point.

Proof The proof of this theorem is similar to that of our existence and uniqueness theorem, Theorem 3.1, in Sect. 3.3. Surprisingly the current proof is shorter and simpler.

As in Sect. 3.3 we choose some continuous function $y_0(x)$ that is defined in the interval $a \leq x \leq b$ and construct the iterations

$$y_{n+1}(x) = y_0 + \int_{x_0}^{x} f(s, y_n(s)) \, ds \, , \qquad a \leq x \leq b \, .$$

In Step 1 of Sect. 3.3 it was necessary to prove that all iterations $y_n(x)$ are located in a finite rectangle B. Here we are exempt from proving such property, since we work in the infinite strip S, and no bounds for $y_n(x)$ are required when $a \leq x \leq b$. Hence, the restriction $x_0 - h \leq x \leq x_0 + h$ that appears in Sect. 3.3 is not needed here. The rest of the discussion is true in the entire interval $a \leq x \leq b$.

Next, we follow the proof of Step 2 of Sect. 3.3. By assumption (3.26), $|\partial f(x, y)/\partial y| \leq K$ in all S and analogously with inequality (3.9),

$$|y_2(x) - y_1(x)| \leq \int_{x_0}^{x} K |y_1(s) - y_0(s)| \, ds$$

for all $a \leq x \leq b$. Again, there is no good estimate of the difference $y_1(x) - y_0(x)$. All we know is that these are two continuous functions in the interval $a \leq x \leq b$, so the difference has a maximal value there, $|y_1(x) - y_0(x)| \leq A$. Therefore,

$$|y_2(x) - y_1(x)| \leq \int_{x_0}^{x} K A \, ds = A K (x - x_0) \, , \qquad a \leq x_0 \leq x \leq b \, .$$

In the following iterations, we will get that

$$|y_{n+1}(x) - y_n(x)| \leq A K^n \frac{(x - x_0)^n}{n!} \, .$$

From here on the proof continues to its end as in Sects. 3.3 and 3.4 and the global existence on the whole interval $[a, b]$ is verified.

3.8 Global Existence Theorem

The seemingly innocent assumption (3.26) that $|\partial f(x, y)/\partial y| \leq K$ for all y, $-\infty < y < \infty$, is actually very restrictive and few equations meet it. According to (3.7),

$$f(x, y) - f(x, y_0) = \frac{\partial f(x, \tilde{y})}{\partial y}(y - y_0);$$

hence, $|f(x, y) - f(x, y_0)| \leq K|y - y_0|$ for all y. So

$$|f(x, y)| \leq K|y| + |f(x, y_0)| + K|y_0|.$$

Select some fixed y_0 and replace $|f(x, y_0)|$ on the right-hand side by its maximal value for $a \leq x \leq b$. We receive that

$$|f(x, y)| \leq K|y| + N$$

for all y, where K, N are some constants. Therefore, the global existence theorem cannot be valid for a differential equation where the rate of growth of $f(x, y)$ as a function of y is greater than linear. This is why the equation $y' = y^2$ does not meet the requirements of Theorem 3.7. In contrast, any linear equation $y' = p(x)y + q(x)$ meets the assumptions of Theorem 3.7 in every interval $a \leq x \leq b$ where $p(x), q(x)$ are continuous and bounded.

Example 3.18 Show that any solution of the nonlinear equation

$$y' = x^3 \sin^7 y + x^7 \cos^3 y \tag{3.27}$$

is defined to all x, $-\infty < x < \infty$.

Let $y(x)$ be some solution of the equation and suppose that at a point $x = x_0$ its value is $y(x_0) = y_0$. Take an arbitrary interval $a \leq x \leq b$ containing the point $x = x_0$. Eq. (3.27) and the initial value condition $y(x_0) = y_0$ satisfy the assumptions of Theorem 3.7. The right-hand side of the equation, $f(x, y) = x^3 \sin^7 y + x^7 \cos^3 y$, is continuous for all $a \leq x \leq b$, $-\infty < x < \infty$, and its partial derivative

$$\frac{\partial f}{\partial y} = 7x^3 \sin^6 y \cos y + 3x^7 \cos^2 y \cdot (-\sin y)$$

is bounded by the constant

$$\left|\frac{\partial f}{\partial y}\right| \leq 7|x|^3 + 3|x|^7 \leq 7\max\{|a|, |b|\}^3 + 3\max\{|a|, |b|\}^7.$$

Therefore, by Theorem 3.7, the solution in question is defined for all $a \leq x \leq b$. But a, b were chosen arbitrarily. If, for example, we want to reason that $y(x)$ is defined for $x = 10^6$, all we have to do is to choose the interval $[a, b] = [-2 \times 10^6, 2 \times 10^6]$. Therefore, the solution $y(x)$ is defined for every value of x.

3.9 The Concept of Stability

We return once again to the logistic equation:

$$y'(t) = ky\left(1 - \frac{y}{L}\right), \quad y(t_0) = y_0.$$

By Example 3.8 (or by the explicit solution in Example 2.7) any solution with $0 < y_0 < L$ or $y_0 > L$ tends, when $t \to +\infty$, to the singular solution $y(t) \equiv L$. On the other hand, a solution with $0 < y_0 < L$ moves away from the solution $y(t) \equiv 0$. When $y_0 < 0$ the corresponding solution descends and also moves away from $y(t) \equiv 0$. Furthermore, by Exercise 2.2, the denominator of the explicit solution (2.16) becomes zero at a certain finite value $t_1 > t_0$ and the solution runs away to $-\infty$ at a finite time. Thus, we cannot ask at all how such solution behaves as $t \to +\infty$. See Fig. 2.5.

To summarize, the solution $y(t) \equiv L$ "attracts" nearby solutions when $t \to +\infty$ and the solution $y(t) \equiv 0$ "repels" other solutions. These phenomena suggest to study *stability* and *instability* of solutions. The precise definitions are as following:[6]

Definition 3.2 A solution $y_1(t)$ of the differential equation $y'(t) = f(t, y)$ is called *stable* when $t \to +\infty$ if:

1. $y_1(t)$ is defined to all t, $t_0 \leq t < \infty$.
2. For all $\varepsilon > 0$ there exists $\delta = \delta(\varepsilon) > 0$

so that for any other solution $y(t)$, the inequality $|y(t_0) - y_1(t_0)| < \delta$ at the initial point $t = t_0$ implies that

$$|y(t) - y_1(t)| < \varepsilon \quad \text{for all} \quad t_0 \leq t < \infty.$$

Otherwise, the solution $y_1(t)$ is called *unstable*.

If in addition $\lim_{t \to +\infty} |y(t) - y_1(t)| = 0$, then the solution $y_1(t)$ is called *asymptotically stable*.

The choice of the initial point t_0 is arbitrary and stability can be discussed in any other interval. Stability for $t \to -\infty$ is defined analogously.

The definition of stability by means of ε, δ reminds the definition of continuity of functions. Stability is a kind of continuous dependence of solutions on initial values.

The concept of stability has great practical importance. Every laboratory measurement is approximate and every numerical calculation that is not done with integers and rational numbers is approximate because computers truncate digits and round the numbers. Hence, measurements and numerical calculations give only approximate solutions to problems and not exact ones. If a theoretical solution is

---

[6] In the context of stability problems it is customary to use the variable t rather than x.

3.9 The Concept of Stability

stable, then a solution that starts close to it will remain close to it in some sense and small calculation errors will not have a dramatic effect. On the other hand, if a solution is unstable, then every small change may take us away from the real solution and the results are not reliable.

Most of the devices that surround us in our daily life are stable in the above sense. For example, the steering wheel of a car "pulls back" to the straight forward direction and reduces the deflection angle. This is precisely the stability of the driving direction. A car without such a feature may increase any error and would be dangerous to use. The well-known "butterfly effect" is an example of the instability of the weather system. The following examples demonstrate some cases of stability and instability.

Example 3.19 Any solution of the equation $y' = 5$ is stable.
Indeed, $y(t) = 5t + c$, $y_1(t) = 5t + c_1$, so

$$y(t) - y_1(t) \equiv y(0) - y_1(0) \equiv c - c_1.$$

With $\delta(\varepsilon) = \varepsilon$, $|y(0) - y_1(0)| < \delta = \varepsilon$ implies $|y(t) - y_1(t)| < \varepsilon$ for all t.

On the other hand, no solution of the equation is asymptotically stable, because although both solutions remain "close" to each other but

$$\lim_{t \to +\infty} |y(t) - y_1(t)| = |c - c_1| \neq 0.$$

Example 3.20 All solutions $y = ce^{-kx}$ of the equation $y' = -ky$ where k is a positive constant are asymptotically stable. All solutions of $y = ce^{kx}$ of the equation $y' = ky$, $k > 0$, are unstable. Explain why.

Example 3.21 What are all the stable solutions of the logistic equation $y'(t) = ky(1 - y/L)$?

We have already seen that the solution $y(t) \equiv L$ is stable, but it is not the only such one. Actually, any solution with initial conditions $0 < y_0 < \infty$ is stable. Any two such solutions approach the value L and therefore the difference between them also approaches 0 when $t \to +\infty$. Therefore, each one of them is asymptotically stable. See Fig. 2.5.

Example 3.22 A stable solution does not have to be bounded at all. Consider the differential equation $y' = \dfrac{1}{y}$ in the upper half-plane $y > 0$, $-\infty < t < \infty$. Its solutions are given by $\int y \, dy = \int dt + \text{const}$, that is, $y = +\sqrt{2t + K}$. See Fig. 3.11.

For every two solutions

$$y_2(t) - y_1(t) = \sqrt{2t + K_2} - \sqrt{2t + K_1} = \frac{K_2 - K_1}{\sqrt{2t + K_2} + \sqrt{2t + K_1}} \to 0$$

Fig. 3.11 Asymptotically stable solutions of the equation $y' = 1/y$

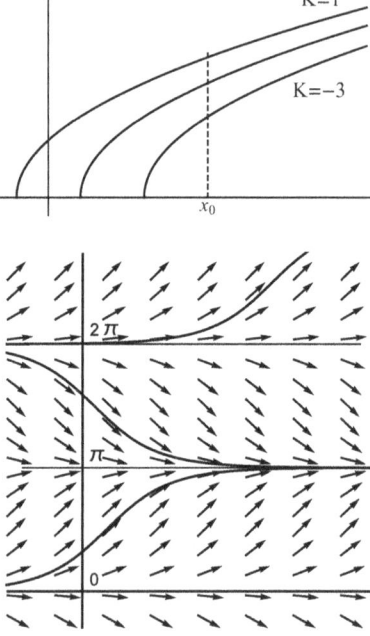

Fig. 3.12 The direction field of $y' = \sin y$

when $t \to +\infty$. Therefore, they all approach each other and every solution is asymptotically stable.

Example 3.23 What are the stable solutions and what are the unstable solutions of the differential equation $y' = \sin(y)$?

The direction field of this equation and some of its solutions are depicted in Fig. 3.12. We analyze Fig. 3.12 similarly to the logistic equation. A solution $y_1(x)$ that starts with an initial condition $y_1(x_0) = y_0$, $0 < y_0 < 2\pi$ is bounded between the singular solutions $y(x) \equiv 0$ and $y(x) \equiv 2\pi$ and tends to the value π when $x \to +\infty$. Hence, for any other solution $y(x)$ which starts sufficiently close to $y_1(x)$,

$$\lim_{x \to +\infty} |y(x) - y_1(x)| = 0 \,. \tag{3.28}$$

Therefore, $y_1(x)$ is asymptotically stable. Note that the solution $y_2(x) \equiv \pi$ is but one of these stable solutions. We emphasize again that the decisive consideration in determining stability of $y_1(x)$ is the limit of the difference (3.28) and not the fact that $y_1(x)$ converges to a constant limit.

On the other hand, the solution $y_3(x) \equiv 0$ is unstable when $x \to +\infty$, since every solution that starts near it moves away from it when $x \to +\infty$. The same consideration shows that also the solution $y_4(x) \equiv 2\pi$ is unstable, etc.

Example 3.24 All the solutions of $y' = y^2 - x$ mentioned in Example 3.15 are stable. By Fig. 3.10, the solution which was found in Example 3.16 is unstable.

3.10 Existence and Uniqueness for Higher-Order Equations

Let us look at a normalized differential equation of order n:

$$y^{(n)}(x) = f(x, y(x), y'(x), \ldots, y^{(n-1)}(x)) . \tag{3.29}$$

It is natural to fit to this differential equation n initial value conditions, all of them located at the same initial point $x = x_0$:

$$y(x_0) = \alpha_0, \ y'(x_0) = \alpha_1, \ \ldots, \ y^{n-1}(x_0) = \alpha_{n-1} . \tag{3.30}$$

Later we will refer to $f(x, z_0, z_1, \ldots, z_{n-1})$ as a function of $n + 1$ independent variables $x, z_0, z_1, \ldots, z_{n-1}$. In the differential equation there appear in the places designated for these variables the expressions $x, y, y', \ldots, y^{(n-1)}$.

Theorem 3.8 *Suppose that the function $f(x, z_0, z_1, \ldots, z_{n-1})$ and its n partial derivatives*

$$\frac{\partial f}{\partial z_0}(x, z_0, z_1, \ldots), \ \frac{\partial f}{\partial z_1}(x, z_0, z_1, \ldots), \ \ldots, \ \frac{\partial f}{\partial z_{n-1}}(x, z_0, z_1, \ldots, z_{n-1})$$

are continuous with respect to all its $(n + 1)$ variables $x, z_0, z_1, \ldots, z_{n-1}$ in a neighborhood of the given initial point $(x_0, \alpha_0, \alpha_1, \ldots, \alpha_{n-1})$. Then the differential equation (3.29) and the initial value conditions (3.30) have a solution $y(x)$ in a certain neighborhood $[x_0 - h, x_0 + h]$ of x_0; it has there n continuous derivatives $y'(x), \ldots, y^{(n)}(x)$ and it is unique.

Note that the theorem assumes nothing about the partial derivative $\partial f/\partial x$. For a first-order equation uniqueness means that no two different solutions satisfy $y_1(x_0) = y_2(x_0)$, that is, the graphs of two solutions do not meet. For an equation of order n, uniqueness means that two different solutions $y_2(x), y_1(x)$ cannot have all their n derivatives of orders $0, 1, \ldots, n - 1$ to be equal at a certain point $x = x_0$:

$$y_1^{(i)}(x_0) = y_2^{(i)}(x_0) , \quad i = 0, 1, \ldots, n - 1 .$$

It is possible, of course, that some of these conditions are met, while the others are not and this does not contradict the existence and uniqueness theorem. For example, a 2nd-order equation cannot have two different solutions so that

$$y_1(x_0) = y_2(x_0) , \quad y_1'(x_0) = y_2'(x_0) ,$$

but it is certainly possible that one of the equalities holds while the other does not, for example,

$$y_1(x_0) = y_2(x_0), \quad y_1'(x_0) \neq y_2'(x_0).$$

The geometric meaning is that the graphs of $y_1(x)$, $y_2(x)$ can intersect without being tangent each to the other.

Remark (Boundary Values) In the context of the existence and uniqueness theorem (and in fact throughout this book), we discuss only problems with initial value conditions that are given at a single point. There exists, of course, other problems in which additional conditions are given at several distinct points. A common case is an equation that is defined in an interval $a \leq x \leq b$; some conditions are given at the endpoint $x = a$ and other conditions at the endpoint $x = b$. Such problem is called a *boundary value problem* (in contrast to an initial value problem). Ordinary differential equations with boundary value conditions are a useful tool in the study of partial differential equations. Boundary value problems are not discussed in this text.

Problems

3.1 Given the initial value problem $y'(t) = ty^\alpha$, $y(0) = y_0$, $y_0 > 1$.
What is the domain of definition of the solution if $0 < \alpha < 1$? And what if $\alpha > 1$? Show that in the second case, the domain of definition gets smaller as α grows.

3.2 Show that every positive solution of the equation $y' = ky^\alpha$ with $k > 0$, $\alpha > 1$, has a vertical asymptote.

3.3 Show that the solution of the initial value problem $y' = y^8 - 1$, $y(0) = y_0$, with $y_0 > 1$, has a vertical asymptote. Hint: Show that $y^8 - 1 \geq ky^8$ for $y \geq y_0 > 1$ and a suitably selected $k < 1$ and use Theorem 3.6.

3.4 Find all solutions of the differential equation $y' = -xy^3$ and draw them. Determine the domain of definition of each solution.

3.5 Given the initial value problem $y'(t) = ty^\alpha$, $y(0) = 0$.
How many solutions do the problem have if $\alpha > 1$? And how many if $0 < \alpha < 1$? And if $\alpha < 0$? Explain the difference between the cases.

3.6 The same question as above for the initial value problem for the Bernoulli equation $y' + p(x)y = q(x)y^\alpha$, $y(0) = 0$.

3.7 Find four different solutions of the initial value problem $y' = xy^{1/3}$, $y(0) = 0$ which are defined for all x and which are different from 0 at any point $x \neq 0$. Hint: When branches are tangent, one can pass from one to the other.

3.8

(a) Find all the solutions of the equation $y' = \dfrac{2y}{x} - 1$ explicitly and draw their graphs.

(b) The assumptions of the existence and uniqueness theorem obviously do not hold at $(x, y) = (0, 0)$. Explain how the solutions of the equation behave in the neighborhood $(0, 0)$.

(c) How many solutions do the initial value condition $y(2) = 3$ have in a small neighborhood of $(2, 3)$? And how many solutions do the same initial value problem have in the whole plane?

3.9 The problem $y'(t) = t\sqrt{1 - y^2}$, $y(0) = -1$ (where $\sqrt{}$ indicates the positive root) has the solution $y(t) \equiv -1$. Find another solution of the problem. Why does the existence of two solutions not contradict the existence and uniqueness theorem?

What is the total number of solutions that satisfy $y(0) = -1$?
And how many solutions satisfy the initial value condition $y(0) = 1$?
Drawing the direction field of the equation helps to answer the questions.

3.10 Show that the equation $y' = |y| - 1$ does not fulfill some of the assumptions of the existence and uniqueness theorem (Theorem 3.1) at points of the form $(x_0, 0)$. Nevertheless, the initial value problem $y(x_0) = 0$ has exactly one solution. Write this solution explicitly for all values of x.

3.11

(a) Draw the direction field of the equation $y' = -\sqrt{xy}$.

(b) Find the solution of the initial value problem $y' = -\sqrt{xy}$, $y(0) = 1$, in all its domain of definition.

(c) Find $\lim_{x \to \infty} y(x)$.

3.12

(a) Find the solution of the initial value problem $y' = +\sqrt{y}$, $y(0) = 4$ in its maximal domain of definition. Is this solution unique?

(b) Answer the same questions for the initial value problem $y' = +\sqrt{|y|}$, $y(4) = 0$ and explain the difference.

3.13

(a) Draw the direction field of the equation $y' = -xy^{1/3}$.

(b) Where do the assumptions of the existence and uniqueness theorem hold and where do they not?

(c) Find the solutions of the initial value problem $y(1) = 0$ for all $-\infty < x < \infty$. How many solutions are there? How does the answer comply with the theorem of existence and uniqueness?

(d) Find the solutions of the initial value problem $y(0) = 0$ for all $-\infty < x < \infty$. How many solutions are there? How does the answer comply with the existence and uniqueness theorem?

3.14 Given the equation $y' = (y^2 - 1)/y$.

(a) Find the solutions of the initial conditions $y(0) = 2$ and $y(0) = 1/2$. Determine the maximal domain of definition of each solution and draw the solutions on the direction field.
(b) Explain how the solutions of the equation behave at the endpoints of their existence.

3.15 Draw the direction field of the equation $y' = y^3 - x$. Show that the solution that satisfies the initial value condition $y(0) = 0$ has a local maximum at $x = 0$.

3.16 Draw the direction field for the equation $y' = xy(y - 1)$ and sketch some approximate solutions.
Find the general solution and determine the domains of definition of the solutions for different initial values.

3.17 In Problem 2.4 we studied the initial value problem $(x^2 - 1)y' + 2xy = f(x)$, $y(4) = 1$, with

$$f(x) = \begin{cases} x, & x \leq 2, \\ x + 2, & x > 2. \end{cases}$$

This differential equation does not fulfill the assumptions of the existence and uniqueness theorem because the function $f(x)$ is discontinuous at $x = 2$. Check which of the conclusions of the existence and uniqueness theorem remains valid for this problem and which do not.

3.18 Find a solution $y(x)$ of the initial value problem

$$y' + y = \begin{cases} 0, x \leq 3, \\ 1, x > 3, \end{cases} \quad y(0) = 1,$$

for each x, so that $y(x)$ will be continuous at $x = 3$. Which of the assumptions of the existence and uniqueness theorem hold? Which conclusions of the theorem hold and which do not?

3.19 In Chap. 2, Problem 2.29 we found all the solutions of the initial value problem $y' = (y + x)/(y - x)$, $y(0) = 0$. Explain how the result fits the existence and uniqueness theorem.

3.20

(a) In which domain and for which initial conditions does the equation $y' = 1/y$ satisfy the assumptions of the existence and uniqueness theorem? Find all

3.10 Existence and Uniqueness for Higher-Order Equations

solutions of this differential equation and plot them. Determine the domain of definition of each solution $y(x)$ and its "endpoint" in the (x, y)-plane.

(b) The function $y(x) = \sqrt{2x}$ satisfies the differential equation above and also satisfies $y(0) = 0$. Explain why, despite this, it does not make sense to call the function $y(x) = \sqrt{2x}$ the solution of the initial value problem $y' = 1/y$, $y(0) = 0$, but it makes sense to say that it satisfies $\lim_{x \to 0} y(x) = 0$.

3.21 Solve the two initial value problems:

$$y' = y, \quad y(0) = 5,$$
$$y' = y^3, \quad y(0) = 5.$$

In both cases the function $f(x, y)$ on the right-hand side and its derivative $\partial f/\partial y$ are continuous for all values of x and y, but in the first case the solution is defined on the whole x-axis while in the second case only in part of the x-axis. Explain this phenomenon.

3.22

(a) Solve by a suitable substitution the two initial value problems:

$$y' = \sqrt[3]{y - 2x} + 1, \quad y(3) = 6,$$
$$y' = \sqrt[3]{y - 2x} + 2, \quad y(3) = 6.$$

(b) In both cases, the function $f(x, y)$ on the right side is continuous at every point, and the derivative $\partial f/\partial y$ is not continuous at the initial point $(3, 6)$. But in the first case there is exactly one solution while in the second case there are more than one. Explain how this complies with the existence and uniqueness theorem.

3.23

(a) For the initial value problem $y' = y - x + 1$, $y(0) = 0$, calculate the sequence of recursive iterations $y_{n+1}(x) = y_0 + \int_{x_0}^{x} f(x, y_n(x)) \, dx$ with the first approximation $y_0(x) \equiv 0$.

(b) Verify that the resulting sequence of functions $\{y_n(x)\}$ converges to a limit when $n \to \infty$.

(b) Compare the limit of the sequence of iterations with the solution that is calculated directly.

3.24 Calculate recursive iterations for the initial value problem $y' = 2xy$, $y(0) = 1$ with first iteration $y_0(x) \equiv 1$.

3.25 Given a sequence of iterations $y_0(x) = x$, $y_{n+1}(x) = \dfrac{\pi}{6} + \int_0^x \tan(y_n(x)) \, dx$. Find, by any method, the limit $\lim_{n \to \infty} y_n(x)$. For which values of x the limit exists?

3.26 In Example 3.8 we saw that if a solution of the equation $y' = ky\left(1 - \frac{y}{L}\right)$ starts between $y = 0$ and $y = L$, then $\lim_{t \to +\infty} y(t) = L$. Here is another alleged "proof" for this claim:

"If $0 < y(t) < L$ then $y' > 0$, so $y(t)$ increases and is bounded from above and therefore it tends to some limit. If a function $y(t)$ tends monotonically to a finite limit then its derivative $y'(t)$ tends to 0. But in our equation $y' = 0$ only if $y = 0$ or $y = L$; hence, the limit must be L."
Where is the error in this "proof?"

3.27 Given the differential equation $y' + 2y = y^5 + y^{55}$.

(a) Prove that if $y(x_1) > 0$ for some x_1, then $y(x) > 0$ in the whole domain of definition of the solution.
(b) Prove that if $|y(x_1)| < 1$, then $|y(x)| < 1$ in its whole domain of definition. What is $\lim_{x \to +\infty} y(x)$?

3.28 Given the differential equation $y' + x^6|y| = 2x^6$. Prove that no solution of this equation passes through both $(-3, 3)$ and $(3, -3)$.

3.29 Let $y(x)$ be the solution of the problem $y' = x^2 + y^2 - 1$, $y(0) = 0$.

(a) Prove that $y(x)$ is an odd function.
(b) Show that the solution $y(x)$ has exactly one local minimum point and one local maximum point. Show that $y(x)$ has an inflection point at $(0, 0)$.

3.30 Given the differential equation $y' = xy^3$.

(a) Show that if $y(x)$ is a solution of this equation, then both $u(x) = y(-x)$ and $v(x) = -y(-x)$ are solutions of the same equation. Can we conclude from this that any solution $y(x)$ is an even function? Or can it be concluded that every $y(x)$ is an odd function?
(b) Find all solutions of the equation and plot them. The figure may help to answer the previous question.

3.31 Prove that all the solutions of the differential equation $y' = x(y^{12} + y)$ are even functions. Plot the solutions with initial values $y(0) = \frac{1}{2}, -\frac{1}{2}, -2$.

3.32

(a) Draw schematically the direction field of the equation $y' = |x| - y$.
(b) With the help of the direction field explain how the solution of the initial value problem $y' = |x| - y$, $y(0) = 1$ looks like for $x \geq 0$.
(c) Solve explicitly the initial value problem for all x. Is the solution continuous at $x = 0$? Is it differentiable at $x = 0$? Compare how the explicit solution fits the qualitative geometric discussion.

3.33 Investigate the existence of a solution of the initial value problem $y' = x^4 - y^8$, $y(2) = 1$ for $2 \leq x \leq \infty$.

3.10 Existence and Uniqueness for Higher-Order Equations

3.34 Prove that the initial value problem $\dfrac{dy}{dx} = \dfrac{y+y^7}{x+x^7}$, $y(2) = 1$ has a solution $y = y(x)$ that is defined for all $0 < x < \infty$. What is the limit of this solution as x tends to 0 from the right-hand side?

3.35 Prove that the initial value problem $\dfrac{dy}{dx} = e^{-y} - e^{y}$, $y(0) = 1$ has a solution $y = y(x)$ for all $0 < x < \infty$. What is the limit of this solution when $x \to +\infty$?

3.36 Given the differential equation $y' = x^4 - y^4$.

(a) Prove that $\alpha(x) = x - 1$, $\beta(x) = x$, $1 \leq x < \infty$ form a funnel of this differential equation.
(b) Prove that $\alpha(x) = -(x+1)$, $\beta(x) = -x$, $0 \leq x < \infty$ are an inverse funnel of the equation.

3.37

(a) Draw schematically the direction field of the equation $y' = (1 - xy)^7$.
(b) Show that the solution of any initial value problem $y(0) = y_0$ is extendable to all $0 \leq x < \infty$. How do these solutions behave when $x \to +\infty$?

3.38 In Chap. 2, Problem 2.48, we drew the direction field and schematic solutions of the equation $y' = (y-1)(y-2)(y-3)$. We add several more features of the same equation:

(a) Show that the solution of the initial value problem $y(x_0) = y_0$, $2 < y_0 < 3$ satisfies $2 < y(x) < 3$ and $\lim_{x \to +\infty} y(x) = 2$.
(b) Which solutions of this equation are stable when $x \to +\infty$?
(c) Find the unstable solutions of this equation.

3.39 Given the initial value problem $y' = (y-1)(y-2)(y-3)(x^2-4)$, $y(0) = 1.5$. Without finding the solution $y(x)$ explicitly, determine where it has its maximal value and where its minimal value. Find $\lim_{x \to +\infty} y(x)$.

3.40

(a) Draw the direction field of the equation $y' = x \cos y$.
(b) Which solutions of this equation are stable when $x \to +\infty$?
(c) Which solutions of this equation are unstable?

3.41 Show that the solution of the initial value problem $y'(t) = +\sqrt{1-y^2}$, $y(0) = 0$, that appears in Example 3.11 is asymptotically stable.

3.42 Given the equation $y'(x) = +\sqrt{1-|y|}$.

(a) What is the domain of definition of this equation in the (x, y) plane?
(b) In which domain are the assumptions of the existence and uniqueness theorem fulfilled?

(c) Write explicitly the solution of the initial value problem $y(0) = 0$ in all of its domain of definition. Note that the solution may be given by different expressions in different places.

3.43

(a) Plot the direction field for the equation $y' = +\sqrt{|y|}$.
(b) Where is this equation defined and where does it fulfill the assumptions of the existence and uniqueness theorem?
(c) Solve the initial value problem $y(2) = 1$. Show that this problem has a unique solution in some neighborhood of the initial point $x = 2$ but its continuation for every x is not unique.
(d) Is there a solution that passes through the two points $(10, 1)$ and $(-10, -1)$? If there is, write it. If not, explain why not.

3.44

(a) Draw the direction field of the differential equation $y' = -\sqrt{y}$ in the half-plane $y \geq 0$, $-\infty < t < \infty$ and plot approximately the solution of the initial value problem $y(0) = 1$.
(b) Find explicitly the solution of the initial value problem and determine its maximal domain of definition. Explain how it complies with the figure above.
(c) How many solutions do the initial value problem have? Where does the equation fulfill the assumptions of the existence and uniqueness theorem?
(d) How will the previous answers change if we replace the differential equation by $y' = -\sqrt{|y|}$ in the whole (x, y) plane?

3.45

(a) Solve the initial value problem $y' = 2\sqrt{1-y}$, $y(0) = 0$. Where is the solution defined?
(b) Check the stability of the solution when $x \to +\infty$ and when $x \to -\infty$.

3.46

(a) Draw schematically the direction field of the equation $y' = 4y + |y|^3$ and solve the initial value problem $y(0) = -1$.
(b) Solve explicitly the initial value problem $y(0) = 1$ and determine its domain of definition.
(c) Does this equation satisfy the conditions of the existence and uniqueness theorem in a neighborhood of the point $(0, 0)$? Is the solution which starts from the point $(0, 0)$ stable when $x \to +\infty$?

3.47 Draw schematically the solution of the initial value problem $y' = \dfrac{x(1-y^2)}{x^2+y^2}$, $y(0) = 3$. What is the domain of definition of this solution? Where does the solution increase and where does it decrease? Is the solution bounded for all x? Does the solution have a minimal value, and if so, where? Does it have a maximal value; if so, where?

3.10 Existence and Uniqueness for Higher-Order Equations

3.48 Given the initial value problem $y' = \dfrac{(y^2 - 1) \sin y}{\sin^2 x + y^2 + 1}$, $y(0) = 2$.

(a) Show that the solution of this initial value problem is bounded from above and from below.
(b) Show that the solution of the problem is monotone increasing.
(c) Explain why the solution of the initial value problem is defined for every x.
(d) To what limit does the solution tend when $x \to +\infty$ and what is its limit when $x \to -\infty$?
(e) Which solutions of this equation are stable and which of them are unstable?

3.49 Solve the equation $y' = x/y$ and prove that every solution in the upper half-plane is asymptotically stable.

3.50 Solve the equation $y' = \dfrac{1}{x^2 + 1} y$ and prove that every solution is stable but not asymptotically stable.

3.51 Solve the initial value problem $xy'/y + 2xy \ln x = -1$, $y(1) = \dfrac{1}{2}$, and prove that the solution is asymptotically stable when $x \to +\infty$.

3.52 Show that every solution of an equation of the form $y' = f(x)$ is stable but not asymptotically stable. Hint: Denote $F(x) = \int f(x)\,dx$.

3.53 Given the differential equation $y'(x) = \dfrac{y^{26} + y^{16}}{y^6 + 1}$. Do not try to solve it explicitly! Prove:

(a) If $y(x)$ is a solution, then $y(x + c)$ is also a solution.
(b) If $y(x)$ is a solution, then $-y(-x)$ is also a solution. What do these two properties mean for drawing the graphs of the solutions?
(c) Let $y_0(x)$ be the solution that satisfies the initial condition $y(2) = 3$. Show that when x increases, the solution $y_0(x)$ is unbounded, and when $x \to -\infty$, then $y_0(x) \to 0$.
(d) (More difficult) Prove that every solution of the equation, except $y(x) \equiv 0$, is obtained from the solution $y_0(x)$ by the two transformations mentioned in (a) and (b).
Hint: To reach any solution, it is enough to reach every initial value condition.

3.54 Read the following explanation:
"In order to solve the initial value problem $y' = y^{1/2}$, $y(0) = 1$, we separate the variables as $\int y^{-1/2}\,dy = \int dx + C$ and get $2y^{1/2} = x + C$; thus, $y = \left(\dfrac{x+C}{2}\right)^2$.
The initial condition $y(0) = 1$ implies $1 = \left(\dfrac{0+C}{2}\right)^2$, i.e., namely, $C = \pm 2$. We obtain two different solutions $y = \left(\dfrac{x}{2} \pm 1\right)^2$.

On the other hand, the existence and uniqueness theorem applies to this differential equation in a neighborhood of the initial point $(x, y) = (0, 1)$ and the initial value problem must have a unique solution."

Which of the statements above is true and which is not? How many solutions do the initial value problem really have? Where is an error hiding?

3.55 Given a problem and its "solution":

"Solve the equation $t^2 y' + 4ty = y^2$, $t > 0$, and show that all its solutions, except one, tend to 0 when $t \to +\infty$.

Solution: This is a Bernoulli's equation (see Problem 2.20), it has the general solution $y(t) = \dfrac{5t}{1 + ct^5}$ and a singular solution $y(t) \equiv 0$. For all $c \neq 0$, $\lim_{t \to \infty} \dfrac{5t}{1 + ct^5} = 0$ and only for $c = 0$ the solution $y_0(t) = 5t$ does not tend to 0."

(a) Find where is an error hiding in the "solution" above and why the claim is false.

Hint: There is no error in the formula $y(t) = \dfrac{5t}{1 + ct^5}$.

(b) Explain how the different solutions of the equation behave as t grows.

(c) Is the solution $y(t) \equiv 0$ stable for $1 \leq t < \infty$? Is it asymptotically stable?

3.56 Fish in a pond multiply according to the logistic equation $y' = ky\left(1 - \dfrac{y}{\ell}\right)$, $k, \ell > 0$. If we are fishing a fixed amount, say h, per unit of time, then the appropriate model is

$$y' = ky\left(1 - \dfrac{y}{\ell}\right) - h, \qquad k, \ell, h > 0.$$

(a) Prove that if $h < k\ell/4$, the equation has two constant solutions:

$$y_1(t) = \text{const}, \qquad y_2(t) = \text{const},$$

the bigger is stable, and the smaller is unstable when $t \to +\infty$. Draw schematically the graphs of the solutions.

(b) Prove that if $h > k\ell/4$, then every positive solution tends to zero in finite time. What is the ecological meaning?

Hint: $y' = ky\left(1 - \dfrac{y}{\ell}\right) - h = -\dfrac{k}{\ell}\left(y - \dfrac{\ell}{2}\right)^2 + \left(\dfrac{k\ell}{4} - h\right) \leq \dfrac{k\ell}{4} - h < 0.$

3.57

(a) Solve the initial value problem $y' = ay^2$, $y(x_0) = y_0$. When $a > 0$, $y_0 > 0$, find the domain of definition of the solution. What happens to the solution when x approaches the right endpoint of the domain of definition? And when x approaches the left endpoint of the domain?

(b) It is given that the function $z(x)$ satisfies a differential inequality and an initial condition:

$$z' > az^2, \quad z(x_0) = y_0, \quad a > 0.$$

Explain why the domain of definition of $z(x)$ cannot be larger than that of the solution $y(x)$ of the initial value problem (a).

Hint: Look at the slopes of the corresponding direction fields at each point.

Chapter 4
Linear Equations of Higher Order

4.1 Linear Algebra Reminder

The study of linear differential equations resembles in some aspects that of linear algebraic equations. We summarize, without proofs, some basic facts from linear algebra that will be used.

Theorem 4.1 *Given a system of n linear equations with n unknown,*

$$a_{11}x_1 + a_{12}x_2 + \cdots + a_{1n}x_n = b_1,$$
$$a_{21}x_1 + a_{22}x_2 + \cdots + a_{2n}x_n = b_2,$$
$$\cdots$$
$$a_{n1}x_1 + a_{n2}x_2 + \cdots + a_{nn}x_n = b_n,$$

which is written in vector-matrix notation as $A\mathbf{x} = \mathbf{b}$, *where* $A = (a_{ij})_{i,j=1}^n$. *The system has a solution*[1] $\mathbf{x} = (x_1, \ldots, x_n)$ *for every* $\mathbf{b}$ *if and only if one of the following equivalent conditions holds:*

1. *The matrix A is invertible.*
2. *A has rank n.*
3. $\det(A) \neq 0$.
4. *All eigenvalues of A are different from 0.*
5. *The rows of A (the columns of A) are linearly independent.*

Under these conditions, the solution is unique and is given by $\mathbf{x} = A^{-1}\mathbf{b}$.

---

[1] All the vectors that are mentioned from here on are column vectors, but for convenience of printing they are written inside the text as row vectors.

In contrast with the general system $A\mathbf{x} = \mathbf{b}$, the homogeneous system $A\mathbf{x} = \mathbf{0}$ always has at least one solution, namely, the trivial solution $\mathbf{x} = \mathbf{0} = (0, \ldots, 0)$. According to Theorem 4.1, if the matrix A is invertible (and all equivalent conditions are met), then the trivial solution is the only solution of the system. Since we are interested mainly in nontrivial solutions, for a homogeneous system we ask whether there are other, nontrivial solutions

Theorem 4.2 *The system of n homogeneous linear equations with n variables, $A\mathbf{x} = \mathbf{0}$, has a nontrivial solution $\mathbf{x} = (x_1, \ldots, x_n)$ if and only if one of the following equivalent conditions holds:*

1. *The matrix A is not is invertible.*
2. *The rank of A is less than n.*
3. $\det(A) = 0.$
4. *At least one of the eigenvalues of A is equal to 0.*
5. *The rows of A (columns A) are linearly dependent.*

The set of nontrivial solution constitute a linear subspace.

We mention some ideas of linear algebra about the structure of the vector space R^n of n-tuples that will be needed later on.

The vectors $\mathbf{v}_1, \mathbf{v}_2, \ldots, \mathbf{v}_m$ are called *linearly dependent* if there exist constants $c_1, c_2, \ldots, c_m$, not all zero, and a *linear combination* so that

$$c_1 \mathbf{v}_1 + c_2 \mathbf{v}_2 + \cdots + c_m \mathbf{v}_m = \mathbf{0}.$$

If there are no such $(c_1, \ldots, c_m) \neq (0, \ldots, 0)$, then the vectors are called *linearly independent*. A collection of vectors is said to *span R^n* if every given vector $\mathbf{v}$ of R^n can be written as their linear combination.

It turns out that in R^n there may be at most n independent vectors and there are required at least n vectors to span R^n. A collection of vectors that are both linearly independent and span the space is called a *basis* of the space. By the above observations a basis of R^n consists of n vectors. The number of elements in a basis is called the *dimension* of the vector space.

4.2 Existence and Uniqueness Theorem for Linear Equations

Recall that an nth-order differential equation is called *linear* if the unknown function y and its derivatives $y', y'', \ldots, y^{(n)}$ appear in the equation only linearly. The independent variable x appears in the coefficients of the derivatives, so the general form of a linear equation is

$$a_0(x)y^{(n)} + a_1(x)y^{(n-1)} + \cdots + a_{n-1}(x)y' + a_n(x)y = b(x).$$

4.2 Existence and Uniqueness Theorem for Linear Equations

Assuming that $a_0(x)$, the coefficient of $y^{(n)}$, is never 0, we divide the equation by $a_0(x)$ and write the equation in a *normalized* form:

$$y^{(n)} + p_1(x)y^{(n-1)} + \cdots + p_{n-1}(x)y' + p_n(x)y = q(x). \tag{4.1}$$

The appropriate initial conditions for an nth-order equation are of the form

$$y(x_0) = \alpha_0, \quad y'(x_0) = \alpha_1, \quad \ldots, \quad y^{(n-1)}(x_0) = \alpha_{n-1}. \tag{4.2}$$

The homogeneous equation with $q(x) \equiv 0$ deserves a special discussion.

Under what conditions can the existence and uniqueness theorem for equations of order n (Theorem 3.8) be applied to a linear equation of order n and the n initial value conditions? We will rewrite Eq. (4.1) as

$$y^{(n)} = q(x) - p_n(x)y - p_{n-1}(x)y' - \cdots - p_1(x)y^{(n-1)}$$

and denote its right-hand side by $f(x, y, y', \ldots, y^{(n-1)})$:

$$f(x, y, y', \ldots, y^{(n-1)}) \equiv q(x) - p_n(x)y - p_{n-1}(x)y' - \cdots - p_1(x)y^{(n-1)}.$$

The assumptions of the existence and uniqueness theorem for an nth-order equation (Theorem 3.8) are that both $f(x, z_0, z_1, \ldots, z_{n-1})$ and its partial derivatives with respect to the variables $z_0, z_1, \ldots, z_{n-1}$,

$$\frac{\partial f}{\partial z_0}, \frac{\partial f}{\partial z_1}, \ldots, \frac{\partial f}{\partial z_{n-1}}$$

(but not with respect to x!), are all continuous. Our function

$$f(x, z_0, z_1, \ldots, z_n) = q(x) - p_n(x)z_0 - p_{n-1}(x)z_1 - \cdots - p_1(x)z_{n-1}$$

is obviously continuous in variables $z_0, z_1, \ldots, z_n$ which appear linearly. To ensure the continuity of f in x it is required that the coefficients $p_1(x), \ldots, p_n(x), q(x)$ are continuous in x. In addition, the partial derivatives of f are

$$\frac{\partial f}{\partial z_0} = -p_n(x), \quad \frac{\partial f}{\partial z_1} = -p_{n-1}(x), \quad \ldots, \quad \frac{\partial f}{\partial z_n} = -p_1(x),$$

and it was already assumed that these functions are continuous. The details are summarized in the following theorem, which is stated without proof.

Theorem 4.3 *If the functions $q(x), p_1(x), \ldots, p_n(x)$ are continuous in a common interval $[a, b]$ and x_0 is a point in this interval, then the linear differential equation (4.1) and the n initial value conditions (4.2) given at the point x_0 have a solution,* and it is unique and is defined in the entire interval $[a, b]$. The solution has continuous derivatives up to order n.

Note that unlike the general existence and uniqueness theorem (Theorem 3.8) where the existence of the solution is guaranteed only locally in a certain neighborhood $[x_0 - h, x_0 + h]$ of x_0, for a linear equation it is possible to prove that the solution exists globally in the entire interval $[a, b]$ where the coefficients of the equation are continuous. At the endpoints of the closed interval $[a, b]$ the solution has only one-sided derivatives.

We do not prove Theorem 4.3 for nth-order linear equations but we will use it frequently in this chapter. The particular case for first-order linear equations was concluded from its explicit solution (Theorem 2.1). Global existence for other first-order differential equation was verified in Theorem 3.7.

The following example demonstrates an application of the existence and uniqueness theorem.

Example 4.1 Given the equation $y'' + p_1(x)y' + p_2(x)y = 0$ where $p_1(x)$ is a continuous, odd function and $p_2(x)$ is a continuous, even function in the same interval $[-a, a]$, prove that the solution of the initial value conditions:

$$y(0) = y_0, \quad y'(0) = 0,$$

is an even function in $[-a, a]$.

Solution According to the existence theorem, the solution $y(x)$ of the initial value problem is defined throughout the interval $[-a, a]$. To argue that $y(-x) \equiv y(x)$ for all x in $[-a, a]$, we define $z(x) = y(-x)$ and prove that the function $z(x)$ is a solution of the same differential equation and satisfies the same initial conditions. For all x in the interval

$$z'(x) = -y'(-x), \quad z''(x) = y''(-x)$$

and it is given that $p_1(-x) = -p_1(x)$, $p_2(-x) = p_2(x)$. We put all this into the given differential equation and get

$$\begin{aligned} z''(x) &+ p_1(x)z'(x) + p_2(x)z(x) \\ &= y''(-x) + (-p_1(-x))(-y'(-x)) + p_2(-x)y(-x) \\ &= (y'' + p_1 y' + p_2 y)(-x) = 0, \end{aligned}$$

since $y(x)$ satisfies the equation at every point of $[-a, a]$, in particular at $(-x)$. At the point $x = 0$, the solution $z(x)$ has the initial values $z(0) = y(0) = y_0$, $z'(0) = -y'(0) = 0$. Hence, $z(x)$ satisfies the same differential equation and the same initial value conditions as $y(x)$, and according to the existence and uniqueness theorem they are identically equal. Thus, we proved that $y(-x) = z(x) \equiv y(x)$, namely, $y(x)$ is an even function.

Exercise 4.1 Prove under the same assumptions as in the Example 4.1 that the solution of the initial value problem

$$y(0) = 0, \quad y'(0) = y_1,$$

is an odd function.
Hint: To show that $y(-x) = -y(x)$, consider the function $z(x) = -y(-x)$.

4.3 Homogeneous Linear Differential Equations

Our discussion starts with homogeneous linear equations where the right-hand side is zero:

(H) $\quad y^{(n)} + p_1(x)y^{(n-1)} + \cdots + p_{n-1}(x)y' + p_n(x)y = 0, \quad a \leq x \leq b.$ (4.3)

Later we will deal with nonhomogeneous equations (4.1).

The collection of operations that are applied to the function $y(x)$ on the left-hand side of Eq. (4.3) is

$$y^{(n)} + p_1(x)y^{(n-1)} + \cdots + p_{n-1}(x)y' + p_n(x)y$$

$$= \left(\frac{d^n}{dx^n} + p_1(x)\frac{d^{n-1}}{dx^{n-1}} + \cdots + p_{n-1}(x)\frac{d}{dx} + p_n(x) \right) y.$$

This will be denoted for short and convenience by L:

$$L[y] = \left(\frac{d^n}{dx^n} + p_1(x)\frac{d^{n-1}}{dx^{n-1}} + \cdots + p_{n-1}(x)\frac{d}{dx} + p_n(x) \right) y.$$

With this notation, the left-hand side of the equation is written as $L[y]$ and the homogeneous equation (H) is written briefly as

$$L[y] = 0.$$

L is an *operator* that maps functions to functions. L is a *linear operator*, that is, for every two functions $u(x)$, $v(x)$ on which L acts and for constants c_1, c_2, the action of L on their linear combination is

$$L[c_1u + c_2v] = c_1L[u] + c_2L[v].$$

This happens because each of the derivatives $\frac{d}{dx}, \ldots, \frac{d^n}{dx^n}$ is a linear operator that satisfies

$$\frac{d}{dx}[c_1 u + c_2 v] = c_1 \frac{d}{dx} u + c_2 \frac{d}{dx} v,$$

etc. The operator L may be applied to functions with n continuous derivatives in a certain interval $[a, b]$. This is the domain of definition of L and it is usually denoted by $C^n[a, b]$. It is a linear space because a linear combination of differentiable functions is also a differentiable function.

Theorem 4.4 *If $y_1(x)$, $y_2(x)$ satisfy the homogeneous differential equation (H) (without reference to initial value conditions), then any linear combination with constant coefficients, $c_1 y_1(x) + c_2 y_2(x)$, satisfies the same equation. Consequently, the collection of all solutions of (H) is a linear space.*

Addition of solutions of Eq. (H) and their multiplication by constants keeps us within the set of the solutions. For if $L[y_1] = L[y_2] = 0$, then $L[c_1 y_1 + c_2 y_2] = c_1 L[y_1] + c_2 L[y_2] = 0 + 0 = 0$. This feature is also called *"the principle of superposition."*

In particular, $y(x) \equiv 0$ is a solution of any linear, homogeneous equation and is called, as expected, *the trivial solution*. This solution corresponds to the n initial value conditions:

$$y(x_0) = 0, \quad y'(x_0) = 0, \quad \ldots, \quad y^{(n-1)}(x_0) = 0.$$

Example 4.2 Verify that $y_1(x) = \cos x$, $y_2(x) = \sin x$ are solutions of the linear, homogeneous, second-order equation $y'' + y = 0$. According to the comments above, every linear combination $y(x) = c_1 \cos x + c_2 \sin x$ is a solution of the same equation. Check by direct calculation that this is the solution which satisfies the initial value conditions $y(0) = c_1$, $y'(0) = c_2$.

In analogy with linear algebra we define also *linear dependence of functions*.

Definition 4.1 The functions $f_1(x)$, $f_2(x)$, ..., $f_m(x)$ which are defined in the same interval $a \leq x \leq b$ are called *linearly dependent* on $[a, b]$ if there exist constants $c_1, c_2, \ldots, c_m$, not all zero, so that

$$c_1 f_1(x) + c_2 f_2(x) + \cdots + c_m f_m(x) \equiv 0 \quad \text{for all } x \text{ in } [a, b].$$

If such $c_1, \ldots, c_m$, not all zero, do not exist, then the functions are called *linearly independent* on the interval $[a, b]$.

For example, the three functions $2 + 3x$, $4 + 7x$, and $1 + x$ are linearly dependent on every interval because $3(2 + 3x) - (4 + 7x) - 2(1 + x) \equiv 0$.

4.3 Homogeneous Linear Differential Equations

Definition 4.2 A collection of functions is said to *span* a linear space of functions if every given function in the space can be written as their linear combination. A collection of functions that are both linearly independent and span the space is called a *basis* of the linear space. The number of elements in a basis is called the *dimension* of the space.

Example 4.3 The $m + 1$ functions $1, x, x^2, \ldots, x^m$ are linearly independent in every interval $[a, b]$ and span the linear space of the polynomials of degree at most m.

Indeed, a linear combination $c_0 \cdot 1 + c_1 x + \cdots + c_m x^m$ is a polynomial of degree m, so it has at most m roots. Hence, this expression may be equal to zero at most for m different values of x and cannot be zero for all values of x in an interval unless it is identically zero and all its coefficients are 0. This shows the linear independence of our set for any integer m. This set of functions obviously spans all the polynomials of degree at most m.

Warning The definition of linear dependence of functions relies heavily on the choice of the interval under consideration. Let us look, for example, at the two functions:

$$f_1(x) = \begin{cases} x^3, & x \leq 0, \\ 0, & 0 \leq x, \end{cases} \qquad f_2(x) = \begin{cases} x^3, & x \leq 0, \\ -x^3, & 0 \leq x. \end{cases}$$

The functions f_1, f_2 are linearly dependent on $(-\infty, 0]$ since $f_1(x) - f_2(x) \equiv 0$ in this interval. They are also linearly dependent on $[0, \infty)$ since $1 \cdot f_1(x) + 0 \cdot f_2(x) \equiv 0$ there. But these functions are linearly independent of $(-\infty, \infty)$ because there are no constants $(c_1, c_2) \neq (0, 0)$ so that $c_1 f_1(x) + c_2 f_2(x) \equiv 0$ for all x.

Our goal in this section is to describe the solution space of the homogeneous equation $L[y] = 0$ without trying to calculate the elements of the base explicitly. We want to know how many linearly independent solutions it contains, how many solutions are required to span it, how to characterize a basis of the solution space, and what is its dimension. As a motivation, recall that according to the existence and uniqueness theorem, a solution of an nth-order equation is determined by n initial values:

$$y(x_0) = \alpha_0, \quad y'(x_0) = \alpha_1, \quad \ldots, \quad y^{(n-1)}(x_0) = \alpha_{n-1}.$$

As the solutions are determined by n arbitrary parameters $\alpha_0, \alpha_1, \ldots, \alpha_{n-1}$, we are facing a problem with n degrees of freedom. This hints that the solution space is n-dimensional. The central theorem on this subject is

Theorem 4.5 *If an nth-order linear homogeneous equation $L[y] = 0$ fulfills the conditions of the existence and uniqueness theorem in a certain interval, then the collection of its solutions in that interval is an n-dimensional linear space.*

Proof To prove the theorem, we introduce n solutions that:

- Span the solution space (that is, there are enough solutions)
- Are linearly independent (meaning there are not too many solutions)

The investigation of the problem is divided into three steps:

A. What is the condition that the solutions $y_1(x), \ldots, y_n(x)$ span the solution space?
B. How to find n solutions that meet this condition?
C. What is the condition that $y_1(x), \ldots, y_n(x)$ will be independent?

We will consider each step separately. The combination of the steps will complete the proof of the theorem about the dimension of the solution space.

A. Spanning of the Solution Space

Given n solutions $y_1(x), \ldots, y_n(x)$ of the homogeneous equation $L[y] = 0$, what will guarantee that these solutions span the entire solution space? Let us start by analyzing the problem and see how it inevitably directs us toward the answer. At the end of the process, we will summarize the conclusion formally.

A collection of solutions $y_1(x), \ldots, y_n(x)$ spans the solution space if for each given solution $u(x)$ there exist constants $c_1, c_2, \ldots, c_n$ so that $u(x)$ can be written as a linear combination:

$$u(x) \equiv c_1 y_1(x) + c_2 y_2(x) + \cdots + c_n y_n(x)$$

for all $a \le x \le b$. We have to show that such constants $c_1, c_2, \ldots, c_n$ and such combination exist and to find them.

At this point the importance of the theorem of existence and uniqueness is revealed. Since usually we do not know to write the solutions of a differential equation explicitly (and we will not be able to do this in the future either), the only way to characterize a solution is by the initial conditions it satisfies at some point. Take a fixed point x_0 in the interior of the interval $[a, b]$. By the existence and uniqueness theorem, the n initial values $y(x_0), y'(x_0), \ldots, y^{(n-1)}(x_0)$ identify a solution $y(x)$ unequivocally. Our given solution $u(x)$ has at the point x_0 certain initial values, say

$$u(x_0) = \alpha_0, \quad u'(x_0) = \alpha_1, \quad \ldots, \quad u^{(n-1)}(x_0) = \alpha_{n-1}$$

4.3 Homogeneous Linear Differential Equations

and the n numbers $\alpha_0, \alpha_1, \ldots, \alpha_{n-1}$ characterize $u(x)$. A linear combination $c_1 y_1(x) + c_2 y_2(x) + \cdots + c_n y_n(x)$ will be identical with the given solution $u(x)$ if (and only if) the combination has at the point x_0 the same initial value conditions as $u(x)$:

$$c_1 y_1(x_0) + c_2 y_2(x_0) + \cdots + c_n y_n(x_0) = \alpha_0 ,$$
$$c_1 y_1'(x_0) + c_2 y_2'(x_0) + \cdots + c_n y_n'(x_0) = \alpha_1 ,$$
$$\ldots \qquad (4.4)$$
$$c_1 y_1^{(n-1)}(x_0) + c_2 y_2^{(n-1)}(x_0) + \cdots + c_n y_n^{(n-1)}(x_0) = \alpha_{n-1} .$$

(4.4) is a system of n linear, nonhomogeneous algebraic equations, with n unknown $c_1, c_2, \ldots, c_n$. It has a solution $\{c_1, c_2, \ldots, c_n\}$ for all values of $\alpha_0, \alpha_1, \ldots, \alpha_{n-1}$ provided that the determinant of the system is nonzero (Theorem 4.1), that is,

$$\begin{vmatrix} y_1(x_0) & y_2(x_0) & \cdots & y_n(x_0) \\ y_1'(x_0) & y_2'(x_0) & \cdots & y_n'(x_0) \\ \cdots & & & \\ y_1^{(n-1)}(x_0) & y_2^{(n-1)}(x_0) & \cdots & y_n^{(n-1)}(x_0) \end{vmatrix} \neq 0 . \qquad (4.5)$$

If this condition is met, we obtain from (4.4) n constants $c_1, c_2, \ldots, c_n$. Using these constants we define the linear combination $c_1 y_1(x) + c_2 y_2(x) + \cdots + c_n y_n(x)$. This linear combination is of course a solution, and by the definition of the constants in (4.4) it satisfies the same initial value conditions at $x = x_0$ as $u(x)$. According to the existence and uniqueness theorem, these two solutions must be identical in their whole domain of definition $[a, b]$:

$$u(x) \equiv c_1 y_1(x) + c_2 y_2(x) + \cdots + c_n y_n(x) , \qquad a \leq x \leq b .$$

Thus, the given solution $u(x)$ is spanned by $y_1(x), \ldots, y_n(x)$.

Why had the coefficients $c_1, \ldots, c_n$ been defined with the aid of the system (4.4) which depends on the values at a certain point x_0 and only later the linear dependence was extended and proved for the entire interval by the existence and uniqueness theorem? Why did we not define $c_1, \ldots, c_n$ directly by the system

$$c_1 y_1(x) + \cdots + c_n y_n(x) = u(x) ,$$
$$c_1 y_1'(x) + \cdots + c_n y_n'(x) = u'(x) ,$$
$$\ldots$$
$$c_1 y_1^{(n-1)}(x) + \cdots + c_n y_n^{(n-1)}(x) = u^{(n-1)}(x)$$

for all x at once? This is because the solutions of the last system will probably depend on x and will not be constant numbers, as required.

The determinant in (4.5) often appears in the study of linear differential equations and is called *Wronskian* named after Józef Wronski.[2] From here on we denote

$$W(x) = W\big(y_1(x), \ldots, y_n(x)\big) = \begin{vmatrix} y_1(x) & y_2(x) & \cdots & y_n(x) \\ y_1'(x) & y_2'(x) & \cdots & y_n'(x) \\ \cdots & & & \\ y_1^{(n-1)}(x) & y_2^{(n-1)}(x) & \cdots & y_n^{(n-1)}(x) \end{vmatrix}.$$

Sometimes, when it is necessary to emphasize the role of the point x, we also use the notation $W(y_1, \ldots, y_n)(x)$.

An interim summary of what was proved so far is:

Theorem 4.6 *Given a homogeneous linear equation* (H) *which fulfills the conditions of the existence and uniqueness theorem in the interval* $[a, b]$ *and n of its solutions* $y_1(x), \ldots, y_n(x)$. *If* $W(y_1, \ldots, y_n)(x_0) \neq 0$ *at some point x_0 in* $[a, b]$, *then* $y_1(x), \ldots, y_n(x)$ *span the solution space of* (H).

The formulation of Theorem 4.6 naturally raises the question of whether and how it depends on the choice of the point x_0. The answer will be given later.

Example 4.4 We have already seen that the two functions $y_1 = \cos x$, $y_2 = \sin x$ are solutions of the differential equation $y'' + y = 0$. Their Wronskian is

$$W(\cos x, \sin x) = \begin{vmatrix} \cos x & \sin x \\ -\sin x & \cos x \end{vmatrix} \equiv 1 \neq 0 \, ;$$

therefore, these two solutions span the solution space of the equation.

B. Existence of a Nonzero Wronskian

In Step A we saw that a sufficient condition for n solutions $y_1(x), \ldots, y_n(x)$ to span the solution space is that their Wronskian is nonzero at one point. Do such solutions exist at all?

The answer is positive and we will select n solutions that meet this requirement. As before, we cannot present explicitly the required solutions. The only way to choose a certain solution is to define it with the help of initial value conditions and to rely on the conclusions of the existence and uniqueness theorem. We define a solution $y_1(x)$ by n initial values:

$$y_1(x_0) = 1 \, , \; y_1'(x_0) = 0 \, , \; y_1''(x_0) = 0 \, , \; \ldots, \; y_1^{(n-1)}(x_0) = 0 \, .$$

---

[2] Józef M. Wronski, 1776–1853.

4.3 Homogeneous Linear Differential Equations

According to the theorem of existence and uniqueness, such a solution exists. The second solution $y_2(x)$ is defined by initial value conditions:

$$y_2(x_0) = 0, \; y_2'(x_0) = 1, \; y_2''(x_0) = 0, \; \ldots, \; y_2^{(n-1)}(x_0) = 0.$$

This solution also exists by the same reasoning. We will continue similarly up to the nth solution $y_n(x)$ which is defined by the initial conditions

$$y_n(x_0) = 0, \; y_n'(x_0) = 0, \; y_n''(x_0) = 0, \; \ldots, \; y_n^{(n-1)}(x_0) = 1.$$

For these n particular solutions the value of the Wronskian at the initial point x_0 is

$$W(y_1, \ldots, y_n)(x_0) = \begin{vmatrix} 1 & 0 & 0 & \ldots & 0 \\ 0 & 1 & 0 & \ldots & 0 \\ 0 & 0 & 1 & \ldots & 0 \\ & & \ldots & & \\ 0 & 0 & 0 & \ldots & 1 \end{vmatrix} = 1 \neq 0.$$

Thus, this particular collection of solutions spans the solution space. The two solutions in Example 4.4 are of this type for $n = 2$ and $x_0 = 0$. Obviously, our choice of initial conditions is arbitrary and many other initial conditions could have been chosen. This situation is natural because every linear space has many bases.

We have reached the third and final step of the discussion.

C. Linear Dependence and Independence of Solutions

So far we have seen a sufficient condition that solutions $y_1(x), \ldots, y_n(x)$ span the solution space. Now we look for a condition that n solutions will also be linearly independent.

Theorem 4.7

(i) If n arbitrary functions (not necessarily related to any differential equation!) $y_1(x), \ldots, y_n(x)$ with $n - 1$ continuous derivatives are linearly dependent in the interval $[a, b]$, then

$$W\big(y_1(x), \ldots, y_n(x)\big) \equiv 0$$

in the whole interval $[a, b]$.

(ii) Conversely, if the linear homogeneous differential equation $L[y] = 0$ satisfies the conditions of the existence and uniqueness theorem in the interval $[a, b]$ and its solutions $y_1(x), \ldots, y_n(x)$ satisfy $W(y_1, \ldots, y_n)(x_0) = 0$ for one point x_0 in the interval, then the solutions $y_1(x), \ldots, y_n(x)$ are linearly dependent on this interval.

Proof Claim (i) is the easy part of the proof. If the functions $y_1(x), \ldots, y_n(x)$ are linearly dependent in an interval, then there exist constants $c_1, \ldots, c_n$, not all zero, so that $c_1 y_1(x) + c_2 y_2(x) + \cdots + c_n y_n(x) \equiv 0$ identically in the interval. We differentiate this identity $n - 1$ times and get finally

$$c_1 y_1(x) + c_2 y_2(x) + \cdots + c_n y_n(x) \equiv 0,$$
$$c_1 y_1'(x) + c_2 y_2'(x) + \cdots + c_n y_n'(x) \equiv 0,$$
$$\cdots$$
$$c_1 y_1^{(n-1)}(x) + c_2 y_2^{(n-1)}(x) + \cdots + c_n y_n^{(n-1)}(x) \equiv 0,$$

for every x in the interval. This may be considered a system of n linear, homogeneous algebraic equations with n unknown $c_1, \ldots, c_n$. But it is given that a nontrivial solution $(c_1, \ldots, c_n) \neq (0, \ldots, 0)$ exists. This is only possible if the determinant of the system is 0. In our case, this is precisely the requirement that

$$W(y_1(x), y_2(x), \ldots, y_n(x)) \equiv 0$$

for all x in the interval, which is the first claim. Note that until here it was not assumed that $y_1(x), \ldots, y_n(x)$ are solutions of some differential equation and claim (i) holds for any differentiable functions.

(ii) In the opposite direction, suppose that $y_1(x), \ldots, y_n(x)$ are solutions of (H) and they satisfy

$$W(y_1, y_2, \ldots, y_n)(x_0) = 0$$

for one arbitrary point x_0. Consider the system of algebraic, linear, homogeneous equations with n unknown $c_1, \ldots, c_n$:

$$c_1 y_1(x_0) + c_2 y_2(x_0) + \cdots + c_n y_n(x_0) = 0,$$
$$c_1 y_1'(x_0) + c_2 y_2'(x_0) + \cdots + c_n y_n'(x_0) = 0,$$
$$\cdots \qquad (4.6)$$
$$c_1 y_1^{(n-1)}(x_0) + c_2 y_2^{(n-1)}(x_0) + \cdots + c_n y_n^{(n-1)}(x_0) = 0.$$

The determinant of this system is precisely $W(y_1, y_2, \ldots, y_n)(x_0)$, and as given, it is 0. Hence, system (4.6) has a nontrivial solution $(c_1, \ldots, c_n) \neq (0, \ldots, 0)$. From here on we will use these n constants $c_1, \ldots, c_n$ and define

$$U(x) = c_1 y_1(x) + c_2 y_2(x) + \cdots + c_n y_n(x).$$

Being a linear combination of solutions, $U(x)$ is a solution of the homogeneous differential equation (H). In addition, thanks to the choice of $c_1, \ldots, c_n$ by the system (4.6), $U(x)$ satisfies at x_0 the n initial value conditions

$$U(x_0) = 0, \quad U'(x_0) = 0, \quad \ldots, \quad U^{(n-1)}(x_0) = 0.$$

On the other hand, the identically zero function $V(x) \equiv 0$ is, of course, a solution of the linear homogeneous equation (H) and it obviously fulfills the conditions

$$V(x_0) = 0, \quad V'(x_0) = 0, \quad \ldots, \quad V^{(n-1)}(x_0) = 0 .$$

It turns out that $U(x)$, $V(x)$ are two solutions of the same initial value problem and they must be identical: $U(x) \equiv V(x)$ for all x in $[a, b]$. It means that

$$c_1 y_1(x) + c_2 y_2(x) + \cdots + c_n y_n(x) \equiv 0$$

in $[a, b]$. This completes the proof of the linear dependence of $y_1(x), \ldots, y_n(x)$.

Part (i) and part (ii) of Theorem 4.7 together lead to an important conclusion about solutions of linear homogeneous equations. According to part (ii), vanishing of the Wronskian of solutions at one point ensures their linear dependence in the whole interval in the discussion. But according to part (i), their linear dependence guarantees that the Wronskian is identically zero in the whole interval $[a, b]$. Hence, the conclusion follows.

Theorem 4.8 *Suppose that the homogeneous equation* (H) *satisfies the conditions of the existence and uniqueness theorem in* $[a, b]$. *Then the Wronskian of solutions* $W(y_1, \ldots, y_n)$ *is zero at one point* x_0 *of* $[a, b]$ *if and only if it is identically zero in the whole interval* $[a, b]$.

In the three steps above and in Theorems 4.6 and 4.7 we verified the existence of a basis of solutions and determined its dimension. Hence, the proof of a Theorem 4.5 is completed. Furthermore, we also characterized the elements of a base. We summarize:

Theorem 4.9 *Given the linear, homogeneous differential equation* (H) *which fulfills the assumptions of the existence and uniqueness theorem (that is, the coefficients* $p_1(x), \ldots, p_n(x)$ *of the equation are continuous) in the interval* $[a, b]$, *then n solutions* $y_1(x), \ldots, y_n(x)$ *of Eq.* (H) *form a basis for the solution space if and only if the Wronskian* $W(y_1(x), \ldots, y_n(x))$ *is nonzero at some point of the interval* $[a, b]$ *(and equivalently at all points of the interval).*

4.4 Wronskian and Abel's Formula

We have seen that it is of crucial importance whether the Wronskian of solutions is zero or nonzero. The following feature sheds more light on Wronskians.

Theorem 4.10 (Abel's Formula)[3] *Given the normalized differential equation*

(H) $\quad y^{(n)} + p_1(x)y^{(n-1)} + \cdots + p_{n-1}(x)y' + p_n(x)y = 0$

in which the coefficients $p_1(x), \ldots, p_n(x)$ *are continuous in the interval* $[a, b]$ *and* x_0 *is some fixed point in this interval, if* $y_1(x), \ldots, y_n(x)$ *are any n solutions of* (H), *then*

$$W(y_1, \ldots, y_n)(x) = W(y_1, \ldots, y_n)(x_0) e^{-\int_{x_0}^{x} p_1(t)\, dt} . \qquad (4.7)$$

Proof We prove formula (4.7) for $n = 2$ and the normalized equation

$$(H_2) \qquad y'' + p_1(x)y' + p_2(x)y = 0 .$$

Let $y_1(x), y_2(x)$ be two solutions of (H_2) and their Wronskian

$$W\bigl(y_1(x), y_2(x)\bigr) = \begin{vmatrix} y_1 & y_2 \\ y_1' & y_2' \end{vmatrix} = y_1 y_2' - y_1' y_2 .$$

The derivative of this determinant is

$$\frac{d}{dx} W(y_1, y_2) = (y_1 y_2' - y_1' y_2)'$$
$$= (y_1' y_2' + y_1 y_2'') - (y_1'' y_2 + y_1' y_2')$$
$$= y_1 y_2'' - y_1'' y_2 .$$

y_1, y_2 are solutions of (H_2), namely, $y_1'' + p_1 y_1' + p_2 y_1 = 0$, $y_2'' + p_1 y_2' + p_2 y_2 = 0$. We take y_1'', y_2'' and substitute them in the above calculation. It goes on as

$$\frac{d}{dx} W(y_1, y_2) = y_1(-p_1 y_2' - p_2 y_2) - y_2(-p_1 y_1' - p_2 y_1)$$
$$= -p_1(y_1 y_2' - y_1' y_2) = -p_1 W(y_1, y_2) .$$

This is a first-order differential equation for W which may be solved by separation of variables:

$$\int_{x_0}^{x} \frac{W'(x)}{W(x)}\, dx = \int_{x_0}^{x} -p_1(x)\, dx ,$$

$$\ln W(x) - \ln W(x_0) = -\int_{x_0}^{x} p_1(x)\, dx ;$$

---

[3] Niels Henrik Abel, 1829–1802.

4.4 Wronskian and Abel's Formula

hence,

$$W(x) = W(x_0) e^{-\int_{x_0}^{x} p_1 \, dx},$$

as claimed by the theorem. Note that $W(x_0)$ on the right-hand side is a fixed number. Do not forget that Abel's formula was proved for a normalized equation.

The proof of identity (4.7) for $n > 2$ is analogous but differentiation of a determinant of order $n \times n$ is more complicated and involves longer calculations.

Abel's formula has a direct implication on the question of Wronskian having a zero. Since the coefficients of the equation are continuous, the integral $\int_{x_0}^{x} p_1 \, dx$ exists, and its exponential is positive and certainly different from zero. Therefore, $W(x)$ is zero for any value of x if and only if $W(x_0) = 0$ for one x_0, as already proved in the Theorem 4.8.

Why did we assume that all the coefficients of the equation are continuous when only one of them, $p_1(x)$, really appears in the formula? The answer is that unless the assumption of continuity, the existence and uniqueness theorem does not hold and very existence of the solutions is not guaranteed. The following three examples demonstrate some possibilities.

Example 4.5 Take, for example, the equation

$$y'' - \frac{4}{x} y' + \frac{6}{x^2} y = 0,$$

whose coefficients $p_1(x) = -4/x$, $p_2(x) = 6/x^2$ are continuous in $(0, \infty)$ but not at $x = 0$. It is easy to verify by direct calculation that $y_1(x) = x^2$, $y_2(x) = x^3$ are solutions of this equation. Their Wronskian is

$$W(x^2, x^3) = \begin{vmatrix} x^2 & x^3 \\ 2x & 3x^2 \end{vmatrix} = x^4.$$

Here $W(x) \neq 0$ at all points of $(0, \infty)$; however, $W(0) = 0$. Indeed, Abel's formula with $x_0 = 0$ is meaningless because the integral $\int_{x_0}^{x} p_1 \, dx = \int_0^x (-4/x) \, dx$ is undefined due to its lower limit.

On the other hand, the equation $y'' + (6/x) y' + (6/x^2) y = 0$ whose coefficients are also discontinuous at $x = 0$ has the solutions $y_1(x) = x^{-2}$, $y_2(x) = x^{-3}$ (check!), and for them $W(x^{-2}, x^{-3}) = -6x^{-6}$. Also here $W(x) \neq 0$ in $(0, \infty)$ but W is not defined at all at $x = 0$.

Finally, for the equation $y'' - (6/x^2) y = 0$, whose solutions are x^{-2}, x^3, the Wronskian is $W(x^{-2}, x^3) \equiv 5 \neq 0$ in the entire domain of definition.

We present some applications of the Wronskian. In all the following examples we assume that the coefficients of the equations are continuous and the existence and uniqueness theorem is in effect.

Example 4.6 Prove that if $y_1(x)$, $y_2(x)$ are two solutions of a second-order homogeneous equation $y'' + p_1(x)y' + p_2(x)y = 0$ which are zero at the same point, then they are a fixed multiple of each other.

Suppose, for example, that y_1, y_2 are zero at $x = \alpha$, i.e., $y_1(\alpha) = y_2(\alpha) = 0$. Then

$$W(y_1, y_2)(\alpha) = \begin{vmatrix} y_1(\alpha) & y_2(\alpha) \\ y_1'(\alpha) & y_2'(\alpha) \end{vmatrix} = \begin{vmatrix} 0 & 0 \\ y_1'(\alpha) & y_2'(\alpha) \end{vmatrix} = 0.$$

Therefore, y_1, y_2 are linearly dependent, that is, $c_1 y_1(x) + c_2 y_2(x) \equiv 0$ for some constants c_1, c_2, not both zero, in their interval of definition. Accordingly, $y_2(x) = (-c_1/c_2) y_1(x)$.

Example 4.7 If a solution $y_1(x)$ of a second-order homogeneous equation is zero at two points $x = \alpha, \beta$, then any solution $y_2(x)$ which is not a multiple of $y_1(x)$ must have a zero between α and β.

Let $y_1(\alpha) = y_1(\beta) = 0$ and suppose, on the contrary, that $y_2(x) \neq 0$ in the open interval (α, β). $y_2(x)$ must be nonzero also at $x = \alpha$ because $y_2(\alpha) = 0 = y_1(\alpha)$ would imply, according to Example 4.6, that y_2 is a multiple of y_1, contrary to our assumption. For the same reason $y_2(\beta) \neq 0$ too. Overall, $y_2(x) \neq 0$ in the whole closed interval $[\alpha, \beta]$. In particular, $y_1(x)$ and $y_2(x)$ are linearly independent.

The quotient of solutions $h(x) = \dfrac{y_1(x)}{y_2(x)}$ is defined in the closed interval $[\alpha, \beta]$ because its denominator is not zero there. By our assumption that $y_1(\alpha) = y_1(\beta) = 0$, also $h(\alpha) = h(\beta) = 0$. Therefore, by Rolle's theorem there exists an intermediate point γ, $\alpha < \gamma < \beta$, so that $h'(\gamma) = 0$. On the other hand,

$$h'(x) = \left(\frac{y_1}{y_2}\right)' = \frac{y_1' y_2 - y_1 y_2'}{y_2^2} = \frac{W(y_2, y_1)}{y_2^2} \neq 0$$

in the whole interval, since the solutions $y_2(x)$, $y_1(x)$ are linearly independent and the numerator $W(y_2, y_1)$ is never 0. This contradiction proves that the assumption that $y_2(x) \neq 0$ in (α, β) is impossible and $y_2(x)$ must have a zero between the two zeros of $y_1(x)$.

Of course, the roles of $y_1(x)$ and $y_2(x)$ can be interchanged and it follows that also $y_1(x)$ has a zero between two zeros of $y_2(x)$. Combining these two results leads to the conclusion that for a second-order equation, the zero points of two linearly independent solutions must separate each other. This result is called the *separation theorem of Sturm*.[4] For example, $y_1(x) = \cos x$, $y_2(x) = \sin x$ are two solutions of $y'' + y = 0$ and between any two zeros of $\cos x$ there is a zero of $\sin x$ and vice versa.

---

[4] Jacques Charles Francois Sturm, 1803–1855.

4.4 Wronskian and Abel's Formula

Example 4.8 Prove that the function x^2 cannot be a solution of any normalized equation $y'' + p_1(x)y' + p_2(x)y = 0$ with continuous coefficients $p_1(x)$, $p_2(x)$ in an neighborhood of $x = 0$.

Suppose that $y_1(x) = x^2$ and $y_2(x)$ are two independent solutions in some neighborhood of $x = 0$. If the coefficients of the equation are continuous and the existence and uniqueness theorem holds, then their Wronskian is nonzero there. But

$$W(y_1, y_2) = \begin{vmatrix} x^2 & y_2 \\ 2x & y_2' \end{vmatrix}$$

is zero at $x = 0$, a contradiction.

Another possible approach is to notice that the supposed solution $y_1(x) = x^2$ fulfills the initial value conditions

$$y_1(0) = 0, \quad y_1'(0) = 0.$$

But the trivial solution $z(x) \equiv 0$ also fulfills the same initial value conditions $z(0) = z'(0) = 0$. If the equation has continuous coefficients and the existence and uniqueness theorem holds, it follows that $y_1(x) \equiv z(x) \equiv 0$. But $y_1 = x^2 \not\equiv 0$, a contradiction.

In fact, we have shown that if $y(x)$ satisfies

$$y(x_0) = 0, \quad y'(x_0) = 0, \tag{4.8}$$

it cannot be a solution of a 2nd-order linear, homogeneous, normalized equation with continuous coefficients, unless it is the trivial solution $y(x) \equiv 0$. Recall that when a function $y(x)$ satisfies (4.8), we say that $x = x_0$ is a *double zero (a zero of multiplicity 2)* of $y(x)$.

Example 4.9 Given two functions $y_1(x)$, $y_2(x)$ having two continuous derivatives in a certain interval and linearly independent there, find a linear, homogeneous differential equation of the second order which is satisfied by these two functions. What is the condition that it will be possible to write this differential equation as a normalized equation with continuous coefficients? Why did we limit the question only to linearly independent functions?

First Solution This solution is based on a clever guessing. Let us consider the equality

$$\begin{vmatrix} y(x) & y_1(x) & y_2(x) \\ y'(x) & y_1'(x) & y_2'(x) \\ y''(x) & y_1''(x) & y_2''(x) \end{vmatrix} = 0. \tag{4.9}$$

The expansion of the determinant according to its first column is

$$y'' \begin{vmatrix} y_1(x) & y_2(x) \\ y_1'(x) & y_2'(x) \end{vmatrix} - y' \begin{vmatrix} y_1(x) & y_2(x) \\ y_1''(x) & y_2''(x) \end{vmatrix} + y \begin{vmatrix} y_1'(x) & y_2'(x) \\ y_1''(x) & y_2''(x) \end{vmatrix} = 0 \,.$$

This is a linear, homogeneous second-order differential equation for $y(x)$ which is not normalized. When $y(x)$ is replaced by $y_1(x)$ in the left-hand side of (4.9), the determinant is zero because two of its columns are equal, so $y_1(x)$ satisfies Eq. (4.9). So does $y_2(x)$, and Eq. (4.9) that we found is the required equation.

The coefficient of y'' is the minor

$$\begin{vmatrix} y_1(x) & y_2(x) \\ y_1'(x) & y_2'(x) \end{vmatrix},$$

namely, $W(y_1, y_2)$. After normalization (dividing by the coefficient of y''), the coefficients of the resulting equation will be continuous if and only if $W(y_1, y_2) \neq 0$ in the suitable interval. It was assumed that y_1, y_2 are linearly independent; otherwise, $W(y_1, y_2)$ would be identically zero.

This solution has a certain drawback. Equation (4.9) was found by guessing, and it is natural to ask whether this is the only equation that meets our requirements, or perhaps another suitable equation may be found by using other methods. The next solution will answer this question in the negative.

Second Solution Suppose that the required equation is $y'' + p_1(x)y' + p_2(x)y = 0$, where $p_1(x)$, $p_2(x)$ are still unknown. We place the given functions $y_1(x)$, $y_2(x)$ in the supposed equation. At any point x, the result may be considered as a system of two algebraic equations for the unknown $p_1(x)$, $p_2(x)$:

$$p_1(x) y_1'(x) + p_2(x) y_1(x) = -y_1''(x) \,,$$
$$p_1(x) y_2'(x) + p_2(x) y_2(x) = -y_2''(x) \,.$$

Wherever the determinant of this system, $\begin{vmatrix} y_1(x) & y_2(x) \\ y_1'(x) & y_2'(x) \end{vmatrix}$, is different from zero, the system can be solved and we get unique $p_1(x)$, $p_2(x)$.

Conclusion *If the Wronskian of two functions $y_1(x)$, $y_2(x)$ differs from zero in an interval, there exists one and only one linear, homogeneous, normalized differential equation of second order that these functions are its solutions.*

It is easy to apply these methods and conclusion to equations of any order.

4.5 Reduction of the Order of an Equation

So far we have not solved any linear equation of order higher than one (and in the future we will know to solve explicitly only a few equations). We demonstrate a method that sometimes may help us to find solutions. Let us start with an equation of second order.

Suppose that one nontrivial solution $y_1(x)$ of the normalized equation

$$L[y] \equiv y'' + p_1(x)y' + p_2(x)y = 0$$

is known in some way. How can this information be used to find another solution of the equation?

We suggest to look for another solution of the form

$$y(x) = y_1(x)v(x) ,$$

when $v(x)$ is a yet unknown function that has to be found. Two questions arise about this suggestion: Why is it possible to present another solution in this way? And why is it profitable to try to present it so? The answer to the first question is simple: any function $f(x)$ can be presented as $f(x) = y_1(x)v(x)$ wherever $y_1(x) \neq 0$. All to be done is to choose $v(x) = f(x)/y_1(x)$. As for the question of usefulness, it will be seen later that the method is convenient to use, and one cannot argue with success.

Differentiate $y = y_1 v$ twice and substitute the results into the given differential equation:

$$y' = y_1'v + y_1 v', \qquad y'' = y_1''v + 2y_1'v' + y_1 v'',$$
$$L[y] = L[y_1 v] = (y_1''v + 2y_1'v' + y_1 v'') + p_1(x)(y_1'v + y_1 v') + p_2(x)(y_1 v)$$
$$= y_1 v'' + (2y_1' + p_1 y_1)v' + (y_1'' + p_1 y_1' + p_2 y_1)v = 0 .$$

But y_1 is a solution of $L[y] = 0$; therefore, the coefficient of v, $y_1'' + p_1 y_1' + p_2 y_1$, is identically zero and we remain with the equation

$$y_1 v'' + (2y_1' + p_1 y_1)v' = 0 .$$

We divide it by y_1 (wherever $y_1 \neq 0$), denote $v' = z$, $v'' = z'$, and receive

$$z' + \left(2\frac{y_1'}{y_1} + p_1\right) z = 0 . \tag{4.10}$$

Thus, we passed from a second-order homogeneous, linear equation to a first-order homogeneous, linear equation. For this reason, the method is called *reduction of the order of an equation*.

The first-order linear equation (4.10) can be solved by separation of variables:

$$\int \frac{z'(x)}{z(x)} \, dx = -\int \left(2 \frac{y_1'(x)}{y_1(x)} + p_1(x) \right) dx ,$$

$$\ln z(x) = -2 \ln y_1(x) - \int p_1(x) \, dx + \ln C ,$$

$$z(x) = C y_1^{-2}(x) e^{-\int p_1 \, dx} .$$

After a further integration of $v' = z$,

$$v(x) = C \int \frac{e^{-\int p_1 \, dx}}{y_1^2} \, dx + D .$$

Finally,

$$y = y_1(x) v(x) = C y_1(x) \int \frac{e^{-\int p_1 \, dx}}{y_1^2} \, dx + D y_1(x) .$$

We aimed to find one more solution, but we found a solution that is a linear combination with two arbitrary constants, C and D. The second term $D y_1(x)$ is already known as a solution. Since the solution space is spanned by two solutions, it turns out that also the first expression, which is multiplied by C, is a solution too. We denote it as

$$y_2(x) = y_1(x) \int \frac{e^{-\int p_1 \, dx}}{y_1^2} \, dx . \qquad (4.11)$$

The solutions $y_1(x)$, $y_2(x)$ are linearly independent. Otherwise there would exist $c_1 y_1(x) + c_2 y_2(x) \equiv 0$ with nonzero c_1, c_2 and $y_2(x)/y_1(x) \equiv -c_1/c_2$. But this is impossible, since $y_2(x)/y_1(x) = \int e^{-\int p_1}/y_1^2 \, dx$ cannot be constant (explain why!). So we have two independent solutions and they span the entire solution space.

Formula (4.11) can be found in a completely different way, without the reduction of order method. For two solutions $y_1(x)$, $y_2(x)$ consider the identity

$$\left(\frac{y_2}{y_1} \right)' = \frac{y_2' y_1 - y_2 y_1'}{y_1^2} = \frac{W(y_1, y_2)}{y_1^2} .$$

4.5 Reduction of the Order of an Equation

(See Example 4.7.) According to Abel's formula, $W(y_1, y_2) = Ke^{-\int p_1 \, dx}$ where K is a constant, our expression is also equal to

$$\left(\frac{y_2}{y_1}\right)' = \frac{Ke^{-\int p_1 \, dx}}{y_1^2}.$$

By integration, we get

$$\frac{y_2}{y_1} = K \int \frac{e^{-\int p_1 \, dx}}{y_1^2} \, dx$$

and the formula (4.11) for $y_2(x)$ follows (multiplied by the constant K).

How does formula (4.11) behave at a point where $y_1(x)$ is zero? If $y_1(c) = 0$, then the integral $\int e^{-\int p_1 \, dx}/y_1^2 \, dx$ diverges at $x = c$ due to y_1^2 in its denominator and (4.11) becomes an expression of the form $0 \cdot \infty$. We rewrite it in the form of ∞/∞ and calculate its limit by l'Hospital's rule:

$$\lim_{x \to c} \frac{\int \left(e^{-\int p_1 \, dx}/y_1^2\right) dx}{1/y_1(x)} = \lim_{x \to c} \frac{e^{-\int p_1 \, dx}/y_1^2}{-y_1'(x)/y_1^2(x)} = -\lim_{x \to c} \frac{e^{-\int p_1 \, dx}}{y_1'(x)}.$$

If $y_1(c) = 0$, then it must be that $y_1'(c) \neq 0$; otherwise, y_1 would be the trivial solution. Therefore, the last limit exists and is finite and the solution $y_2(x)$ which is given by (4.11) behaves well even at a point where $y_1(x)$ is zero.

Exercise 4.2 One solution of $(1 - x^2)y'' + 2xy' - 2y = 0$ is $y_1(x) = x$ (check!). Find all the solutions of the equation. Remember that (4.11) had been developed for a normalized equation.

We move on to reduce the order of a linear equation of any order. Let be given a normalized equation of order n:

$$L[y] \equiv y^{(n)} + p_1(x)y^{(n-1)} + \cdots + p_{n-1}(x)y' + p_n(x)y = 0$$

and suppose that one of its nontrivial solutions, say $y_1(x)$, is known. We try again a substitution of the form $y(x) = y_1(x)v(x)$ which replaces $y(x)$ by a new unknown $v(x)$. First, we apply the operator $L[\ ]$ to $y_1(x)v(x)$, namely,

$$L[y_1 v] = (y_1 v)^{(n)} + p_1(x)(y_1 v)^{(n-1)} + \cdots + p_{n-1}(x)(y_1 v)' + p_n(x)(y_1 v).$$
(4.12)

When the derivatives of the products are calculated, we arrange the terms according to the derivatives $v, \ldots, v^{(n)}$. The right-hand side becomes a linear combination of $v, \ldots, v^{(n)}$ with certain coefficients:

$$L[y_1 v] = q_0(x)v^{(n)} + q_1(x)v^{(n-1)} + \cdots + q_{n-1}(x)v' + q_n(x)v . \qquad (4.13)$$

The last two expressions are valid for any function v with n derivatives. The explicit form of all the coefficients $q_0(x), \ldots, q_n(x)$ is not important at this moment. All we need is that $q_0(x), \ldots, q_n(x)$ are independent of v and are expressed by $y_1, y_1', \ldots, y_1^{(n)}$ and $p_1, \ldots, p_n$ only. By the way, the derivative $v^{(n)}$ can only appear from the expression

$$(y_1 v)^{(n)} = y_1 v^{(n)} + \cdots$$

and therefore $q_0(x) = y_1(x)$.

The critical fact is that $q_n(x) \equiv 0$. By choosing $v(x) \equiv 1$ in (4.13), we get that

$$L[y_1 \times 1] = q_0(x) \times 0 + q_1(x) \times 0 + \cdots + q_{n-1}(x) \times 0 + q_n(x) \times 1 .$$

On the other hand, $y_1(x)$ is a solution of the homogeneous equation, so the left-hand side is $L[y_1] = 0$. Consequently, $0 \equiv q_n(x)$. Since the coefficients $q_0(x), \ldots, q_n(x)$ are independent of v, this conclusion is true regardless of the specific choice of v and we remain with the identity

$$L[y] = L[y_1 v] = q_0(x)v^{(n)} + q_1(x)v^{(n-1)} + \cdots + q_{n-1}(x)v' . \qquad (4.14)$$

The equation $L[y] = 0$ becomes

$$q_0(x)v^{(n)} + q_1(x)v^{(n-1)} + \cdots + q_{n-1}(x)v' = 0 .$$

An additional substitution $z = v'$ finally brings us to

$$q_0(x)z^{(n-1)} + q_1(x)z^{(n-2)} + \cdots + q_{n-1}(x)z = 0 .$$

Thus, we passed from a nth-order linear equation to another one of order $n - 1$, which justifies the name of the method, "reduction of the order of an equation."

The motivation for the reduction of order is the assumption that the lower the order of an equation is, the easier it is to solve. This hope is not realized in general, but there are certain cases where the reduction of order brings us closer to a solution.

Exercise 4.3 The third-order equation

$$x^3 y^{(3)} - 3x^2 y'' + x(6 - x^2)y' - (6 - x^2)y = 0$$

has a solution $y_1(x) = x$ (check!). Reduce the problem to a second-order equation. The resulting second-order equation will be solved by methods that we will learn later.

4.6 Equations with Constant Coefficients

Until now we have not solved systematically any differential equation of order higher than one and this is not accidental. They are very few equations that we know to solve explicitly and even these cases are often based on a successful guess.

A family of equations that we know to solve systematically are the linear equations with constant coefficients:

$$L[y] = a_0 y^{(n)} + a_1 y^{(n-1)} + \cdots + a_{n-1} y' + a_n y = 0, \quad -\infty < x < \infty, \quad (4.15)$$

where $a_0, \ldots, a_n$ are real constants. We assume that $a_0 \neq 0$, so this equation will be exactly of order n.

From the general theory it is known that this equation has exactly n linearly independent solutions which form a basis of the solution space. If we manage to find somehow n independent solutions, they will be a basis for the solution space and the problem will be fully solved.

We look for solutions of the form $y(x) = e^{rx}$, where r is an unknown constant. The logic behind this choice is that all derivatives of the exponential function e^{rx} are multiples of e^{rx} and all the terms on the left-hand side of the equation will have the common factor e^{rx}. Application of the operator L to $y(x) = e^{rx}$ is

$$\begin{aligned} L[e^{rx}] &= a_0(r^n e^{rx}) + a_1(r^{n-1} e^{rx}) + \cdots + a_{n-1} r e^{rx} + a_n e^{rx} \\ &= e^{rx} \left(a_0 r^n + a_1 r^{n-1} + \cdots + a_{n-1} r + a_n \right) \quad (4.16) \\ &= e^{rx} P(r), \end{aligned}$$

where $P(r) = a_0 r^n + a_1 r^{n-1} + \cdots + a_{n-1} r + a_n$. So the differential equation $L[y] = 0$ becomes by the substitution $y = e^{rx}$

$$L[e^{rx}] = P(r) e^{rx} = 0.$$

Hence, e^{rx} is a solution of the (4.15) if and only if r satisfies

$$P(r) = a_0 r^n + a_1 r^{n-1} + \cdots + a_n = 0.$$

The polynomial $P(r) = a_0 r^n + a_1 r^{n-1} + \cdots + a_{n-1} r + a_n$ is called the *characteristic polynomial* of the operator L and the equation is $P(r) = 0$ is called the *characteristic equation* of the homogeneous differential equation (4.15). According to the fundamental theorem of algebra, the characteristic equation has exactly n roots (solutions), different or equal, real or complex valued. Our goal is to obtain a basis of solutions of Eq. (4.15) for each of these cases.

Case A All the roots of the characteristic polynomial are real numbers, different from each other.

We mark the roots of the polynomial by $r_1, \ldots, r_n$, and assume that $r_i \neq r_j$ for all $i \neq j$. In this case the n functions $e^{r_1 x}, \ldots, e^{r_n x}$ are different from each other and each of them is a solution. To check whether these n solutions are linearly independent for all x, we consider their Wronskian determinant:

$$W\left(e^{r_1 x}, e^{r_2 x}, \ldots, e^{r_n x}\right) = \begin{vmatrix} e^{r_1 x} & e^{r_2 x} & \cdots & e^{r_n x} \\ r_1 e^{r_1 x} & r_2 e^{r_2 x} & \cdots & r_n e^{r_n x} \\ \vdots & & & \\ r_1^{n-1} e^{r_1 x} & r_2^{n-1} e^{r_2 x} & \cdots & r_n^{n-1} e^{r_n x} \end{vmatrix}.$$

From each of its columns we take outside a common factor $e^{r_i x}$ to get

$$e^{(r_1 + r_2 + \cdots + r_n) x} \begin{vmatrix} 1 & 1 & \cdots & 1 \\ r_1 & r_2 & \cdots & r_n \\ \vdots & & & \\ r_1^{n-1} & r_2^{n-1} & \cdots & r_n^{n-1} \end{vmatrix}.$$

The last determinant is called the *determinant of Vandermonde*.[5] It is shown in linear algebra that its value is

$$\prod_{i<j}(r_j - r_i) = (r_2 - r_1)(r_3 - r_1) \ldots (r_n - r_1)(r_3 - r_2) \ldots (r_n - r_{n-1}) \neq 0. \quad (4.17)$$

Therefore, $e^{r_1 x}, \ldots, e^{r_n x}$ are linearly independent for all x and the general solution of the Eq. (4.15) is

$$y(x) = c_1 e^{r_1 x} + c_2 e^{r_2 x} + \ldots + c_n e^{r_n x}.$$

Exercise 4.4 Verify (4.17) for the Vandermonde determinant of order $n = 3$.

---

[5] Alexandre-Theophile Vandermonde, 1735–1796.

4.6 Equations with Constant Coefficients

Example 4.10 The equation $y'' - 5y' + 6y = 0$ has the characteristic polynomial $r^2 - 5r + 6 = 0$ whose roots are $r_1 = 2$, $r_2 = 3$. Therefore, the general solution of the differential equation is

$$y(x) = c_1 e^{2x} + c_2 e^{3x} \, .$$

Find a solution that satisfies the initial conditions $y(0) = 3$, $y'(0) = 5$.

Because of the importance of the topic we will prove the linear independence of different exponential functions by an additional method.

Theorem 4.11 *Every n different exponential functions are linearly independent on each interval.*

The proof is by induction on n. For $n = 1$ there is nothing to prove since $c_1 e^{r_1 x} = 0$ is possible only when $c_1 = 0$. We assume now that the claim is true for $n - 1$ exponential functions and will prove its validity for n functions.

Suppose that $r_i \neq r_j$ for all $i \neq j$ and the identity

$$c_1 e^{r_1 x} + \ldots + c_{n-1} e^{r_{n-1} x} + c_n e^{r_n x} \equiv 0 \tag{4.18}$$

holds for all x. We divide it by $e^{r_n x} \neq 0$ and get

$$c_1 e^{(r_1 - r_n)x} + \cdots + c_{n-1} e^{(r_{n-1} - r_n)x} + c_n \equiv 0 \, .$$

The derivative with respect to x is

$$c_1 (r_1 - r_n) e^{(r_1 - r_n)x} + \cdots + c_{n-1}(r_{n-1} - r_n) e^{(r_{n-1} - r_n)x} \equiv 0 \, ,$$

which is a linear combination of $n - 1$ different exponential functions. According to the assumption of the induction, these functions are linearly independent; therefore, each coefficient of the linear combination must be 0:

$$c_1 (r_1 - r_n) = c_2 (r_2 - r_n) = \ldots = c_{n-1}(r_{n-1} - r_n) = 0 \, .$$

Due to the assumption $r_i \neq r_j$ it follows that $c_1 = c_2 = \ldots = c_{n-1} = 0$. By identity (4.18) it follows that also $c_n = 0$. This verifies that all n different exponential functions are linearly independent for all x.

Exercise 4.5 Prove that if $r_1, \ldots, r_m$ are m real numbers which are all different from each other, then the m functions $x^{r_1}, x^{r_2}, \ldots, x^{r_m}$ are linearly independent in the interval $(0, \infty)$.

Case B All the roots of the characteristic polynomial are different from each other but some of them are complex numbers.

Suppose, for example, that one of the complex roots is $r_1 = \alpha + i\beta$. Since the coefficients $a_0, \ldots, a_n$ of the differential equation are real (we are discussing real-valued differential equations), the coefficients of the characteristic polynomial are real, as well. As we know, the complex roots of a real polynomial appear in pairs of conjugate complex numbers, and together with $r_1 = \alpha + i\beta$ also the conjugate complex $r_2 = \alpha - i\beta$ is a root the characteristic polynomial.

What is the meaning of the function $e^{r_1 x} = e^{(\alpha + i\beta)x}$, namely, how is the exponential of a complex number defined? The natural place of this question is in a course about functions of a complex variable and not in the study of differential equations. Therefore, our discussion will be as restricted as possible.

Two principles direct us: We want a complex exponential function to satisfy the same identities as the real equivalent and will have a Taylor series similar to the real series. The source of these two requirements is anchored in the concept of "analytic continuation" of complex functions.

First, we require that $e^{(\alpha + i\beta)x} = e^{\alpha x} e^{i\beta x}$. The real factor $e^{\alpha x}$ is known to us and the factor $e^{i\beta x}$ will be defined by a Taylor series. The Taylor series of the real function e^t is

$$e^t = 1 + t + \frac{t^2}{2!} + \frac{t^3}{3!} + \cdots, \quad -\infty < t < \infty.$$

Therefore, we define e^{it} by the series

$$e^{it} = 1 + it + \frac{(it)^2}{2!} + \frac{(it)^3}{3!} + \frac{(it)^4}{4!} + \cdots = 1 + it - \frac{t^2}{2!} - i\frac{t^3}{3!} + \frac{t^4}{4!} + \cdots$$

$$= \left(1 - \frac{t^2}{2!} + \frac{t^4}{4!} - + \cdots\right) + i\left(t - \frac{t^3}{3!} + \frac{t^5}{5!} - + \cdots\right)$$

$$= \cos t + i \sin t.$$

This equality is called *Euler's formula*.[6] For $t = \pi$, the identity $e^{i\pi} + 1 = 0$ connects the five most fundamental constants of mathematics, 0, 1, i, e, π and it is considered by some to be the most beautiful mathematical equation.

According to Euler's formula for all real βx, $e^{i\beta x} = \cos \beta x + i \sin \beta x$ and

$$e^{r_1 x} = e^{(\alpha + i\beta)x} = e^{\alpha x}(\cos \beta x + i \sin \beta x). \tag{4.19}$$

In the same way also

$$e^{r_2 x} = e^{(\alpha - i\beta)x} = e^{\alpha x}(\cos \beta x - i \sin \beta x). \tag{4.20}$$

---

[6] Leonhard Euler, 1707–1783.

4.6 Equations with Constant Coefficients

Differentiation of complex functions of a real variable of the form $u(x) + iv(x)$ does not pose any problem. We refer to i as to a constant and get that the derivative is $u'(x) + iv'(x)$. Note that we do not discuss here functions of a *complex variable*!

Exercise 4.6 Prove by direct calculation that the derivative of the expression $e^{r_1 x} = e^{(\alpha + i\beta)x}$ which was defined above satisfies $(e^{r_1 x})' = r_1 e^{r_1 x}$.

Thanks to the last observation, there is no difficulty to consider complex functions of a real variable as solutions of a differential equation and properties of real solutions hold correspondingly also to complex-valued solutions. Linear combination of complex-valued solutions will naturally have their coefficients be complex numbers.

While complex-valued solutions are mathematically acceptable, it is natural that in practical problems we wish to have real-valued solutions of real equations. Therefore, we will generate real solutions by appropriate linear combinations of $e^{r_1 x}$, $e^{r_2 x}$. By addition and subtraction of (4.19), (4.20) we get

$$y_1(x) = \frac{1}{2}\left(e^{(\alpha + i\beta)x} + e^{(\alpha - i\beta)x}\right) = e^{\alpha x} \cos \beta x, \tag{4.21}$$

$$y_2(x) = \frac{1}{2i}\left(e^{(\alpha + i\beta)x} - e^{(\alpha - i\beta)x}\right) = e^{\alpha x} \sin \beta x. \tag{4.22}$$

Hereafter we will replace the two conjugate complex solutions $e^{r_1 x}$, $e^{r_2 x}$ by two real solutions $e^{\alpha x} \cos \beta x$, $e^{\alpha x} \sin \beta x$. To show that these solutions are part of a basis, it must be shown that all our operations do not harm linear independence. We skip the proof of this claim and summarize:

Theorem 4.12 *Suppose that all the roots $r_1, r_2, \ldots, r_n$ of the characteristic polynomial are different from each other, and suppose that 2k of them are conjugate complex numbers:*

$$r_{2j-1} = \alpha_j + i\beta_j, \quad r_{2j} = \alpha_j - i\beta_j, \quad j = 1, 2, \ldots, k,$$

and the other $n - 2k$ roots are real. Then the n functions

$$e^{\alpha_1 x} \cos \beta_1 x, \ e^{\alpha_1 x} \sin \beta_1 x, \ldots, e^{\alpha_k x} \cos \beta_k x, \ e^{\alpha_k x} \sin \beta_k x, \ e^{r_{2k+1} x}, \ldots, e^{r_n x}$$

are a basis of the solution space.

Example 4.11 The equation $y'' + y = 0$ has the characteristic polynomial $r^2 + 1 = 0$ and its roots are $r_1 = i$, $r_2 = -i$, which correspond to $\alpha = 0$, $\beta = 1$. Therefore, $e^{ix} = \cos x + i \sin x$, $e^{-ix} = \cos x - i \sin x$ are two complex-valued solutions and $\cos x$, $\sin x$ are two real-valued, linearly independent solutions of this differential equation. Verify this also by direct calculation. This equation was mentioned in Examples 4.2, 4.4, and 4.7.

Example 4.12 The equation $2y''' - y' - y = 0$ has the characteristic equation

$$2r^3 - r - 1 = (r-1)(2r^2 + 2r + 1) = 0$$

and its roots are $1, -\frac{1}{2} + \frac{1}{2}i, -\frac{1}{2} - \frac{1}{2}i$. Hence, a basis for the solution space is

$$\{e^t, e^{-t/2}\cos(t/2), e^{-t/2}\sin(t/2)\}.$$

Remark (Complex Coefficients) So far we have discussed differential equations with real constant coefficients and we have, of course, looked for real solutions for them. Indeed, the differential equations appearing in the applications are often real. But our solution method is suitable for equations whose coefficients are complex constants. In that case, the characteristic polynomial will be a polynomial with complex coefficients and its roots will not necessarily be conjugate to each other. Consider, for example, the equation

$$y'' + (-3 + 4i)y = 0. \tag{4.23}$$

It is clear that this equation has no real solution except the trivial one, because a real-valued $y(x)$ has also a real second derivative y'' and that contradicts the equality $y''/y = 3 - 4i$.

The characteristic polynomial of Eq. (4.23) is $r^2 + (-3 + 4i) = 0$ and its two roots are $r_2 = -1 - 2i$, $r_1 = 1 + 2i$ (Check!). Two complex-valued solutions are

$$y_1(x) = e^{(1+2i)x} = e^x\left(\cos 2x + i\sin 2x\right),$$
$$y_2(x) = e^{(-1-2i)x} = e^{-x}\left(\cos 2x - i\sin 2x\right),$$

and the general solution is

$$y(x) = c_1 e^x\left(\cos 2x + i\sin 2x\right) + c_2 e^{-x}\left(\cos 2x - i\sin 2x\right),$$

with complex c_2, c_1. As mentioned above, no such nontrivial linear combination can be real for all x.

Exercise 4.7 Find all solutions of $y'' - (2+3i)y' + (1+3i)y = 0$. Is there any real solution among its solutions?

Case C The characteristic polynomial has double roots.

If $r_1 = r_2$, then $e^{r_1 x} \equiv e^{r_2 x}$ and these are of course dependent solutions (the same solution!). On the other hand, it is known that there exist n linearly independent solutions. All that remains is to find them.

For example, the characteristic polynomial of the equation $y'' - 2ay' + a^2 y = 0$ has the roots $r_1 = r_2 = a$ and $y_1 = e^{ax}$ is a solution. A second solution can be found by reducing the order of the equation or by the formula (4.11). Since

4.6 Equations with Constant Coefficients

$-\int p_1 \, dx = 2ax$, it follows that

$$y_2(x) = y_1(x) \int \frac{e^{-\int p_1 \, dx}}{y_1^2} \, dx = e^{ax} \int \frac{e^{2ax}}{e^{2ax}} \, dx = xe^{ax}$$

is another linearly independent solution.

To address the general problem, let us assume that exactly k of the n roots of the characteristic polynomial are equal and the others differ from them:

$$r_1 = r_2 = \ldots = r_k \neq r_{k+1}, \ldots, r_n \, .$$

Definition 4.3 If exactly k roots of a polynomial are equal, their common value is said to be *root of multiplicity k* of the polynomial. A root of a polynomial that is different from any other root (its multiplicity is one) is called a *simple root*.

We will discuss the case when r_1 is a root of multiplicity k of the characteristic polynomial. Which functions appear in this case in the solution basis instead of $e^{r_1 x} \equiv e^{r_2 x} \equiv \ldots \equiv e^{r_k x}$?

The characteristic polynomial $P(r)$ that has roots $r_1, \ldots, r_n$ is factorized as $P(r) = a_0(r - r_1)(r - r_2) \ldots (r - r_n)$. If $r_1 = \ldots = r_k$, it is

$$P(r) = a_0(r - r_1)^k (r - r_{k+1}) \ldots (r - r_n) = (r - r_1)^k Q(r) \, ,$$

where $Q(r)$ denotes a polynomial of degree $n - k$ which contains all the other factors. The identity $L[e^{rx}] = P(r)e^{rx}$ (Eq. (4.16)), becomes

$$L[e^{rx}] = (r - r_1)^k Q(r) e^{rx} \tag{4.24}$$

for all r and for every x.

To justify the continuation of the process, we suggest an intuitive motivation. If it were $r_1 \neq r_2$, then also the linear combination

$$\frac{e^{r_2 x} - e^{r_1 x}}{r_2 - r_1}$$

would be a solution. When $r_2 \to r_1$, the quotient above tends to $\left. \frac{\partial}{\partial r} e^{rx} \right|_{r=r_1} = xe^{r_1 x}$. This is completely formal, as in our problem $r_1 = r_2$ are fixed numbers, and we cannot talk about $r_2 \to r_1$. But the structure of the limit suggests that it is useful to consider r as a variable parameter in (4.24) and differentiate it with respect to r. That what we will do. Only later the value of r will be chosen.

Since identity (4.24) depends on two variables, x and r, we apply partial derivatives. The derivative of the right-hand side of (4.24) is

$$\frac{\partial}{\partial r}\left((r-r_1)^k Q(r)e^{rx}\right) = k(r-r_1)^{k-1}Q(r)e^{rx} + (r-r_1)^k \frac{\partial}{\partial r}\left(Q(r)e^{rx}\right). \tag{4.25}$$

Next, we differentiate the left-hand side of (4.24), $L[e^{rx}]$, with respect to r. $L[e^{rx}]$ consists of derivatives with respect to x. The order of the differentiations with respect to the variables x and r can be interchanged:

$$\frac{\partial}{\partial r}\frac{\partial}{\partial x}e^{rx} = \frac{\partial}{\partial x}\frac{\partial}{\partial r}e^{rx},\ \ldots,\ \frac{\partial}{\partial r}\frac{\partial^k}{\partial x^k}e^{rx} = \frac{\partial^k}{\partial x^k}\frac{\partial}{\partial r}e^{rx},\ \ldots.$$

Consequently, it is allowed to change the order of the differential operators $\dfrac{\partial}{\partial r}$ and L, so the derivative of the left-hand side of (4.24) will be

$$\frac{\partial}{\partial r}L[e^{rx}] = L\left[\frac{\partial}{\partial r}e^{rx}\right] = L\left[xe^{rx}\right]. \tag{4.26}$$

From the comparison of (4.25) and (4.26) it follows that

$$L[xe^{rx}] = k(r-r_1)^{k-1}Q(r)e^{rx} + (r-r_1)^k \frac{\partial}{\partial r}\left(Q(r)e^{rx}\right).$$

This identity is true for all values of the variable r. Now we set the value $r = r_1$ and receive

$$L[xe^{r_1 x}] = 0.$$

This shows that the function $xe^{r_1 x}$ is a solution of the homogeneous differential equation $L[y] = 0$.

We may repeat the differentiation process with respect to r once again. On the left-hand side will appear this time $L[x^2 e^{rx}]$ and on the right-hand side the lowest power of $r - r_1$ will drop from $(k-1)$ to $(k-2)$. For $r = r_1$ it follows that

$$L[x^2 e^{r_1 x}] = 0;$$

hence, also $x^2 e^{r_1 x}$ is a solution of our equation. This process can be repeated as long as on the right-hand side there remains a factor $r - r_1$, namely, $k - 1$ times. To sum up, we see that when r_1 is a root of multiplicity k of the characteristic polynomial, then the k functions

$$e^{r_1 x},\ xe^{r_1 x},\ \ldots,\ x^{k-1}e^{r_1 x}$$

are k solutions of the linear homogeneous differential equation $L[y] = 0$.

4.6 Equations with Constant Coefficients

Example 4.13 The equation $y'' + 4y' + 4y = 0$ has the characteristic polynomial

$$r^2 + 4r + 4 = (r+2)^2 = 0$$

whose roots are $r_1 = r_2 = -2$. The general solution of the equation is $y = c_1 e^{-2x} + c_2 x e^{-2x}$. Find a solution that satisfies the initial conditions $y(1) = 3$, $y'(1) = 5$.

So far we have dealt with the case where the characteristic equation has a single root of high multiplicity. It is of course possible that there is also another, different root from a certain multiplicity, for example,

$$r_1 = r_2 = \ldots = r_k \neq r_{k+1} = r_{k+2} = \ldots = r_{k+l} \neq \ldots, r_n.$$

The second group of equal roots is treated exactly as the first group. Are the solutions we got linearly independent? The affirmative answer is given in the following theorem.

Theorem 4.13 *If r_1 is a root of multiplicity k_1 of the characteristic polynomial, r_2 is a root of multiplicity k_2, and so on, then the $k_1 + k_2 + \cdots = n$ solutions*

$$e^{r_1 x}, \; xe^{r_1 x}, \; \ldots, \; x^{k_1-1}e^{r_1 x}, \; e^{r_2 x}, \; xe^{r_2 x}, \; \ldots, \; x^{k_2-1}e^{r_2 x}, \; e^{r_3 x}, \; \ldots$$

are independent and establish a basis of the solution space of the equation $L[y] = 0$.

Proof Linear combinations of the solutions listed above are sums of products of exponential functions, different from each other, by polynomials. Therefore, we will prove a more general claim: If the numbers $r_1, \ldots, r_l$ are different from each other and $P_{m_1}(x), \ldots, P_{m_l}(x)$ are any polynomials of degrees $m_1, \ldots, m_l$, respectively, then the l functions

$$P_{m_1}(x)e^{r_1 x}, \; \ldots, \; P_{m_l}(x)e^{r_l x}$$

are linearly independent for all x.

The proof is by induction on the number of different exponentials and it is analogous to the proof of Theorem 4.11 for exponential functions only. For $l = 1$ there is nothing to prove since $c_1 P_{m_1}(x)e^{r_1 x} \equiv 0$ for all x only if $c_1 = 0$. We assume that the claim is true for $l - 1$ functions of this type and we prove its validity for l functions. Let

$$c_1 P_{m_1}(x)e^{r_1 x} + \ldots + c_l P_{m_l}(x)e^{r_l x} \equiv 0 \tag{4.27}$$

for all x. We divide by $e^{r_l x} \neq 0$ and receive

$$c_1 P_{m_1}(x)e^{(r_1-r_l)x} + \cdots + c_{l-1} P_{m_{l-1}}(x)e^{(r_{l-1}-r_l)x} + c_l P_{m_l}(x) \equiv 0$$

in which all exponents $r_i - r_l$ are different from zero. To eliminate the last polynomial in this identity, we differentiate it $m_l + 1$ times according to x. Before carrying out the differentiation, note that for every polynomial and every exponential function,

$$\left(P(x)e^{\alpha x}\right)' = \left(\alpha P(x) + P'(x)\right)e^{\alpha x},$$

and if $\alpha \neq 0$, then $\alpha P(x) + P'(x)$ is a polynomial of exactly the same degree as the polynomial $P(x)$. In our case after $m_l + 1$ differentiations of (4.27) we have

$$c_1 Q_{m_1}(x)e^{(r_1-r_l)x} + \cdots + c_{l-1} Q_{m_{l-1}}(x)e^{(r_{l-1}-r_l)x} \equiv 0$$

where $Q_{m_1}(x), \ldots, Q_{m_{l-1}}(x)$ are polynomials of the same degree as $P_{m_1}(x), \ldots, P_{m_{l-1}}(x)$, respectively. But according to the induction assumption these $l - 1$ functions are linearly independent; therefore, $c_1 = \ldots = c_{l-1} = 0$. Finally, by (4.27), we accept that $c_l = 0$ as well. This verifies the linear independence of the l functions in (4.27) and completes the claim of the induction for l.

It is also possible that the roots of the characteristic polynomial will be both complex-valued and also of multiplicity higher than one, for example,

$$r_1 = r_2 = \alpha + i\beta, \quad r_3 = r_4 = \alpha - i\beta.$$

In this case, the corresponding four solutions of the differential equation will be

$$e^{\alpha x}\cos\beta x, \ e^{\alpha x}\sin\beta x, \ xe^{\alpha x}\cos\beta x, \ xe^{\alpha x}\sin\beta x.$$

4.7 Qualitative Behavior and Stability

We will ask several questions about the behavior of the solutions of the linear homogeneous equation $L[y] = 0$ with constant coefficients when $x \to +\infty$.

Example 4.14 Under what conditions all solutions of the equation $L[y] = 0$ with constant coefficients tend to 0 when $x \to +\infty$?

All solutions consist of terms of the forms e^{rx}, $x^l e^{rx}$, $e^{\alpha x}\cos\beta x$, $x^l e^{\alpha x}\cos\beta x$ and their combinations, where $r, \alpha + i\beta$ are roots of the characteristic equation. Such expressions tend to zero as $x \to +\infty$ only when the exponential functions have negative exponents. This occurs when $r < 0$ (if r is real) or when $\alpha = \mathrm{Re}\{r\} < 0$ (if r is a complex number). The second condition includes the first one, and therefore, the condition we need is

$$\mathrm{Re}\{r_i\} < 0$$

4.7 Qualitative Behavior and Stability

for each root of the characteristic polynomial. Note that in this case not just every solution y tends to zero but also all its derivatives $y, y', \ldots, y^{(n)}$.

Example 4.15 Under what conditions all solutions of the equation $L[y] = 0$ with constant coefficients tend to 0 both when $x \to +\infty$ and when $x \to -\infty$?

It is easy to see that this situation is impossible.

Example 4.16 Under what conditions all solutions of the equation $L[y] = 0$ are bounded when $x \to +\infty$?

Recall that a solution $y(x)$ is bounded as $x \to +\infty$ if there is a constant M so that $|y(x)| \leq M$ for all x from a certain point and to the right.

Obviously a positive exponent r or a positive $\alpha = \text{Re}\{r\}$ are not suitable; therefore, we start with the requirement that $\text{Re}\{r_i\} \leq 0$. Moreover, if $r = 0$ or $\alpha = \text{Re}\{r\} = 0$, a positive power x^l should not appear in any solution. That is, the multiplicity of such root of the characteristic polynomial must be 1. To sum up, we require that for each root of the characteristic polynomial

$$\text{Re}\{r_i\} \leq 0$$

and every root r with $\text{Re}\{r\} = 0$ must be of multiplicity 1 (simple root).

Under these conditions not only every solution is bounded but also all its derivatives.

Exercise 4.8 Under what conditions all solutions of the equation $L[y] = 0$ with constant coefficients will be bounded when $x \to +\infty$ and also when $x \to -\infty$?

The previous examples lead naturally to the idea of stability of solutions. This concept has already been defined for first-order equations $y' = f(x, y)$. Now we define it for equations of any order n.

Definition 4.4 A solution $y_1(x)$ of the equation $y^{(n)} = f(x, y, y', \ldots, y^{(n-1)})$ is called *stable* when $x \to \infty$ if:

1. $y_1(x)$ is defined for all x, $x_0 \leq x < \infty$.
2. For all $\varepsilon > 0$ there exists $\delta = \delta(\varepsilon) > 0$ so that if any solution $y(x)$ of the equation satisfies at the initial point $x = x_0$

$$|y(x_0) - y_1(x_0)|, |y'(x_0) - y_1'(x_0)|, \ldots, |y^{(n-1)}(x_0) - y_1^{(n-1)}(x_0)| < \delta,$$

then also

$$|y(x) - y_1(x)|, |y'(x) - y_1'(x)|, \ldots, |y^{(n-1)}(x) - y_1^{(n-1)}(x)| < \varepsilon$$

for all x, $x_0 \leq x < \infty$. Otherwise, the solution $y_1(x)$ is called *unstable*.

If, in addition to the above, also

$$\lim_{x \to +\infty} |y(x) - y_1(x)|, \quad \lim_{x \to +\infty} |y'(x) - y_1'(x)|, \ldots,$$

$$\lim_{x \to +\infty} |y^{(n-1)}(x) - y_1^{(n-1)}(x)| = 0,$$

then the solution $y_1(x)$ is called *is asymptotically stable*.

The choice of the starting point $x = x_0$ is arbitrary and it is possible to discuss stability in any other interval $[a, \infty)$.

The meaning of stability is that if the initial value conditions of any solution $y(x)$ are sufficiently close to those of $y_1(x)$, then the solution $y(x)$ and its derivatives remain close to $y_1(x)$ and its derivatives, respectively, from here on. We emphasize that it is necessary to demand closeness not only for the solution but also for each of its derivatives separately because this is not at all self-evident. See the following exercise:

Exercise 4.9 Give an example of two functions $u(x)$, $v(x)$ so that $|u(x) - v(x)| < \varepsilon$ to all $0 \le x < \infty$ but $|u'(x) - v'(x)|$ accepts as large values as we wish. Hint: Think about a graphic example.

Examples 4.14 and 4.16 can be interpreted in terms of stability. Example 4.14 means that the trivial solution $y(x) \equiv 0$ is asymptotically stable, since all solutions tend to 0 when $x \to \infty$. In fact, much more is true.

Theorem 4.14 *Every solution of a linear differential equation with constant coefficients (4.15) is asymptotically stable when $x \to +\infty$ if and only if $\mathrm{Re}\{r_i\} < 0$ for all n roots of the characteristic equation.*

Indeed, all the solutions and their derivatives tend to zero and thus for any two solutions $y_1(x)$, $y_2(x)$,

$$\lim_{x \to +\infty} |y_2^{(i)}(x) - y_1^{(i)}(x)| = 0, \qquad i = 1, 2, \ldots, n-1.$$

Theorem 4.15 *Any solution of Eq. (4.15) is stable when $x \to +\infty$ if and only if $\mathrm{Re}\{r_i\} \le 0$ for all the roots of the characteristic equation, and the roots on the imaginary axis, $\mathrm{Re}\{r_i\} = 0$, (if there are any) are simple.*

Example 4.17 The equation of motion for a mass attached to a spring and moving without friction is

$$my'' + ky = 0, \qquad m, k > 0.$$

See Example 1.6. If on the mass there acts friction which is proportional to the speed and its direction is opposite to the direction of motion, then the equation of motion is

$$my'' + ky = -hy', \qquad h > 0.$$

This is a linear, homogeneous differential equation with constant coefficients whose characteristic equation is $mr^2 + hr + k = 0$ and its roots are $r_{1,2} = (-h \pm \sqrt{h^2 - 4km})/2m$.

If $h^2 < 4km$, then r_1, r_2 are complex valued and the motion of the mass is given by

$$y(t) = e^{\alpha t}(A \cos \beta t + B \sin \beta t),$$

$\alpha = \text{Re}\{r_i\} = -h/(2m) < 0$. Therefore, the motion decays to zero through infinite back-and-forth oscillations.

If $h^2 \geq 4km > 0$, then r_1, r_2 are real and negative (since $|h| > \sqrt{h^2 - 4km}$), the equation of motion is of the form

$$y(t) = Ae^{r_1 t} + Be^{r_1 t}, \qquad r_1, r_2 < 0.$$

This motion eventually decays to zero monotonically.

In any case, all solutions of the differential equation are asymptotically stable. In particular, each solution tends to the trivial solution $y(t) \equiv 0$ which describes the state of rest.

A similar analysis holds for the current equation of a simple electric circuit consisting of a resistor, coil, and capacitor, $LI''(t) + RI'(t) + C^{-1}I(t) = 0$. See Example 1.9.

4.8 Euler's Equation

A second-order Euler equation is the equation

$$L[y] \equiv ax^2 y'' + bxy' + cy = 0,$$

in which a, b, c are constants. Analogously, the Euler equation of order n is

$$a_0 x^n y^{(n)} + a_1 x^{n-1} y^{(n-1)} + \ldots + a_{n-1} xy' + a_n y = 0$$

with constant $a_0, a_1, \ldots, a_n$.

We discuss this equation for several reasons:

1. Second-order Euler equation naturally appears in the solution of Laplace's equation (a partial differential equation!) in polar coordinates.
2. Euler equation is a main motivation for solving ordinary differential equations by power series near certain singular points. This is studied in Chap. 7.
3. As we have already mentioned, we know to solve explicitly very few differential equations. Euler's equation will serve us as another solvable example. It was already used in Example 4.5 without mentioning the name of the equation.

We will settle for studying Euler equations of order two. The treatment of higher-order equations is very similar. After normalization the equation is

$$y'' + \frac{b/a}{x}y' + \frac{c/a}{x^2}y = 0$$

which is undefined at $x = 0$. Therefore, we restrict its domain of definition to $(0, \infty)$ or to $(-\infty, 0)$.

The solution of the equation is based on the observation that a change of independent variables $x = e^t$ converts Euler equation into an equation with constant coefficients that we already know to solve. Let $x = e^t$, namely, $t = \ln x$, and mark the unknown function by $y(x) = y(e^t) = v(t)$. According to the chain rule

$$\frac{dy}{dx} = \frac{dv}{dt}\frac{dt}{dx} = \frac{dv}{dt}\frac{1}{x}.$$

To prevent confusion between derivatives with respect to different variables we mark the derivatives by $\frac{d}{dx}, \frac{d}{dt}$ and do not use the notation $(\ )'$. The second derivative with respect to x is

$$\frac{d^2y}{dx^2} = \frac{d}{dx}\left(\frac{dy}{dx}\right) = \frac{d}{dx}\left(\frac{dv}{dt}\frac{1}{x}\right).$$

Here the factor dv/dt depends on t and the second factor, $1/x$, on x. Therefore, we differentiate the first factor according to the chain rule and the second one directly:

$$= \left[\frac{d}{dt}\left(\frac{dv}{dt}\right)\frac{dt}{dx}\right]\frac{1}{x} + \frac{dv}{dt}\left(-\frac{1}{x^2}\right)$$

$$= \frac{d^2v}{dt^2}\frac{1}{x}\frac{1}{x} - \frac{dv}{dt}\frac{1}{x^2} = \frac{1}{x^2}\left(\frac{d^2v}{dt^2} - \frac{dv}{dt}\right).$$

The derivatives are placed into the original equation:

$$ax^2\frac{d^2y}{dx^2} + bx\frac{dy}{dx} + cy(x)$$

$$= ax^2\frac{1}{x^2}\left(\frac{d^2v}{dt^2} - \frac{dv}{dt}\right) + bx\left(\frac{1}{x}\frac{dv}{dt}\right) + cv(t)$$

$$= a\frac{d^2v}{dt^2} + (b-a)\frac{dv}{dt} + cv(t) = 0.$$

4.8 Euler's Equation

Table 4.1 Solutions of Euler equation

	Equation with constant coefficients, $-\infty < t < \infty$	Euler equation $0 < x < \infty$
$r_1 \neq r_2$	$e^{r_1 t}$, $e^{r_2 t}$	x^{r_1}, x^{r_2}
$r_1 = r_2$	$e^{r_1 t}$, $t e^{r_1 t}$	x^{r_1}, $x^{r_1} \ln x$
$\bar{r}_1 = r_2$, $\text{Im}\{r_1\} \neq 0$	$e^{\alpha t} \cos \beta t$, $e^{\alpha t} \sin \beta t$	$x^\alpha \cos(\beta \ln x)$, $x^\alpha \sin(\beta \ln x)$

This equation with constant coefficients has a characteristic polynomial:

$$P(r) = ar^2 + (b-a)r + c = 0, \tag{4.28}$$

and when $r_1 \neq r_2$, its solutions are $v_1(t) = e^{r_1 t}$, $v_2(t) = e^{r_2 t}$. When we return to the original variable $x = e^t$, we find that the original Euler equation has two solutions $y_1(x) = x^{r_1}$, $y_2(x) = x^{r_2}$.

If $r_1 = r_2$, the solution $v_2(t) = te^{r_1 t}$ corresponds to $y_2(x) = x^{r_1} \ln x$. For complex conjugates $r_2 = \bar{r}_1$ the solutions $v_1(t) = e^{\alpha t} \cos \beta t$, $v_2(t) = e^{\alpha t} \sin \beta t$ turn into

$$y_1(x) = x^\alpha \cos(\beta \ln x), \quad y_2(x) = x^\alpha \sin(\beta \ln x) .$$

The solutions are summarized in Table 4.1.

Once the form of the solution of an Euler equation is already known, there is no practical need to change the variables. It is enough to substitute a supposed solution of the form $y = x^r$ into the given differential equation and find for which values of r it is a solution. In fact, the substitution $y = x^r$ into the equation $L[y] \equiv ax^2 y'' + bxy' + cy = 0$ yields

$$L[x^r] = ax^2 \cdot r(r-1)x^{r-2} + bx \cdot rx^{r-1} + cx^r$$
$$= x^r [ar(r-1) + br + c] = x^r P(r) .$$

It turns out, as expected, that x^r is a solution when r is a root of the equation

$$P(r) = ar(r-1) + br + c = 0 .$$

This polynomial is the same as the characteristic polynomial (4.28) of the equation with the constant coefficients that was obtained by the change of variables. If two roots r_1, r_2 are different from each other, then the two solutions of the Euler equation are x^{r_1}, x^{r_2}. What should be done if $r_1 = r_2$? The two roots of the characteristic equation

$$r_{1,2} = \frac{-(b-a) \pm \sqrt{(b-a)^2 - 4ac}}{2a}$$

are equal only when the discriminant is zero and then $r_1 = r_2 = -\dfrac{b-a}{2a}$. In this case the first solution is $x^{r_1} = x^{(b-a)/2a}$ and the second one is $x^{r_1} \ln x$. This conclusion is obtained also by using the reduction of order method and formula (4.11). See also the next exercise.

Exercise 4.10 It is given that $r_1 = r_2 = -\dfrac{b-a}{2a}$ and $y_1(x) = x^{r_1}$. Verify that in this case the second solution

$$y_2(x) = y_1(x) \int y_1^{-2} e^{-\int p_1 \, dx} \, dx$$

of the Euler equation is $x^{r_1} \ln x$. Note that $p_1(x) = b/ax$.

4.9 Nonhomogeneous Linear Equations

In the previous sections we dealt with homogeneous linear equations of the form

$$\text{(H)} \quad y^{(n)} + p_1(x) y^{(n-1)} + \cdots + p_{n-1}(x) y' + p_n(x) y = 0 \, .$$

We determined the structure of its solution space and found several properties of the solutions. Now we study the nonhomogeneous equation:

$$\text{(NH)} \quad y^{(n)} + p_1(x) y^{(n-1)} + \cdots + p_{n-1}(x) y' + p_n(x) y = q(x) \, .$$

Usually the homogeneous equation represents the free self-motion of a certain model, whereas the nonhomogeneous equation describes the behavior of the same model under the influence of an external force. For example, the equation $mx''(t) + kx(t) = 0$ represents the free self-oscillations of a spring while

$$mx''(t) + kx(t) = F(t)$$

describes the motion of the same spring when an external force $F(t)$ is applied to it. Similarly, the homogeneous equation

$$LI''(t) + RI'(t) + C^{-1} I(t) = 0$$

represents the natural response of a simple resistor-inductor-capacitor circuit (RLC) while

$$LI''(t) + RI'(t) + C^{-1} I(t) = E_0 \cos \omega t$$

4.9 Nonhomogeneous Linear Equations

yields the response of the circuit to an external periodic potential. In general, it is possible to refer to $q(x)$ as the input of the system (NH) and to the solution $y(x)$ as the system's response—its output.

Suppose that $y_1(x)$, $y_2(x)$ are two solutions of (NH):

$$L[y_1] \equiv y_1^{(n)} + p_1(x)y_1^{(n-1)} + \cdots + p_n(x)y_1 = q(x),$$

$$L[y_2] \equiv y_2^{(n)} + p_1(x)y_2^{(n-1)} + \cdots + p_n(x)y_2 = q(x).$$

Let us subtract them from each other:

$$L[y_2 - y_1] \equiv (y_2 - y_1)^{(n)} + p_1(x)(y_2 - y_1)^{(n-1)} + \cdots + p_n(x)(y_2 - y_1) = 0.$$

Therefore, $y_2 - y_1$ is a solution of the corresponding homogeneous equation (H). But the structure of the solution space of (H) is known: Every solution is given by a linear combination of a basis of solutions, say $u_1(x), \ldots, u_n(x)$. So

$$y_2(x) - y_1(x) = c_1 u_1(x) + \cdots + c_n u_n(x).$$

From here on we decide to fix y_1 as one particular solution of (NH) and name it y_p, and take y_2 to be any solution of (NH), marked as $y_{\text{NH}}(x)$. Thus,

$$y_{\text{NH}}(x) = y_p(x) + c_1 u_1(x) + \cdots + c_n u_n(x).$$

We summarize:

Theorem 4.16 *The general solution of* (NH) *can be written as one particular solution of* (NH) *plus the linear combinations of the solutions of* (H). *The choice of the particular solution is not unique. The collection of all solutions of* (NH) *depends on n arbitrary parameters* $c_1, \ldots, c_n$, *but it does not constitute a linear subspace.*

Assuming that we know the solutions of (H) (a theoretical assumption, since we know about their existence and the structure of the solution space, but usually do not know them explicitly), we are left with the problem to find one particular solution of (NH). This problem will be solved in the next section, Sect. 4.10

We close this section with some examples.

Example 4.18 Solve the equation $y'' - y = x$.

We guess that $y_p(x) = -x$ is a particular solution of the given equation. The solutions of the corresponding homogeneous equation $u'' - u = 0$ are $u_1(x) = e^x$, $u_2(x) = e^{-x}$; therefore, a general solution is

$$y(x) = -x + c_1 e^x + c_2 e^{-x}.$$

The same general solution can be written also as

$$y(x) = (-x + 5e^x) + (c_1 - 5)e^x + c_2 e^{-x}.$$

This time we identify $-x + 5e^x$ as a particular solution of (NH) and c_1, c_2 represent arbitrary coefficients of the linear combinations of solutions of (H).

If we add two initial value conditions to the equation, say $y(1) = 3$, $y'(1) = 5$, the values of c_1, c_2 are determined from the two equations:

$$y(1) = -1 + c_1 e + c_2 e^{-1} = 3,$$

$$y'(1) = -1 + c_1 e - c_2 e^{-1} = 5.$$

Example 4.19 Given that three functions $f_1(x)$, $f_2(x)$, $f_3(x)$ are three different solutions of a linear, nonhomogeneous equation of the second order, $y'' + p_1 y' + p_2 y = q$, what are all the solutions of the equation?

As said above, the differences

$$u_1(x) = f_2(x) - f_1(x), \quad u_2(x) = f_3(x) - f_1(x)$$

are two solutions of the corresponding homogeneous equation. Assume that they are linearly independent. Take, for example, $f_1(x)$ as a particular solution of the given nonhomogeneous equation. Thus, the general solution can be written in the form

$$y_{\text{NH}}(x) = f_1(x) + c_1 \big[f_2(x) - f_1(x)\big] + c_2 \big[f_3(x) - f_1(x)\big],$$

when c_1, c_2 are two arbitrary constants.

This choice is not the only possible one. In the same way, we can also write

$$y_{\text{NH}}(x) = f_3(x) + d_1 \big[f_1(x) - f_2(x)\big] + d_2 \big[f_2(x) - f_3(x)\big].$$

Find the connection between c_1, c_2, and d_1, d_2.

Exercise 4.11 Given that $y_1(x)$ is a solution of the nonhomogeneous equation $L[y] = q_1(x)$ and that $y_2(x)$ is a solution of $L[y] = q_2(x)$. Prove that $y_1(x) + y_2(x)$ is a solution of the equation $L[y] = q_1(x) + q_2(x)$.

4.10 The Method of Variation of Parameters

We have already mentioned that if we know the solutions of the homogeneous linear equations, it remains to find one particular solution of the nonhomogeneous equation. How that can be done? We present a method that was developed by Lagrange[7] and answers this question.

Here is the fundamental idea: All solutions of the homogeneous equation (H) are of the form $c_1 u_1(x) + \cdots + c_n u_n(x)$, where $u_1(x), \ldots, u_n(x)$ are a basis of

---

[7] Joseph Lagrange, 1736–1813.

4.10 The Method of Variation of Parameters

solutions of (H) and $c_1, \ldots, c_n$ are n arbitrary constants. For the nonhomogeneous equation (NH) we try to find a particular solution of the form

$$y(x) = c_1(x)u_1(x) + \cdots + c_n(x)u_n(x) \qquad (4.29)$$

when $u_1(x), \ldots, u_n(x)$ are as above and $c_1(x), \ldots, c_n(x)$ are n unknown functions. Due to the idea of replacing the constant parameters $c_1, \ldots, c_n$ by functions, the method is called the *method of variation of parameters* or *variation of constants*. A similar technique was used in Chap. 2, where for a linear, nonhomogeneous first-order equation we looked for a solution of the form $c(x)e^{-\int p(x)\,dx}$, with an unknown function $c(x)$.

It is not obvious a priori why are we looking for a particular solution precisely of the form (4.29) and why this is indeed possible. At the end of the process, it will become clear that the method is successful and is also simple and effective.

We have n unknown functions $c_1(x), \ldots, c_n(x)$ and only one restriction, namely, (NH). Therefore, it is likely that along the way we can and will impose $n-1$ more conditions (equations) on $c_1(x), \ldots, c_n(x)$. During the process we differentiate (4.29) n times and substitute all the derivatives into (NH). The first step is, after a rearrangement,

$$y' = (c_1'u_1 + c_1u_1') + (c_2'u_2 + c_2u_2') + \ldots + (c_n'u_n + c_nu_n')$$
$$= (c_1'u_1 + c_2'u_2 + \ldots + c_n'u_n) + (c_1u_1' + c_2u_2' + \ldots + c_nu_n') .$$

As a first of the $n-1$ requirements we alluded to, let us choose that the first group of terms equals 0, that is,

$$c_1'u_1 + c_2'u_2 + \ldots + c_n'u_n = 0 . \qquad (4.30)$$

Then y' will consist only of the second group of terms:

$$y' = c_1u_1' + c_2u_2' + \ldots + c_nu_n' .$$

The logic and benefit of this choice will become clear in a short time. Next, we differentiate this y':

$$y'' = (c_1'u_1' + c_1u_1'') + (c_2'u_2' + c_2u_2'') + \ldots + (c_n'u_n' + c_nu_n'')$$
$$= (c_1'u_1' + c_2'u_2' + \ldots + c_n'u_n') + (c_1u_1'' + c_2u_2'' + \ldots + c_nu_n'') .$$

Our second condition is that the first group of terms in y'' equals 0:

$$c_1'u_1' + c_2'u_2' + \ldots + c_n'u_n' = 0. \qquad (4.31)$$

As a result, y'' is transformed into

$$y'' = c_1 u_1'' + c_2 u_2'' + \ldots + c_n u_n''.$$

The purpose of our conditions begins to clarify: We ensure that despite the repeated differentiations, no derivatives of the unknown $c_1(x), \ldots, c_n(x)$ of order higher than one appear explicitly.

At the third differentiation, we accept

$$y''' = (c_1' u_1'' + c_2' u_2'' + \ldots + c_n' u_n'') + (c_1 u_1''' + c_2 u_2''' + \ldots + c_n u_n''').$$

This time we demand that

$$c_1' u_1'' + c_2' u_2'' + \ldots + c_n' u_n'' = 0 \tag{4.32}$$

and there remains

$$y''' = c_1 u_1''' + c_2 u_2''' + \ldots + c_n u_n''',$$

and so on. After the $(n-1)$th differentiation we assume that

$$c_1' u_1^{(n-2)} + c_2' u_2^{(n-2)} + \ldots + c_n' u_n^{(n-2)} = 0 \tag{4.33}$$

and there remains only

$$y^{(n-1)} = c_1 u_1^{(n-1)} + c_2 u_2^{(n-1)} + \ldots + c_n u_n^{(n-1)}.$$

So far we have specified $n-1$ conditions, as announced in advance. The last step, the nth differentiation, is

$$y^{(n)} =$$
$$\left(c_1' u_1^{(n-1)} + c_2' u_2^{(n-1)} + \ldots + c_n' u_n^{(n-1)} \right) + \left(c_1 u_1^{(n)} + c_2 u_2^{(n)} + \ldots + c_n u_n^{(n)} \right). \tag{4.34}$$

This time we demand no further condition, but rather substitute the derivatives $y^{(n)}, \ldots, y', y$ which were calculated above into the original differential equation:

(NH) $\quad y^{(n)} + p_1(x) y^{(n-1)} + \cdots + p_{n-1}(x) y' + p_n(x) y = q(x).$

4.10 The Method of Variation of Parameters

Note that the equation is normalized and the coefficient of $y^{(n)}$ is 1. The result of the substitution is

$$\left(c_1' u_1^{(n-1)} + c_2' u_2^{(n-1)} + \ldots + c_n' u_n^{(n-1)}\right) + \left(c_1 u_1^{(n)} + c_2 u_2^{(n)} + \ldots + c_n u_n^{(n)}\right)$$

$$+ p_1(x)\left(c_1 u_1^{(n-1)} + c_2 u_2^{(n-1)} + \ldots + c_n u_n^{(n-1)}\right)$$

$$\vdots$$

$$+ p_n(x)\left(c_1 u_1 + c_2 u_2 + \ldots + c_n u_n\right) = q(x) \,.$$

We add the terms, except for the first group which contains $c_1', \ldots, c_n'$, in the vertical direction and reorganize them by pulling $c_1, \ldots, c_n$ out of the corresponding groups:

$$\left(c_1' u_1^{(n-1)} + \ldots + c_n' u_n^{(n-1)}\right) + c_1 \left(u_1^{(n)} + p_1 u_1^{(n-1)} + \ldots + p_n u_1\right)$$

$$+ c_2 \left(u_2^{(n)} + p_1 u_2^{(n-1)} + \ldots + p_n u_2\right)$$

$$\vdots$$

$$+ c_n \left(u_n^{(n)} + p_1 u_n^{(n-1)} + \ldots + p_n u_n\right) = q(x) \,.$$

But u_i is a solution of (H), that is, $u_i^{(n)} + p_1 u_i^{(n-1)} + \ldots + p_n u_i = 0$ for all $i = 1, \ldots, n$. Due to the zeroing of these sums, there remains only the equation

$$c_1' u_1^{(n-1)} + c_2' u_2^{(n-1)} + \ldots + c_n' u_n^{(n-1)} = q(x) \,. \tag{4.35}$$

We have finished the calculation of n derivatives and put them into the original nonhomogeneous differential equation (NH). On the way there accumulated the n Eqs. (4.30), (4.31), ..., (4.33), and (4.35):

$$\begin{aligned} c_1' u_1 + c_2' u_2 + \ldots + c_n' u_n &= 0 \,, \\ c_1' u_1' + c_2' u_2' + \ldots + c_n' u_n' &= 0 \,, \\ &\vdots \\ c_1' u_1^{(n-2)} + c_2' u_2^{(n-2)} + \ldots + c_n' u_n^{(n-2)} &= 0 \,, \\ c_1' u_1^{(n-1)} + c_2' u_2^{(n-1)} + \ldots + c_n' u_n^{(n-1)} &= q(x) \,. \end{aligned} \tag{4.36}$$

For each value of x this is a system of n linear, nonhomogeneous algebraic equations, of the n unknown $c_1'(x), \ldots, c_n'(x)$. The determinant of the system is exactly Wronskian $W(u_1, \ldots, u_n)$ which is different from zero. Therefore, the system (4.36) has one and only one solution $c_1'(x), \ldots, c_n'(x)$.

The original unknown are the functions $c_1(x), \ldots, c_n(x)$ and they are obtained by integration as $c_i(x) = \int c_i'(x)\,dx + K_i$, where K_i are n constants of integration. Hence, a solution of (NH) that we looked for (and found) is

$$y(x) = c_1(x)u_1(x) + \ldots + c_n(x)u_n(x)$$

$$= \left(\int c_1'(x)\,dx + K_1\right) u_1(x) + \ldots + \left(\int c_n'(x)\,dx + K_n\right) u_n(x)$$

$$= \left(u_1(x) \int c_1'(x)\,dx + \ldots + u_n(x) \int c_n'(x)\,dx\right)$$

$$+ \Big(K_1 u_1(x) + \ldots + K_n u_n(x)\Big).$$

Although we looked only for one particular solution, in practice we got an expression that includes a general solution of (H), $K_1 u_1(x) + \ldots + K_n u_n(x)$, with arbitrary constants $K_1, \ldots, K_n$. Therefore, we have a general solution of (NH), whose first group of terms can be identified as a particular solution $y_p(x)$.

Remark Do not forget that in this process we assumed that the equation (NH) is written in a normalized form.

Example 4.20

$$y'' - 2y + y = \frac{e^x}{x^2}.$$

The corresponding homogeneous equation has constant coefficients, its characteristic polynomial is $r^2 - 2r + 1 = (r-1)^2 = 0$ whose roots are $r_1 = r_2 = 1$, and therefore, a basis of solutions of (H) is $u_1(x) = e^x$, $u_2(x) = xe^x$. Next, we look for a solution of the nonhomogeneous equation of the form

$$y = c_1(x)\,e^x + c_2(x)\,xe^x.$$

There is no need, of course, to repeat the process of finding the system (4.36). Having proved it once, we will use it for our convenience as needed. In the current example, the system of algebraic equations (4.36) for $c_1'(x)$, $c_2'(x)$ is

$$c_1' e^x + c_2' xe^x = 0,$$

$$c_1' e^x + c_2'(e^x + xe^x) = \frac{e^x}{x^2}.$$

We subtract one equation from the other and get $c_2'(x) = 1/x^2$; hence, it follows that $c_1'(x) = -1/x$. By integrations, $c_2(x) = -1/x + K_2$, $c_1(x) = -\ln x + K_1$. To

4.10 The Method of Variation of Parameters

summarize,

$$\begin{aligned} y_{\text{NH}}(x) &= c_1(x)u_1(x) + c_2(x)u_2(x) \\ &= (-\ln x + K_1)e^x + (-1/x + K_2)xe^x \\ &= (-\ln x \cdot e^x - e^x) + (K_1 e^x + K_2 xe^x) \,. \end{aligned}$$

At this point we considered the term $(-\ln x \cdot e^x - e^x)$ as a particular solution, and all the other terms as a general solution of the homogeneous equation. However, it may be more convenient to write the same result as

$$y_{\text{NH}}(x) = -\ln x \cdot e^x + (K_1 - 1)e^x + K_2 xe^x$$

and $y_p(x) = -\ln x \cdot e^x$ is identified as a particular solution.

Exercise 4.12 Solve the equation $x^2 y'' - 2xy' + 2y = 4x^7$ with the two initial value conditions $y(1) = 3$, $y'(1) = 5$. Note that this equation is not in normalized form.

So far we have presented the variation of parameters method as an algorithm that ultimately provides a particular solution to a nonhomogeneous equation. In addition, an explicit formula for the solution can also be derived from the method. We will do this, for simplicity, only for second-order equations.

For the second-order nonhomogeneous equation $y'' + p_1(x)y' + p_2(x)y = q(x)$, the requirements (4.36) of the variation of parameters are expressed by the two equations:

$$c_1' u_1(x) + c_2' u_2(x) = 0 \,,$$
$$c_1' u_1'(x) + c_2' u_2'(x) = q(x) \,.$$

The explicit solution, given by Cramer's formula or by direct calculation, is

$$c_1'(x) = \frac{\begin{vmatrix} 0 & u_2 \\ q & u_2' \end{vmatrix}}{\begin{vmatrix} u_1 & u_2 \\ u_1' & u_2' \end{vmatrix}} = \frac{-u_2(x)q(x)}{W(u_1, u_2)(x)} \,,$$

$$c_2'(x) = \frac{\begin{vmatrix} u_1 & 0 \\ u_1' & q \end{vmatrix}}{\begin{vmatrix} u_1 & u_2 \\ u_1' & u_2' \end{vmatrix}} = \frac{u_1(x)q(x)}{W(u_1, u_2)(x)} \,.$$

In the integration of c_1', c_2' we disregard the arbitrary constants of integration and the lower limits of the integrals, since they only generate solution $K_1 u_1(x) + K_2 u_2(x)$

of the homogeneous equation. The result is

$$y_p(x) = c_1(x)u_1(x) + c_2(x)u_2(x)$$
$$= u_1(x)\int^x \frac{-u_2(t)q(t)}{W(u_1,u_2)(t)}\,dt + u_2(x)\int^x \frac{u_1(t)q(t)}{W(u_1,u_2)(t)}\,dt\ .$$

Here we denoted the variable of integration within the integral by t to distinguish it from the variable x outside the integral. The choice of the variable of integration is completely arbitrary as long as we take care to place x as the upper limit of the integral. After removing the concern of confusion between the different variables, we push $u_1(x)$, $u_2(x)$ into the integrals (since these are constants with respect to the integration according to t) and unite the two integrals. The result is

$$y_p(x) = \int^x \frac{-u_1(x)u_2(t) + u_2(x)u_1(t)}{u_1(t)u_2'(t) - u_2(t)u_1'(t)}\,q(t)\,dt$$

$$= \int^x \frac{\begin{vmatrix} u_1(t) & u_2(t) \\ u_1(x) & u_2(x) \end{vmatrix}}{\begin{vmatrix} u_1(t) & u_2(t) \\ u_1'(t) & u_2'(t) \end{vmatrix}}\,q(t)\,dt\ .$$

For brevity, we write it in the form

$$y_p(x) = \int^x K(x,t)q(t)\,dt\ . \tag{4.37}$$

The function $K(x,t)$ is called *Cauchy kernel*.[8] The kernel function $K(x,t)$ depends solely on the solutions u_1, u_2 of the homogeneous equation, while $q(t)$ represent the right-hand side of the nonhomogeneous equation. This formula illustrates the role of $q(x)$ as the input of the model and $y_p(x)$ as its output. The model acts as a "black box" according to the rule

$$\xrightarrow{\text{input } q}\ \boxed{\int^x K(x,t)q(t)\,dt}\ \xrightarrow{\text{output } y_p} \tag{4.38}$$

Presentation of functions by integrals of the type (4.37) occurs in many areas of mathematics. In Chap. 8 we will meet another kernel function.

Example 4.21

$$y'' + y = q(x)\ .$$

---

[8] Augustin Louis Cauchy, 1789–1857.

The solutions of the homogeneous equation are $u_1(x) = \cos x$, $u_2(x) = \sin x$,

$$K(x,t) = \frac{\begin{vmatrix} \cos t & \sin t \\ \cos x & \sin x \end{vmatrix}}{\begin{vmatrix} \cos t & \sin t \\ -\sin t & \cos t \end{vmatrix}} = \sin(x-t),$$

and finally

$$y_p(x) = \int^x \sin(x-t) \, q(t) \, dt.$$

Expressions of the form $\int_0^x f(x-t) g(t) \, dt$ appear in many uses and are called a *convolution*. See Definition 8.3.

4.11 The Method of Undetermined Coefficients

The variation of parameters method is suitable for any nonhomogeneous linear equation when the solutions of the corresponding homogeneous equation are known. The method involves integrations of the $c_i'(t)$-s, which may be difficult to do. For a certain type of nonhomogeneous equations, there exists a method of intelligent and systematic "guessing" to find a particular solution, requiring only elementary algebra. These are the equations where the left-hand side is a differential operator with constant coefficients and the right-hand side contains only polynomials, exponential functions, trigonometric functions, their sums, and their products. The reason for choosing precisely these functions and not others will become clear later.

First, we will demonstrate the method when the left-hand side is $ay'' + by' + cy$ and on the right side appears a polynomial $Q_m(x) = q_0 x^m + q_1 x^{m-1} + \cdots + q_{m-1} x + q_m$:

$$L[y] \equiv ay'' + by' + cy = q_0 x^m + q_1 x^{m-1} + \cdots + q_{m-1} x + q_m. \tag{4.39}$$

We will deal with equations of this type in great detail so that the solution will serve as an introduction to the following equations. Since the derivatives of a polynomial are polynomials of lower degree, it makes sense to look for a particular solution that is also a polynomial. When $c \neq 0$, we try a polynomial exactly of degree m:

$$y_p(x) = c_0 x^m + c_1 x^{m-1} + \cdots + c_{m-1} x + c_m \tag{4.40}$$

whose coefficients are yet unknown and we wish to determine. This is the origin of the name of the method, *undetermined coefficients*. Before we deal with the general problem of Eq. (4.39), we solve a simple example.

Example 4.22

$$y'' - y' + y = 3x^2 + 4x + 5.$$

We look for a particular solution $y_p(x) = Ax^2 + Bx + C$ and substitute this supposed solution and its derivatives $y' = 2Ax + B$, $y'' = 2A$ in the differential equation:

$$(2A) - (2Ax + B) + (Ax^2 + Bx + C) = 3x^2 + 4x + 5.$$

Let us compare the coefficients of the different powers:

$$\begin{aligned} x^2 : \quad & A & = 3, \\ x^1 : \quad & -2A + B & = 4, \\ x^0 : \quad & 2A - B + C & = 5. \end{aligned}$$

The coefficients are found to be $A = 3$, $B = 10$, $C = 9$, a particular solution is $y_p(x) = 3x^2 + 10x + 9$, and a general solution is

$$y_{\text{NH}}(x) = 3x^2 + 10x + 9 + K_1 e^{r_1 x} + K_2 e^{r_2 x}.$$

K_1, K_2 are determined by two initial value conditions.

Now we substitute the supposed solution (4.40) in Eq. (4.39):

$$\begin{aligned} L[y_p] &= ay_p'' + by_p' + cy_p \\ &= a\left(c_0 m(m-1)x^{m-2} + c_1(m-1)(m-2)x^{m-3} + \cdots\right) \\ &\quad + b\left(c_0 m x^{m-1} + c_1(m-1)x^{m-2} + c_2(m-2)x^{m-3} + \cdots\right) \\ &\quad + c\left(c_0 x^m + c_1 x^{m-1} + c_2 x^{m-2} + \cdots +\right) \\ &= q_0 x^m + q_1 x^{m-1} + \cdots + q_{m-1} x + q_m. \end{aligned}$$

Let us compare the $m+1$ coefficients of the powers $x^m, x^{m-1}, \ldots, x, 1$ on the two sides:

$$\begin{aligned} x^m : \quad & c\, c_0 = q_0, \\ x^{m-1} : \quad & bm\, c_0 + c\, c_1 = q_1, \\ & \cdots \\ x^{m-k} : \quad & a(m-k+2)(m-k+1)c_{k-2} + b(m-k+1)c_{k-1} + c\, c_k = q_k, \end{aligned}$$

4.11 The Method of Undetermined Coefficients

for $k = 2, 3, \ldots, m$. If $c \neq 0$, the unknown $c_0, \ldots, c_m$ are easily calculated due to the simple structure of the equations. From the first equation we get $c_0 = q_0/c$, substitute it into the second equation, and calculate $c_1 = (q_1 - bmc_0)/c = q_1/c - bmq_0/c^2$, and recursively

$$c_k = \big(q_k - a(m-k+2)(m-k+1)c_{k-2} - b(m-k+1)c_{k-1}\big)/c .$$

Thus, if $c \neq 0$, the problem has a unique solution $c_0, \ldots, c_m$, which proves the existence of a particular solution of the form $c_0 x^m + \ldots + c_m$ and justifies the guess (4.40).

If in (4.39) we have $c = 0$ but $b \neq 0$, the differential equation becomes

$$ay'' + by' = q_0 x^m + \cdots + q_m .$$

To have the power x^m on both sides, we look for a particular solution that is a polynomial of degree $m+1$:

$$y_p(x) = c_0 x^{m+1} + c_1 x^m + \cdots + c_m x + c_{m+1} .$$

Its substitution into $ay'' + by' = q(x)$, $b \neq 0$, leads again to a recursive formula for $c_0, \ldots, c_m$ (but c_{m+1} disappears and remains arbitrary!). The differential equation $ay'' + by' = 0$ has the characteristic polynomial $ar^2 + br = 0$ with roots $r_1 = 0$, $r_2 = -b/a \neq 0$. So the general solution of this homogeneous equation is $K_1 + K_2 e^{r_2 x}$ with arbitrary K_1, K_2. The term c_{m+1} of the particular solution is absorbed in the arbitrary K_1 and is not mentioned anymore. So the requested particular solution will be $y_p(x) = x(c_0 x^m + c_1 x^{m-1} + \cdots + c_m)$.

If $c = b = 0$, $a \neq 0$ in (4.39), our differential equation is

$$ay'' = q_0 x^m + \cdots + q_m .$$

This is easily solved by two integrations, but for sake of compatibility we prefer to look again for a particular solution of the form

$$y_p(x) = c_0 x^{m+2} + + \cdots + c_m x^2 + c_{m+1} x + c_{m+2} .$$

The corresponding characteristic polynomial is $ar^2 = 0$, its two roots are $r_1 = r_2 = 0$, and the solutions of the homogeneous equation are $K_1 + K_2 x$. The terms $c_{m+1} x + c_{m+2}$ of $y_p(x)$ are absorbed in $K_1 + K_2 x$, so it is enough to take a particular solution:

$$y_p(x) = x^2(c_0 x^m + c_1 x^{m-1} + \cdots + c_m) .$$

The results of the discussion are summarized in a way that is not the most simple one, but which is advantageous for the further investigation:

Summary *A nonhomogeneous equation* $ay'' + by' + cy = Q_m(x)$ *where* $Q_m(x)$ *is a polynomial of degree m has a particular solution of the form*

$$y_p(x) = x^s(c_0 x^m + c_1 x^{m-1} + \cdots + c_m) , \qquad (4.41)$$

where $s = 0, 1, 2$ *is the multiplicity of the number* $r = 0$ *as a root of the characteristic polynomial* $P(r)$ *of* (NH).

The coefficients $c_0, \ldots, c_m$ *are determined by substitution of* y_p *in the nonhomogeneous equation and the comparison of the coefficients of the different powers on both sides of the equation.*

We move on to nonhomogeneous equations where the right-hand side contains both a polynomial and an exponential function:

$$L[y] \equiv ay'' + by' + cy = e^{\alpha x} Q_m(x) .$$

Since the derivative of $e^{\alpha x}$ is a multiple of $e^{\alpha x}$, we get rid of the exponential function in the equation by the change of variables $y = e^{\alpha x} u$:

$$y' = e^{\alpha x} u' + \alpha e^{\alpha x} u = e^{\alpha x} (u' + \alpha u) ,$$

$$y'' = e^{\alpha x}(u'' + \alpha u') + \alpha e^{\alpha x}(u' + \alpha u) = e^{\alpha x}(u'' + 2\alpha u' + \alpha^2 u) .$$

$$L[e^{\alpha x} u] = a e^{\alpha x} \left(u'' + 2\alpha u' + \alpha^2 u \right) + b e^{\alpha x} (u' + \alpha u) + c e^{\alpha x}$$

$$= e^{\alpha x} \left[au'' + (2a\alpha + b)u' + (a\alpha^2 + b\alpha + c)u \right] = e^{\alpha x} Q_m(x) .$$

After dividing by $e^{\alpha x} \neq 0$ we have

$$au'' + (2a\alpha + b)u' + (a\alpha^2 + b\alpha + c)u = Q_m(x) , \qquad (4.42)$$

an equation with constant coefficients and a polynomial right-hand side. Thus, we returned to an equation of type (4.39) that has been solved above by the method of undetermined coefficients.

The characteristic polynomial of the homogeneous equation $ay'' + by' + cy = 0$ is $P(r) = ar^2 + br + c$ and its derivative is $P'(r) = 2ar + b$. Therefore, Eq. (4.42) may be written as

$$au'' + P'(\alpha)u' + P(\alpha)u = Q_m(x) .$$

According to the solution of (4.39) we distinguish between the case where $P(\alpha) \neq 0$, the case where $P(\alpha) = 0$, $P'(\alpha) \neq 0$, and the case $P(\alpha) = P'(\alpha) = 0$. These three cases are interpreted, respectively, as follows:

1. $P(\alpha) \neq 0 \longleftrightarrow \alpha$ is not a root of the characteristic polynomial $P(r)$.
2. $P(\alpha) = 0, \ P'(\alpha) \neq 0 \longleftrightarrow \alpha$ is a root of multiplicity 1 of $P(r)$.
3. $P(\alpha) = 0, \ P'(\alpha) = 0 \longleftrightarrow \alpha$ is a root of multiplicity 2 of $P(r)$.

4.11 The Method of Undetermined Coefficients

Consequently, according to (4.41), a particular solution u_p of (4.42) is $u_p(x) = x^s(c_0 x^m + \ldots + c_m)$, where $s = 0, 1, 2$ is the multiplicity of $r = \alpha$ as a root of the characteristic polynomial $P(r)$. In terms of the original unknown $y = e^{\alpha x} u$ we have a particular solution

$$y_p(x) = x^s(c_0 x^m + \ldots + c_m) e^{\alpha x} .$$

The coefficients $c_0, \ldots, c_m$ are calculated by substituting it in the differential equation and comparing the coefficients on both sides. To this, we add the general solution of the homogeneous equation.

The following equations demonstrate the different situations that were discussed.

Example 4.23 Consider the nonhomogeneous equation:

$$y'' - 6y' + 9y = e^{3x}(5x + 7) .$$

On the right-hand side there is a polynomial of degree $m = 1$. The roots of the characteristic polynomial $r^2 - 6r + 9 = 0$ are $r_1 = r_2 = 3$. As the exponent also contains $\alpha = 3$, its multiplicity as a root is $s = 2$. Consequently, the form of y_p is

$$y_p(x) = x^2(c_0 x + c_1) e^{3x} .$$

c_0, c_1 are calculated by substitution into the differential equation and comparison of coefficients.

Example 4.24

$$y'' - 6y' + 9y = e^{2x}(5x + 7) .$$

Here $\alpha = 2$, the polynomial is of degree $m = 1$, and its roots are $r_1 = r_2 = 3$. In this case $\alpha \neq r_1, r_2$; therefore, $s = 0$,

$$y_p(x) = (c_0 x + c_1) e^{2x} .$$

Example 4.25

$$y'' - 5y' + 6y = e^{2x}(5x + 7) .$$

This time $\alpha = 2$, $m = 1$, the roots of the characteristic polynomial $r^2 - 5r + 6 = 0$ are $r_1 = 2$, $r_2 = 3$. Since $\alpha = r_1 \neq r_2$, we have $s = 1$ and y_p is of the form

$$y_p(x) = x(c_0 x + c_1) e^{2x} .$$

Finally, we discuss equations in which the right-hand side contains, in addition to a polynomial and an exponential function, also sine or cosine, say

$$L[y] \equiv ay'' + by' + cy = e^{\alpha x} \cos(\beta x) Q_m(x). \tag{4.43}$$

By (4.21) we rewrite Eq. (4.43) as

$$L[y] \equiv ay'' + by' + cy = \frac{1}{2}\left(e^{(\alpha+i\beta)x} + e^{(\alpha-i\beta)x}\right) Q_m(x).$$

We saw in Exercise 4.11 that if $y_1(x)$ is a solution of $L[y] = q_1(x)$ and $y_2(x)$ is a solution of $L[y] = q_2(x)$, then $y_1(x) + y_2(x)$ is the solution of the equation $L[y] = q_1(x) + q_2(x)$. Therefore, in our case, it is sufficient to solve separately each of the two equations:

$$L[y] = \frac{1}{2} Q_m(x) e^{(\alpha+i\beta)x}, \tag{4.44}$$

$$L[y] = \frac{1}{2} Q_m(x) e^{(\alpha-i\beta)x}, \tag{4.45}$$

and finally, add their solutions. We will solve one of them, say (4.44).

If $\alpha + i\beta$ is not a root of the characteristic polynomial, then due to a polynomial and the exponential functions on the right-hand side, there is a particular solution of the form

$$y_1(x) = e^{(\alpha+i\beta)x}(c_0 x^m + \ldots + c_m).$$

The complex number in the exponent causes no difficulty because algebraic operations and differentiations in the presence of complex constants are the same as the corresponding operations for real constants.

The second Eq. (4.45) differs from the previous one only by the replacement of $\alpha + i\beta$ by its complex conjugate $\alpha - i\beta$. This is also the only change in all operations for calculating the unknown $c_0, \ldots, c_m$. Therefore, they will also be replaced by their complex conjugates. In short, the particular solution y_2 of Eq. (4.45) is the complex conjugate of y_1:

$$y_2 = \overline{y_1} = e^{(\alpha-i\beta)x}(\overline{c_0} x^m + \ldots + \overline{c_m}).$$

4.11 The Method of Undetermined Coefficients

A particular solution of the original equation (4.43), $y_1 + y_2$, consists of the sum of two complex conjugate expressions so it is real-valued and can be written as

$$y_p(x) = 2\text{Re}\{y_1\} = 2\text{Re}\left\{e^{(\alpha+i\beta)x}(c_0 x^m + \ldots + c_m)\right\}$$

$$= e^{\alpha x}\text{Re}\Big\{(\cos\beta x + i\sin\beta x)$$

$$\times \big[(d_0 + ie_0)x^m + \cdots + (d_{m-1} + ie_{m-1})x + (d_m + ie_m)\big]\Big\} \quad (4.46)$$

$$= e^{\alpha x}\cos\beta x\big(d_0 x^m + \cdots + d_{m-1}x + d_m\big)$$

$$- e^{\alpha x}\sin\beta x\big(e_0 x^m + \cdots + e_{m-1}x + e_m\big),$$

with $2c_k = d_k + ie_k$. To find the $2(m+1)$ unknown $d_0, \ldots, d_m, e_0, \ldots, e_m$, we substitute $y_p(x)$ into the original differential equation and compare the coefficients of the $2(m+1)$ expressions:

$$x^m \cos\beta x, \; x^m \sin\beta x, \; x^{m-1}\cos\beta x, \; \ldots, \; \cos\beta x, \; \sin\beta x$$

on both sides.

If $\alpha + i\beta$ happens to be equal to a root r_1 of multiplicity s of the characteristic polynomial, its conjugate $\alpha - i\beta$ is equal to the conjugate root $r_2 = \overline{r_1}$ of the same multiplicity. In this case the solution (4.46) that we got above should be multiplied by an additional factor x^s.

Examination of the methods above explains why we can utilize the method of undetermined coefficients to differential equations with constant coefficients and a right-hand side containing polynomials, exponential functions, and trigonometric functions. These are exactly the families of functions that also have their derivatives belonging to the same family. Applying a differential operator with constant coefficients leaves us within the same function space; therefore, the method of undetermined coefficients is possible.

Example 4.26 The equation $y'' + 4y = x\sin 2x$ has on its right-hand side a polynomial of degree $m = 1$, $\alpha = 0$ (no exponential function), $\beta = 2$, so $\alpha + i\beta = 2i$, $\alpha - i\beta = -2i$.

The characteristic polynomial of the left-hand side is $r^2 + 4 = 0$ and its roots are $r_1 = -2i$, $r_2 = 2i$. It turns out that $\alpha \pm i\beta$ are exactly the two roots of the characteristic equation and the multiplicity of each of them is $s = 1$. Therefore, we seek a particular solution of the form

$$y_p(x) = x\big[(A_0 x + A_1)\cos 2x + (B_0 x + B_1)\sin 2x\big].$$

The four unknown A_0, A_1, B_0, B_1 are determined by substituting the required y_p into the differential equation and comparing the coefficients of $\cos 2x$, $x\cos 2x$, $\sin 2x$, $x\sin 2x$ on both sides.

Example 4.27 Solve the equation $y'' + \omega_0^2 y = \cos \omega t$ when $\omega \neq \omega_0$ and when $\omega = \omega_0$.

In this equation ω_0 represents the natural frequency of the homogeneous system (simple harmonic motion) and ω is the frequency of an external driving force, $\cos \omega t$, acting on the system.

According to the notation of our calculations, on the right-hand side $\alpha = 0$, $\beta = \omega$, so $\alpha \pm i\beta = \pm i\omega$. The characteristic polynomial of the left side is $r^2 + \omega_0^2 = 0$ and its roots are $r_1, r_2 = \pm i\omega_0$. Therefore, it is necessary to distinguish between the cases $\omega \neq \omega_0$ and $\omega = \omega_0$.

The Case $\omega \neq \omega_0$ In this case $\alpha \pm i\beta$ are different from r_1, r_2; therefore, a particular solution is of the form

$$y_p(x) = A \cos \omega t + B \sin \omega t \ .$$

Substitute y_p in the differential equation:

$$(-A\omega^2 \cos \omega t - B\omega^2 \sin \omega t) + \omega_0^2 (A \cos \omega t + B \sin \omega t) = \cos \omega t \ .$$

Comparison of the coefficients of $\cos \omega t$, $\sin \omega t$ leads to

$$\cos \omega t : \quad -A\omega^2 + A\omega_0^2 = 1 \ ,$$
$$\sin \omega t : \quad -B\omega^2 + B\omega_0^2 = 0 \ .$$

So $B = 0$, $A = 1/(\omega_0^2 - \omega^2)$ and the general solution is

$$y(t) = \frac{1}{\omega_0^2 - \omega^2} \cos \omega t + K_1 \cos \omega_0 t + K_2 \sin \omega_0 t$$

with arbitrary K_1, K_2.

Note that when the forcing frequency ω approaches the natural frequency ω_0 (but remains different from it), the amplitude of $\cos \omega t$ becomes larger (but independent of time) and the first term becomes the dominant one.

The Case $\omega = \omega_0$ In this case $\alpha \pm i\beta = \pm i\omega$ coalesce with $r_1, r_2 = \pm i\omega_0$ and we must seek a particular solution of the form

$$y_p(t) = t \big[A \cos \omega_0 t + B \sin \omega_0 t \big] \ .$$

Its derivatives are

$$y_p'(t) = \omega_0 t \big[-A \sin \omega_0 t + B \cos \omega_0 t \big] + \big[A \cos \omega_0 t + B \sin \omega_0 t \big] \ ,$$
$$y_p''(t) = -\omega_0^2 t \big[A \cos \omega_0 t + B \sin \omega_0 t \big] + 2\omega_0 \big[-A \sin \omega_0 t + B \cos \omega_0 t \big] \ .$$

4.11 The Method of Undetermined Coefficients

Finally,

$$y_p'' + \omega_0^2 y_p = 2\omega_0 \left[-A \sin \omega_0 t + B \cos \omega_0 t \right] = \cos \omega_0 t.$$

The comparison of coefficients yields $A = 0$, $B = 1/(2\omega_0)$ and the general solution is

$$y(t) = \frac{t \sin \omega_0 t}{2\omega_0} + K_1 \cos \omega_0 t + K_2 \sin \omega_0 t.$$

When $t \to \infty$, the first term "runs wild" and oscillates between $+\infty$ and $-\infty$. This is the *resonance* phenomenon where the forcing frequency is equal to the natural frequency of the system and causes its oscillations to increase.

Note the essential difference between the situation in which "ω is close to ω_0" and the situation in which "ω equals ω_0."

Problems

4.1 Given the equation $y'' + p(x)y' + q(x)y = 0$ with continuous coefficients $p(x)$, $q(x)$, which of the following statements is true and which is not:

(a) Two solutions that satisfy $y(0) = 0$ are linearly dependent.
(b) Two solutions that satisfy $y(0) = 1$ are linearly dependent.
(c) If two solutions y_1, y_2 get their maximal values at the same point, they are linearly dependent.

4.2 Given the equation $y'' + p(x)y' + q(x)y = 0$ with continuous coefficients $p(x)$, $q(x)$, prove that if $y_1(x)$, $y_2(x)$ are the two solutions of the equation and $y_1(x) \neq 0$ in some interval, then the quotient $h(x) = y_2(x)/y_1(x)$ is either a constant or a monotonic function there.

4.3 Given the equation $y'' + p(x)y' + q(x)y = 0$ with continuous coefficients $p(x)$, $q(x)$, it has two solutions $y_1(x)$, $y_2(x)$ which satisfy

$$y_1(a) = y_2(a), \quad y_1(b) = y_2(b)$$

and $y_1(x) \neq 0$ in the interval $[a, b]$. Prove that $y_1(x) \equiv y_2(x)$.
Hint: Use Problem 4.2.
Give an example that without the assumption that $y_1(x) \neq 0$ in $[a, b]$, the claim is false.

4.4

(a) Given a function $f(x) \not\equiv 0$, twice differentiable and satisfying $f(7) = 0$, explain why the function $(f(x))^2$ cannot be a solution of a homogeneous linear equation $y'' + p(x)y' + q(x)y = 0$ with continuous coefficients in a neighborhood of $x = 7$.

(b) Verify that $(x - 7)^{20}$ is a solution of a differential equation of the above type, but with discontinuous coefficients at $x = 7$.

4.5 In each of the following four diagrams there are drawn two functions.

For each pair of functions, explain why they cannot be two solutions of a second-order linear equation $y'' + p(x)y' + q(x)y = 0$, which fulfills the conditions of the existence and uniqueness theorem.

4.6 Calculate the Wronskian of two solutions of a second-order equation of the form $(p(x)y')' + q(x)y = 0$. Assume that $p(x)$ is differentiable and is never 0.

4.7 Let $y_1(x)$, $y_2(x)$ be two solutions of a linear homogeneous differential equation of the second order with continuous coefficients. Prove that if

$$W[y_1, y_2](x_0) = 0, \quad y_1(x_0) = 0,$$

then only one of the following two possibilities may happen: Either $y_1(x) \equiv 0$, or $y_2(x) = \text{const} \cdot y_1(x)$.

4.8 Prove that the two functions $1 - \dfrac{x^2}{2}$, $\cos x$ cannot be solutions of a homogeneous linear equation of order three that satisfies the conditions of the existence and uniqueness theorem everywhere.

4.9 Given that the two functions $y_1(x) = \cos(x^2)$, $y_2(x) = \sin(x^2)$ are solutions of a linear, homogeneous, normalized second-order differential equation. Prove, without finding the differential equation, that its two coefficients cannot be both continuous at the point $x = 0$.

4.10 Given a nth-order differential equation that satisfies the conditions of the existence and uniqueness theorem:

(a) For what values of n can graphs of two different solutions intersect each other?
(b) For what values of n can graphs of two different solutions be tangent to each other?
(c) For what values of n can $y = x^3$ be a solution of the given equation?

4.11 The Method of Undetermined Coefficients

4.11 Given two functions $f_1(x) = x^2$, $f_2(x) = x|x|$, prove:

(a) f_1, f_2 are linearly dependent on $(0, \infty)$. What is the dependency?
(b) f_1, f_2 are linearly dependent on $(-\infty, 0)$. What is the dependency?
(c) f_1, f_2 are not linearly dependent on $(-\infty, \infty)$.
(d) $W(f_1, f_2) \equiv 0$ for all x. Why (d) does not contradict (c)?
(e) Explain why $f_1(x)$, $f_2(x)$ cannot be solutions of any homogeneous, normalized, second-order linear equation with continuous coefficients on the entire line.

4.12

(a) Prove the following statement or find a counterexample:
"If the functions $f_1(x), \ldots, f_n(x)$ are linear independent in a certain interval (a, b), then they are also linearly independent in any larger interval containing (a, b)."
(b) Prove, by some method of your choice, that the functions e^x, e^{2x}, x^3 are linearly independent on the whole line $-\infty < x < \infty$.

4.13

(a) Verify that the four functions $\cos x$, $\sin x$, $\cosh x$, $\sinh x$ are the solutions of the differential equation $y^{(4)} - y = 0$.
(b) What is the solution of the differential equation that satisfies the initial value conditions $y(0) = 1$, $y'(0) = y''(0) = y'''(0) = 0$?
(c) Find the solution that satisfies the initial value conditions

$$y(0) = 0,\ y'(0) = 1,\ y''(0) = y'''(0) = 0.$$

Which solution satisfies $y(0) = y'(0) = 0$, $y''(0) = 1$, $y'''(0) = 0$? And which satisfies $y(0) = y'(0) = y''(0) = 0$, $y'''(0) = 1$?
(d) Find the solution of the initial value conditions

$$y(0) = y'(0) = y''(0) = y'''(0) = 0.$$

(e) Find all solutions that satisfy the two initial value conditions $y(0) = 3$, $y'(0) = 5$. How many such solutions are there?

4.14 Calculate the Wronskian of the four solutions mentioned in Problem 4.13(a) above. Show that this result fits Abel's formula (Theorem 4.10).

4.15 Three functions $\cos x$, $\sin x$, 1 are solutions of a differential, linear, homogeneous equation of order 3. Find the three solutions of the same equation which satisfy, respectively, the three sets of initial conditions:

$$y(0) = 1,\ y'(0) = 0,\ y''(0) = 0,$$
$$y(0) = 0,\ y'(0) = 1,\ y''(0) = 0,$$

and

$$y(0) = 0, \ y'(0) = 0, \ y''(0) = 1 .$$

4.16 For each of the following pairs of functions find an equation $y'' + p(x)y' + q(x)y = 0$ that these are its solutions. Where are $p(x), q(x)$ continuous? What is the Wronskian of a pair solutions?

(a) $\{e^{2x}, e^{5x}\}$.
(b) $\{e^x, xe^x\}$.
(c) $\left\{ \dfrac{\sin x}{\sqrt{x}}, \dfrac{\cos x}{\sqrt{x}} \right\}$.
(d) $\left\{ \dfrac{\sin x}{x}, \dfrac{\cos x}{x} \right\}$.
(e) $\{\sin(x^2), \cos(x^2)\}$. See Problem 4.9 above.
(f) Find a linear, homogeneous equation of order 3 whose solutions are x, e^x, xe^x.

4.17 Find an equation $y'' + p(x)y' + q(x)y = 0$ whose solutions are $\{e^{2x}, xe^x\}$. For this equation solve the initial value problem $y(1) = 0, \ y'(1) = 0$. If you found that this initial value problem has a unique solution, explain why this is the case. If the initial value problem has more than one solution, find all of them and explain why this happens.

4.18 Let $y(x)$ be a nontrivial solution of a differential equation $y'' + p(x)y' + 2y = 0$, and it is given that also $y^2(x)$ is a solution of the same equation. What is the equation and what are its solutions? Find all possible answers to this question.

4.19 In each of the following equations a particular solution is given. Check and confirm. Find all the solutions of each equation. Note that the equations are not normalized!

(a) $(x-1)y'' - xy' + y = 0, \ y_1(x) = e^x$.
(b) $(1-x^2)y'' + 2xy' - 2y = 0, \ y_1(x) = x$.
(c) $x(x+1)y'' - 2y' - 2y = 0, \ y_1(x) = \dfrac{1}{x+1}$. Solve the initial value problem $y(1) = 0, \ y'(1) = 2$.
(d) $xy'' + (x-2)y' - 3y = 0, \ y_1(x) = x^3$. Note that not every integral can be calculated explicitly.
(e) $ty'' - (1+3t)y' + 3y = 0, \ y_1(t) = e^{3t}$.
(f) $(x-1)y''' + (1-2x)y'' + (x+1)y' - y = 0, \ y_1(t) = e^x$

4.20

(a) Verify that the equation $xy'' - (x+N)y' + Ny = 0$, where N is a positive integer, has a solution $y_1(x) = e^x$.
(b) Show that the formula for the second solution is $y_2(x) = e^x \int x^N e^{-x} \, dx$
(c) Calculate y_2 for $N = 1, 2, 3$ and guess what is $y_2(x)$ for every integer N.

4.11 The Method of Undetermined Coefficients

4.21 Check that $1/x$ is the solution of the equation $x^2 y'' - 3xy' - 5y = 0$. Find, using reduction of the order of the equation, all solutions of $x^2 y'' - 3xy' - 5y = x^3$.

4.22 The equation $y'' - xy' + y = 0$ has a solution $y_1(x) = x$.

(a) Explain why another solution of this equation can be written in the form $y_2(x) = x \int_a^x e^{t^2/2} t^{-2} \, dt$ for all $a \neq 0$ but not as $x \int_0^x e^{t^2/2} t^{-2} \, dt$.

(b) Calculate $\lim_{x \to 0} y_2(x)$ and write a second solution that is also defined and continuous at the point $x = 0$.

(c) Find a solution that satisfies the initial conditions $y(0) = 1$, $y'(0) = 0$.

4.23 Given an nth-order differential equation $F(x, y, y', \ldots, y^{(n)}) = 0$ and a function $y_0(x)$ which is not a solution of the given equation, prove that

$$\left(\frac{F(x, y, y', \ldots, y^{(n)})}{F(x, y_0(x), y_0'(x), \ldots, y_0^{(n)}(x))} \right)' = 0$$

is a differential equation of order $n+1$ such that the solutions of the original equation and the additional function $y_0(x)$ are solutions of this new equation.

4.24

(a) The equation $xy'' + (x - 1)y' - y = 0$ has a solution which is a polynomial of the first degree. Find it.
(b) Find another solution of this equation.
(c) Solve the nonhomogeneous equation $xy'' + (x - 1)y' - y = x^2 e^{3x}$.

4.25

(a) The equation $xy'' - (1+x)y' + y = 0$ has the solution $y_1(x) = e^x$. Find another solution of the equation.
(b) Solve the equation $xy'' - (1 + x)y' + y = x^2 e^{2x}$ by variation of parameters.

4.26

(a) The equation $2xy'' + (1 - 4x)y' + (2x - 1)y = 0$, $x > 0$, has a solution $y_1(x) = e^x$. Find another solution of it.
(b) Solve by variation of parameters the nonhomogeneous equation $2xy'' + (1 - 4x)y' + (2x - 1)y = e^x$.

4.27 Solve the following equations with constant coefficients:

(a) $y''' - 5y'' - 5y' + y = 0$.
(b) $y''' + y' = 0$.
(c) $y''' - y = 0$.
(d) $y''' - 3y'' + 3y' - y = 0$.
(e) $y^{(4)} - y = 0$.
(f) $y^{(4)} + 4y = 0$.

4.28 Find a normalized homogeneous linear differential equation with constant coefficients whose general solution is $c_1 + c_2 x + c_3 x^2 + c_4 e^{-x}\sin(2x) + c_5 e^{-x}\cos(2x)$.
Is there another homogeneous linear equation of the same order and the same solutions but with coefficients that are not necessarily constants?

4.29

(a) Find a linear, homogeneous differential equation with constant coefficients of order as low as possible such that the function $\cos x + x^2 \sin x$ is its solution. What are all the solutions of this equation?
(b) Answer the same questions for the function $(x + \cos x)\cos x$. Is the solution $y(x) \equiv 0$ a stable solution of the equation you found?
(c) The same question is asked for the function $(x + \cos x)\sin x$.

4.30 Find a linear, homogeneous equation with constant coefficients exactly of order 6 such that the two functions $x \cos x$ and 1 are its solutions. Write all possible equations that meet the conditions of the question.

4.31

(a) Find a third-order, homogeneous, normalized linear equation with constant coefficients such that $\sin(2x)$ is its solution and all its solutions are bounded for all values of x.
How many answers this question has? If there is more than one answer, write them all.
(b) Find a fourth-order, homogeneous, normalized linear equation with constant coefficients such that $\sin(2x)$ is its solution and all its solutions are bounded for all values of x.
How many answers this question has? If there is more than one answer, write them all.

4.32 Prove that if $y(x)$ is a solution of a homogeneous, linear differential equation with constant coefficients, then $y(x + c)$ is a solution of the same equation as well, for every constant c.

4.33

(a) Let $y(x)$ be a solution of a homogeneous, linear equation with constant coefficients. Prove that if $\lim_{x \to \infty} y(x) = 0$, then also

$$\lim_{x \to \infty} y'(x) = 0, \ \lim_{x \to \infty} y''(x) = 0, \ \ldots.$$

(b) Explain why a function that is not a solution of an equation of this type does not necessarily have this property.

4.11 The Method of Undetermined Coefficients

4.34 A differential equation has a characteristic polynomial $(r^2-1)^2(r^2+1) = 0$.
(a) Show that the collection of solutions tending to zero as $t \to +\infty$ constitutes a subspace. What is the dimension of this subspace?
(b) Prove that the set of solutions that are bounded when $t \to +\infty$ constitutes a subspace. What is the dimension of this subspace?
(c) Explain why the collection of the solutions which are unbounded when $t \to +\infty$ does not constitute a subspace. Hint: Which properties of a subspace do not hold for this collection?

4.35 For which values of b, c, the equation $y'' + by' + cy = 0$ will satisfy, respectively, the following conditions:
(a) All solutions of the equation tend to 0 when $x \to +\infty$.
(b) All solutions are bounded for all $-\infty < x < \infty$.
(c) There exists at least one nontrivial solution that tends to 0 when $x \to +\infty$.
(d) Each solution has the value 0 at an infinite number of points.

Remember that the roots of the characteristic polynomial $r^2 + br + c = 0$ satisfy the formulas of Viete, $r_1 + r_2 = -b$, $r_1 r_2 = c$.

4.36 Conditions that are given at various points, such as $y(x_1) = \alpha$, $y(x_2) = \beta$ are called "boundary value conditions." Show that if the equation $ay'' + by' + cy = 0$ with constant coefficients has real solutions $e^{r_1 x}, e^{r_2 x}$, $r_1 \neq r_2$, then the boundary value problem $y(0) = 0$, $y(1) = 1$ has only one solution:

$$y = \frac{e^{r_1 x} - e^{r_2 x}}{e^{r_1} - e^{r_2}}$$

and the boundary value problem $y(0) = 1$, $y(1) = 0$ has only the solution:

$$y = \frac{e^{r_1(1-x)} - e^{r_2(1-x)}}{e^{r_1} - e^{r_2}}.$$

What is the solution of the boundary value problem $y(0) = \alpha$, $y(1) = \beta$?

4.37 Solve the Euler equation $y'' + \dfrac{a}{x^2} y = 0$ for $a > \dfrac{1}{4}$, $a = \dfrac{1}{4}$, $a < \dfrac{1}{4}$.

4.38
(a) Solve the 3rd-order Euler equation $x^3 y''' + 4x^2 y'' + x y' - y = 0$.
(b) There are given three initial value conditions $y(1) = 0$, $y'(1) = \beta$, $y''(1) = \gamma$. For which initial values β, γ the solution of the initial value problem is bounded for all $1 \leq x < \infty$?

4.39 Show that the differential equations $\left(x^2 \left(\frac{y}{x^3}\right)'\right)' = 0$ and $\left(\frac{y}{x^2}\right)'' = 0$ are two additional ways of writing the Euler equation $y'' - \frac{4}{x}y' + \frac{6}{x^2}y = 0$. Solve these equations using their special forms.

4.40

(a) Show that the differential equation $\left(\frac{1}{x}\left(\frac{1}{x}\left(\frac{y}{x}\right)'\right)'\right)' = 0$ is a 3rd-order Euler equation and solve it.

(b) Show that the differential equation $\left(x^\alpha \left(x^\beta \left(x^\gamma y\right)'\right)'\right)' = 0$, where α, β, γ are constants, is a 3rd-order Euler equation and solve it.

4.41

(a) Prove that the set of solutions of a linear, nonhomogeneous equation $y'' + p(x)y' + q(x)y = g(x)$, $g(x) \not\equiv 0$, is not a linear space.
(b) It is given that the three functions $f(x), g(x), h(x)$ are solutions of the nonhomogeneous second-order equation above. Prove that the expression $c_1 f(x) + c_2 g(x) + c_3 h(x)$ is a solution of this equation if and only if $c_1 + c_2 + c_3 = 1$.

4.42 It is given that the functions $1, x, x^2$ are solutions of the equation $y'' + p(x)y' + q(x)y = g(x)$. Find the equation and all its solutions.

4.43 It is given that the three functions $1 + e^t$, $1 + te^t$, $1 + (1+t)e^t$ are solutions of the equation $y'' + p(t)y' + q(t)y = g(t)$. Find the equation and all its solutions.

4.44 Find all the solutions of the equation

$$y'' - 4y' + 4y = \left(1 + x + x^2 + \ldots + x^{127}\right) e^{2x}.$$

Hint: If we substitute according to the method of undetermined coefficients $y = x^s \left(\sum_{i=0}^{127} a_i x^i\right) e^{2x}$, it will take a long time to calculate the solution. It is useful to get rid first of the exponential function, as in the proof of the method.

4.45 Solve by the variation of parameter method:

(a) $y'' + 4y' + 4y = e^{-2x}/x^2$.
(b) $y'' + 4y' + 4y = x^2 e^{2x}$.
(c) $y'' + 6y' + 9y = t^{5/2} e^{-3t}$.
(d) $y'' - y = \dfrac{1}{1 + e^x}$.
(e) $y'' + y = \sin x + e^x$.

4.11 The Method of Undetermined Coefficients

(f) $x^2 y'' - 3xy' + 4y = 2x^2$.

(g) $y'' + \dfrac{1}{4t^2} y = t^5$.

4.46 Solve the equation $y'' - y = x$ by two methods: by the method of undetermined coefficients and by variation of parameters. Which way is easier?

4.47 The equation $y'' + xy' + y = 0$ has a solution $y_1 = e^{-x^2/2}$. (Check!)

(a) Find a second solution of the equation. (The solution contains an integral that cannot be calculated by elementary functions.)
(b) Solve the nonhomogeneous initial value problem $y'' + xy' + y = 1$, $y(0) = 0$, $y'(0) = 0$ by a method as simple as possible.

4.48 Solve the following equations by the method of undetermined coefficients. From Exercise (e) onwards it is enough to write the form of a particular solution without explicitly calculating its coefficients.

(a) $y'' + y = e^{-x}(x + 1)$, $y(0) = 0$, $y'(0) = 0$.
(b) $y'' - 6y' + 12y = x^2 e^{3x} - 3\cos(2x)$, $y(0) = 2$, $y'(0) = 3$.
(c) $2y''' - y' - y = e^x$.
(d) $y'' - 2y' + y = e^x$.
(e) $y''' + y' = 1 + 2\sin x + 3\cos x$.
(f) $y^{(4)} - 2y^{(2)} + y = e^x$
(g) $y''' - y'' - y' + y = 1$.
(h) $y''' + y' = x^3 + \sin(2x)$.
(i) $y^{(4)} + 8y'' + 16y = x(1 - \sin 2x)$.

4.49 Find all the solutions to the equation $y^{(5)} - y^{(4)} - y' + y = 1$ which are bounded for all values of x.

4.50

(a) Is it possible that a differential equation of the form $y'' + cy' + dy = e^x$, where c, d are constant, will have a solution $xe^{2x} - 4e^{-2x} + Be^x$?
(b) Answer the same question for the function $xe^{-2x} - 4e^{-2x}$.

4.51 The two equations

$$y'' + y = 7\sin(2x), \qquad y'' + y' + y = 7\sin(2x)$$

are similar, but the first equation has a particular solution of the form $A\sin(2x)$ while the second one has a solution of the form $A\sin(2x) + B\cos(2x)$. What is the reason to the difference?

4.52 For which ω and ω_0 the equation $y'' + \omega_0^2 y = \sin \omega t$ has at least one periodic solution? When are all its solutions periodic?

4.53 Consider the following "solution":
"The equation $y'' - xy = 0$ has a characteristic polynomial $r^2 - x = 0$ and its roots are $r_{1,2} = \pm\sqrt{x}$. Therefore, the differential equation has two independent solutions

$$y_1 = e^{\sqrt{x}\cdot x} = e^{x^{3/2}}, \quad y_2 = e^{-\sqrt{x}\cdot x} = e^{-x^{3/2}}."$$

Show that the two functions $e^{\pm x^{3/2}}$ do not satisfy the differential equation. What is wrong with the "solution" above?

Chapter 5
Systems of Differential Equations

5.1 Existence and Uniqueness Theorem for Systems

So far, we have dealt with a single differential equation of first-order or some higher-order n, with one unknown function. Now we turn to systems of several first-order differential equations with the same number of unknown functions. The general form of a first-order system of n differential equations with n unknown functions $x_1(t), \ldots, x_n(t)$ is[1]

$$\begin{aligned} x_1'(t) &= F_1(t, x_1, \ldots, x_n), \\ x_2'(t) &= F_2(t, x_1, \ldots, x_n), \\ &\cdots \\ x_n'(t) &= F_n(t, x_1, \ldots, x_n), \end{aligned} \tag{5.1}$$

and n initial value conditions are

$$x_1(t_0) = \alpha_1, \quad \ldots, \quad x_n(t_0) = \alpha_n. \tag{5.2}$$

Here $F_1, \ldots, F_n$ are functions of $n+1$ variables, and $\alpha_1, \ldots, \alpha_n$ are constants.

We quote (without proof) the theorem of existence and uniqueness for a system of differential equations and initial value conditions:

Theorem 5.1 *Given the system (5.1) of differential equations and the initial value conditions (5.2). Suppose that the n functions*

$$F_1(t, x_1, \ldots, x_n), \ldots, F_n(t, x_1, \ldots, x_n)$$

---

[1] In the context of systems, it is customary to denote the independent variable by t.

and their partial derivatives $\dfrac{\partial F_i}{\partial x_j}$, $i, j = 1, \ldots, n$, are continuous in some $n + 1$ dimensional domain D (there is no assumption about the derivative with respect to t). Assume also that the initial point $(t_0, \alpha_1, \alpha_2, \ldots, \alpha_n)$ is inside D. Then the initial value problem (5.1)–(5.2) has a unique solution $(x_1(t), \ldots, x_n(t))$, and this solution is defined for values of t in a certain neighborhood of t_0. Each component $x_i(t)$ of the solution has a continuous derivative $x_i'(t)$ there.

The existence and uniqueness theorem for one differential equation (Theorem 3.1) is just a special case of this theorem for $n = 1$. We will not prove Theorem 5.1 here, but we will hint at the way to prove it. It is useful to consider the solution $(x_1(t), \ldots, x_n(t))$ as a vector function

$$\mathbf{x}(t) = \big(x_1(t), \ldots, x_n(t)\big),$$

and the n functions $F_1, \ldots, F_n$ as the n components of a vector function

$$\mathbf{F}(t, \mathbf{x}) = \big(F_1(t, x_1, \ldots, x_n), \ldots, F_n(t, x_1, \ldots, x_n)\big)$$

and write the n initial values $\alpha_1, \ldots, \alpha_n$ as a constant vector $\boldsymbol{\alpha} = (\alpha_1, \ldots, \alpha_1)$. So the initial value problem will be written in vector notation as

$$\frac{d}{dt}\mathbf{x}(t) = \mathbf{F}(t, \mathbf{x}), \qquad \mathbf{x}(t_0) = \boldsymbol{\alpha}.$$

The proof of Theorem 5.1 is analogous to that of Theorem 3.1. Here the iterations are performed on vector functions, and absolute values of scalars are replaced by norms of vectors. The choice of the norm $||\mathbf{x}|| = \max\{|x_1|, \ldots, |x_n|\}$ is especially convenient. For example, the rectangle B in Sect. 3.3 will be replaced here by

$$B = \big\{(t, \mathbf{x}) \mid |t - t_0| \le a, \ ||\mathbf{x} - \boldsymbol{\alpha}|| \le b \big\}.$$

and $||\mathbf{x} - \boldsymbol{\alpha}|| = \max\{|x_1 - \alpha_1|, \ldots, |x_n - \alpha_n|\}$. So B is a $n + 1$ dimensional box that may be written in terms of components as

$$t_0 - a \le t \le t_0 + a,$$
$$\alpha_1 - b \le x_1 \le \alpha_1 + b,$$
$$\ldots$$
$$\alpha_n - b \le x_n \le \alpha_n + b.$$

The definition of stability is also naturally generalized to the solutions of a system of differential equations. Since these are vector solutions, the distance between two solutions is measured using the norm of vectors.

5.1 Existence and Uniqueness Theorem for Systems

Definition 5.1 A solution $\mathbf{x}^{(1)}(t)$ of the differential system (5.1) is called *stable* when $t \to +\infty$ if

1. $\mathbf{x}^{(1)}(t)$ is defined for all t, $t_0 \leq t < \infty$.
2. For all $\varepsilon > 0$ there exists $\delta = \delta(\varepsilon) > 0$ such that for any other solution $\mathbf{x}(t)$, the inequality $||\mathbf{x}^{(1)}(t_0) - \mathbf{x}(t_0)|| < \delta$ at t_0 implies $||\mathbf{x}^{(1)}(t) - \mathbf{x}(t)|| < \varepsilon$ for all t, $t_0 \leq t < \infty$. Otherwise, the solution $\mathbf{x}^{(1)}(t)$ is called *unstable*.
   If, in addition to the above, also $\lim_{t \to +\infty} ||\mathbf{x}^{(1)}(t) - \mathbf{x}(t)|| = 0$, then the solution $\mathbf{x}^{(1)}(t)$ is called *asymptotically stable*.

Note the similarity between this definition and Definition 3.2 for the stability of a single differential equation. The only difference between them is the replacement of the absolute value by the norm.

The differential equation (3.29) of order n,

$$y^{(n)} = f(x, y, y', \ldots, y^{(n-1)})$$

may be transformed into a system of n first-order equations by an appropriate substitution. We define n new unknown functions $v_1(x), \ldots, v_n(x)$ by

$$v_1 = y, \ v_2 = y', \ \ldots, \ v_{n-1} = y^{(n-2)}, \ v_n = y^{(n-1)}.$$

So $v_i' = v_{i+1}$, $i = 1, 2, \ldots, n-1$, and $v_n' = y^{(n)} = f(x, y, y', \ldots, y^{(n-1)}) = f(x, v_1, v_2, \ldots, v_n)$. Thus, we get a system of n first-order differential equations,

$$v_1' = v_2,$$
$$v_2' = v_3,$$
$$\vdots$$
$$v_{n-1}' = v_n,$$
$$v_n' = f(x, v_1, v_2, \ldots, v_n).$$

The n initial value conditions $y(x_0) = \alpha_1$, $y'(x_0) = \alpha_2$, $\ldots$, $y^{(n-1)}(x_0) = \alpha_n$ become $v_1(x_0) = \alpha_1$, $v_2(x_0) = \alpha_2$, $\ldots$, $v_n(x_0) = \alpha_n$.

In particular, the linear, homogeneous n-th order equation

$$y^{(n)} + p_1(x) y^{(n-1)} + \ldots + p_{n-1}(x) y' + p_n(x) y = 0$$

is transformed by the abovementioned substitution into the system

$$
\begin{aligned}
v_1' &= v_2, \\
v_2' &= v_3, \\
&\vdots \\
v_{n-1}' &= v_n, \\
v_n' &= -p_n(x)v_1 - p_{n-1}(x)v_2 - \ldots - p_1(x)v_n
\end{aligned}
\tag{5.3}
$$

of n first-order linear equations in the unknown $v_1, \ldots, v_n$.

From this change of variables, it follows that the existence and uniqueness theorem for an n-th order equation (Theorem 3.8) is a special case of the existence and uniqueness theorem for systems (Theorem 5.1).

From now on, we leave the general systems of differential equations and concentrate in the special case of linear differential systems.

5.2 Systems of Linear Equations

A system of differential equations is called *linear* if the functions $F_1, \ldots, F_n$ on the right-hand side of (5.1) are linear in the n unknown $x_1(t), \ldots, x_n(t)$. The general form of a linear system is

$$
\begin{aligned}
x_1' &= p_{11}(t)x_1 + p_{12}(t)x_2 + \ldots + p_{1n}(t)x_n + q_1(t), \\
x_2' &= p_{21}(t)x_1 + p_{22}(t)x_2 + \ldots + p_{2n}(t)x_n + q_2(t), \\
&\ldots \\
x_n &= p_{n1}(t)x_1 + p_{n2}(t)x_2 + \ldots + p_{nn}(t)x_n + q_n(t),
\end{aligned}
\tag{5.4}
$$

that is, $F_i(t, x_1, \ldots, x_n) = p_{i1}(t)x_1 + p_{i2}(t)x_2 + \ldots + p_{in}(t)x_n + q_i(t)$. Each function F_i is obviously continuous in the variables $x_1, \ldots, x_n$ (it is linear). To be continuous also in t, it is required that the coefficients $p_{ij}(t)$ and $q_i(t)$ be continuous in t. This also guarantees the continuity of the derivatives $\partial F_i/\partial x_j = p_{ij}(t)$. Therefore, for a linear system, all the requirements of the existence and uniqueness theorem are reduced to the requirement that all coefficients of the system will be continuous in the independent variable t.

For linear differential systems, it can be proved that every solution exists not only in a certain small neighborhood of the initial point t_0 but in the whole interval where all the coefficients are continuous. This is a global solution. See Theorem 3.7 in Sect. 3.8. We will not prove this but only state:

5.2 Systems of Linear Equations

Theorem 5.2 *If all coefficients $p_{ij}(t)$ of the linear system (5.4) are continuous in the interval $a \leq t \leq b$ and $a \leq t_0 \leq b$, then the initial value problem (5.4), (5.2) has only one solution, and it is defined in the entire interval $[a, b]$.*

Due of the linearity of the system of equations, it is natural to use vector and matrix notation. We will use the following notations:

$$P(t) = \begin{pmatrix} p_{11}(t) & \cdots & p_{1n}(t) \\ \vdots & & \vdots \\ p_{n1}(t) & \cdots & p_{nn}(t) \end{pmatrix}, \quad \mathbf{x}(t) = \begin{pmatrix} x_1(t) \\ \vdots \\ x_n(t) \end{pmatrix}, \quad \mathbf{q}(t) = \begin{pmatrix} q_1(t) \\ \vdots \\ q_n(t) \end{pmatrix}, \quad \boldsymbol{\alpha} = \begin{pmatrix} \alpha_1 \\ \vdots \\ \alpha_n \end{pmatrix}.$$

The terms of the $n \times n$ matrix $P(t)$ are functions of the independent variable t, and $\mathbf{x}(t)$, $\mathbf{q}(t)$ are column vectors that depend on t, i.e., vector functions. $\boldsymbol{\alpha}$ is a constant vector. The derivative of a vector function is the derivative of each and every component separately, i.e.,

$$\mathbf{x}'(t) = \begin{pmatrix} x_1'(t) \\ \vdots \\ x_n'(t) \end{pmatrix}.$$

With these notations, system (5.4) and the corresponding initial value conditions (5.2) will be written in the form

$$\mathbf{x}'(t) = P(t)\mathbf{x}(t) + \mathbf{q}(t), \quad \mathbf{x}(t_0) = \boldsymbol{\alpha}.$$

The corresponding homogeneous system where $q_1(t) \equiv 0, \ldots, q_n(t) \equiv 0$ is

$$\mathbf{x}'(t) = P(t)\mathbf{x}(t). \tag{5.5}$$

When we work with several solutions, the solution vectors are enumerated with superscript indices, while the component indices are written as subscripts. With this notation,

$$\mathbf{x}^{(1)}(t) = \begin{pmatrix} x_1^{(1)}(t) \\ x_2^{(1)}(t) \\ \vdots \\ x_n^{(1)}(t) \end{pmatrix}, \quad \mathbf{x}^{(2)}(t) = \begin{pmatrix} x_1^{(2)}(t) \\ x_2^{(2)}(t) \\ \vdots \\ x_n^{(2)}(t) \end{pmatrix}, \quad \ldots \quad \mathbf{x}^{(k)}(t) = \begin{pmatrix} x_1^{(k)}(t) \\ x_2^{(k)}(t) \\ \vdots \\ x_n^{(k)}(t) \end{pmatrix}, \ldots,$$

that is to say, $x_i^{(k)}$ marks the i-th component of vector solution number k.

Example 5.1 The equation $y'' + y = 0$ is transformed by the substitution $x_1 = y$, $x_2 = y'$ to the system

$$x_1' = x_2,$$
$$x_2' = -x_1,$$

and in vector-matrix notation $\begin{pmatrix} x_1(t) \\ x_2(t) \end{pmatrix}' = \begin{pmatrix} 0 & 1 \\ -1 & 0 \end{pmatrix} \begin{pmatrix} x_1(t) \\ x_2(t) \end{pmatrix}$. Check by direct calculation that $\mathbf{x}^{(1)}(t) = \begin{pmatrix} \cos t \\ -\sin t \end{pmatrix}$ and $\mathbf{x}^{(2)}(t) = \begin{pmatrix} \sin t \\ \cos t \end{pmatrix}$ are two of its solutions.

5.3 Homogeneous System and the Structure of Its Solutions

The properties of linear systems are similar to those of higher-order linear equations.

Theorem 5.3 *If* $\mathbf{x}^{(1)}(t)$, $\mathbf{x}^{(2)}(t)$ *are two solutions of the homogeneous differential system (5.5), then any linear combination of them with constant coefficients* $c_1 \mathbf{x}^{(1)}(t) + c_2 \mathbf{x}^{(2)}(t)$ *is a solution of the same system.*

For

$$\frac{d}{dt}\left[c_1 \mathbf{x}^{(1)} + c_2 \mathbf{x}^{(2)}\right] = c_1 \frac{d}{dt}\mathbf{x}^{(1)} + c_2 \frac{d}{dt}\mathbf{x}^{(2)}$$
$$= c_1 P(t) \mathbf{x}^{(1)} + c_2 P(t) \mathbf{x}^{(2)}$$
$$= P(t) \left[c_1 \mathbf{x}^{(1)} + c_2 \mathbf{x}^{(2)}\right].$$

For example, $c_1 \begin{pmatrix} \cos t \\ -\sin t \end{pmatrix} + c_2 \begin{pmatrix} \sin t \\ \cos t \end{pmatrix} = \begin{pmatrix} c_1 \cos t + c_2 \sin t \\ -c_1 \sin t + c_2 \cos t \end{pmatrix}$ is a solution of the 2×2 system mentioned in Example 5.1.

The zero vector function,

$$\mathbf{x}(t) \equiv \begin{pmatrix} 0 \\ \vdots \\ 0 \end{pmatrix} = \mathbf{0}$$

is a solution of any homogeneous linear differential system and is called the *trivial solution*. From now on, $\mathbf{0}$ will denote the zero vector in the n dimensional space.

An equivalent formulation of Theorem 5.3 is

Conclusion *The collection of all solutions of the homogeneous system (5.5) is a linear space.*

5.3 Homogeneous System and the Structure of Its Solutions

To describe the structure of a linear space of vector functions, we must define the concept of "linear dependence" for elements that are both vectors and functions of a variable t.

Definition 5.2 The m vector functions

$$\mathbf{v}^{(1)}(t) = \begin{pmatrix} v_1^{(1)}(t) \\ v_2^{(1)}(t) \\ \vdots \\ v_n^{(1)}(t) \end{pmatrix}, \ldots, \mathbf{v}^{(m)}(t) = \begin{pmatrix} v_1^{(m)}(t) \\ v_2^{(m)}(t) \\ \vdots \\ v_n^{(m)}(t) \end{pmatrix}$$

are called *linearly dependent on the interval* $[a, b]$ if there are m constants $c_1, c_2, \ldots, c_m$, not all 0, so that

$$c_1 \mathbf{v}^{(1)}(t) + c_2 \mathbf{v}^{(2)}(t) + \cdots + c_m \mathbf{v}^{(m)}(t) \equiv \mathbf{0}$$

for all t in the interval $[a, b]$. If there are no such $(c_1, \ldots, c_m) \neq (0, \ldots, 0)$, then the functions are called *linearly independent* on the interval $[a, b]$.

Since the initial value conditions $\mathbf{x}(t_0) = \boldsymbol{\alpha}$ contains n arbitrary parameters, our problem has n degrees of freedom. This is the motivation to prove that the solution space of the system (5.4) is n dimensional. To prove this, we have to show that there are n linearly independent vector solutions and that they span the entire solution space. In addition, we will also see how these solutions are characterized. The process is similar to that given for equations of order n. Let's start this time, for the sake of variety, with the definition of the Wronskian and the question of linear dependence.[2]

Definition 5.3 The Wronskian of any n vector functions $\mathbf{x}^{(1)}(t), \ldots, \mathbf{x}^{(n)}(t)$ is the determinant of the $n \times n$ matrix, which consists of their columns,

$$\begin{vmatrix} x_1^{(1)}(t) & x_1^{(2)}(t) & \cdots & x_1^{(n)}(t) \\ \vdots & \vdots & & \\ x_n^{(1)}(t) & x_n^{(2)}(t) & \cdots & x_n^{(n)}(t) \end{vmatrix},$$

and it is marked by

$$W(t) = W\big(\mathbf{x}^{(1)}(t), \ldots, \mathbf{x}^{(n)}(t)\big).$$

Sometimes, we prefer for convenience of writing and for emphasizing the role of t to write the Wronskian in the form $W(t) = W\big(\mathbf{x}^{(1)}, \ldots, \mathbf{x}^{(n)}\big)(t)$.

---

[2] Part of the following proof is "copy and paste" of Sect. 4.3 of Chap. 4, with minor necessary changes.

Note the similarities and differences between the definition of Wronskian for an equation of order n and for a system of n first-order equations. While for an n-th order equation in each column appears a solution and its derivatives until order $n-1$, in a Wronskian of a system, there appears a solution with its n components.

Theorem 5.4

(i) *If n vector functions $\mathbf{x}^{(1)}(t), \ldots, \mathbf{x}^{(n)}(t)$ are linearly dependent on an interval $[a, b]$, then their Wronskian is identically zero in the whole interval $[a, b]$,*

$$W(\mathbf{x}^{(1)}(t), \ldots, \mathbf{x}^{(n)}(t)) \equiv 0, \qquad a \leq t \leq b.$$

(ii) *In the opposite direction, if the homogeneous system (5.5), $\mathbf{x}' = P(t)\mathbf{x}$, satisfies the conditions of the existence and uniqueness theorem in the interval $[a, b]$ and its n solutions $\mathbf{x}^{(1)}(t), \ldots, \mathbf{x}^{(n)}(t)$ satisfy*

$$W(\mathbf{x}^{(1)}, \ldots, \mathbf{x}^{(n)})(t_0) = 0$$

for some point t_0 in the interval, then these solutions are linearly dependent on the interval.

Proof

(i) The first part of the proof has nothing to do with differential equations. For every t in the interval, the n vectors $\mathbf{x}^{(1)}(t), \ldots, \mathbf{x}^{(n)}(t)$ are the columns of the Wronskian and a determinant whose columns are dependent equals 0.
Note that it is not assumed that $\mathbf{x}^{(1)}(t), \ldots, \mathbf{x}^{(n)}(t)$ are solutions of some differential system and claim (i) is valid for any n vector functions.

(ii) Now suppose that $\mathbf{x}^{(1)}(t), \ldots, \mathbf{x}^{(n)}(t)$ are solutions of the system (5.5), and they satisfy

$$W(\mathbf{x}^{(1)}, \ldots, \mathbf{x}^{(n)})(t_0) = 0$$

for some point t_0 in the interval $[a, b]$. Let's look at the system of algebraic, linear, homogeneous equations in n unknown $c_1, \ldots, c_n$,

$$c_1 x_1^{(1)}(t_0) + \ldots + c_n x_1^{(n)}(t_0) = 0,$$
$$\vdots \qquad\qquad (5.6)$$
$$c_1 x_n^{(1)}(t_0) + \ldots + c_n x_n^{(n)}(t_0) = 0.$$

The determinant of this system is exactly the Wronskian, and according to the assumption, it is $W(\mathbf{x}^{(1)}(t_0), \ldots, \mathbf{x}^{(n)}(t_0)) = 0$. Therefore, this system has a non-trivial solution $(c_1, \ldots, c_n) \neq (0, \ldots, 0)$. From here on, we will fix these

5.3 Homogeneous System and the Structure of Its Solutions

n constants $c_1, \ldots, c_n$ and use them. For these $c_1, \ldots, c_n$, consider the vector function

$$\mathbf{U}(t) = c_1 \mathbf{x}^{(1)}(t) + \ldots + c_n \mathbf{x}^{(1)}(t) .$$

Being a linear combination of the solutions $\mathbf{x}^{(1)}(t), \ldots, \mathbf{x}^{(n)}(t)$, $\mathbf{U}(t)$ is a solution of the homogeneous differential system (5.5). Moreover, by the choice of $c_1, \ldots, c_n$ in the system (5.6), $\mathbf{U}(t)$ satisfies at the point t_0 the n initial conditions

$$\mathbf{U}(t_0) = \begin{pmatrix} 0 \\ \vdots \\ 0 \end{pmatrix} = \mathbf{0}.$$

On the other hand, take the identically zero vector function, $\mathbf{V}(t) \equiv \mathbf{0}$. It is the trivial solution of the homogeneous system (5.5) (and of any other homogeneous system) and satisfies at the point t_0 the initial value conditions $\mathbf{V}(t_0) = \mathbf{0}$. It turns out that $\mathbf{U}(t), \mathbf{V}(t)$ are two solutions of the same differential system that fulfill the same initial value conditions. By the existence and uniqueness theorem, they are identically equal in the interval in question, $\mathbf{U}(t) \equiv \mathbf{V}(t)$, i.e.,

$$c_1 \mathbf{x}^{(1)}(t) + \ldots + c_n \mathbf{x}^{(1)}(t) \equiv \mathbf{0} \qquad \text{for } a \leq t \leq b .$$

This proves the linear dependence of the solutions $\mathbf{x}^{(1)}(t), \ldots, \mathbf{x}^{(1)}(t)$ in $[a, b]$.

The two parts of the theorem lead together to an important conclusion. According to part (ii), the zeroing of the Wronskian at one point ensures linear dependence in the interval, and according to part (i), the linear dependence guarantees that the Wronskian is identically zero in the whole interval $[a, b]$. This conclusion is summarized in the following theorem.

Theorem 5.5 *If the homogeneous differential system (5.5) satisfies the conditions of the existence and uniqueness theorem in an interval $[a, b]$ and the Wronskian $W(\mathbf{x}^{(1)}(t), \ldots, \mathbf{x}^{(n)}(t))$ equals 0 at some point $t = t_0$ of $[a, b]$, then it is identically 0 in the whole interval $[a, b]$.*

By this, we characterized the linear dependence of solutions of a homogeneous system. Now we turn to the question of spanning the solution space.

Theorem 5.6 *Given the homogeneous system (5.5) that fulfills the assumptions of the theorem of existence and uniqueness in the interval $[a, b]$ and n vector solutions $\mathbf{x}^{(1)}(t), \ldots, \mathbf{x}^{(n)}(t)$ of the system. If*

$$W(\mathbf{x}^{(1)}, \ldots, \mathbf{x}^{(n)})(t_0) \neq 0$$

at some point t_0 in $[a, b]$, then $\mathbf{x}^{(1)}(t), \ldots, \mathbf{x}^{(n)}(t)$ span the solution space of (5.5).

Proof A collection of solutions $\mathbf{x}^{(1)}(t), \ldots, \mathbf{x}^{(n)}(t)$ spans the solution space if for each solution $\mathbf{u}(t)$ that we choose, there exist certain constants $c_1, c_2, \ldots, c_n$ so that $\mathbf{u}(t)$ can be written on $[a, b]$ as a linear combination of the form

$$\mathbf{u}(t) \equiv c_1 \mathbf{x}^{(1)}(t) + \cdots + c_n \mathbf{x}^{(n)}(t) \qquad \text{for } a \le t \le b .$$

We have to check whether there are such constants $c_1, c_2, \ldots, c_n$ and find them.

Take the point t_0 where the Wronskian $W(\mathbf{x}^{(1)}, \ldots, \mathbf{x}^{(n)})(t_0)$ is different from 0. At this point, the given solution $\mathbf{u}(t)$ has certain initial values, say

$$\mathbf{u}(t_0) = \begin{pmatrix} \beta_1 \\ \vdots \\ \beta_n \end{pmatrix} .$$

A linear combination $c_1 \mathbf{x}^{(1)}(t) + \cdots + c_n \mathbf{x}^{(n)}(t)$ will be identical with the given solution $\mathbf{u}(t)$ if both have at the point t_0 the same initial value,

$$c_1 \mathbf{x}^{(1)}(t_0) + \cdots + c_n \mathbf{x}^{(n)}(t_0) = \mathbf{u}(t_0),$$

or component-wise,

$$c_1 x_1^{(1)}(t_0) + \ldots + c_n x_1^{(n)}(t_0) = \beta_1 ,$$
$$\vdots$$
$$c_1 x_n^{(1)}(t_0) + \ldots + c_n x_n^{(n)}(t_0) = \beta_n .$$

This is a system of n linear algebraic non-homogeneous equations for n unknown $c_1, c_2, \ldots, c_n$. The determinant of the system is $W(\mathbf{x}^{(1)}, \ldots, \mathbf{x}^{(n)})(t_0) \ne 0$; therefore, for all values of $\beta_1, \ldots, \beta_n$, the system has a unique solution $c_1, c_2, \ldots, c_n$. We fix the constants $c_1, \ldots, c_n$ and construct a solution $\mathbf{v}(t) = c_1 \mathbf{x}^{(1)}(t) + \cdots + c_n \mathbf{x}^{(n)}(t)$. The given solution $\mathbf{u}(t)$ and the solution $\mathbf{v}(t)$ that we built have the same initial values at the point $t = t_0$. Therefore, according to the theorem of existence and uniqueness, they are identical in their domain of existence:

$$\mathbf{u}(t) \equiv \mathbf{v}(t) = c_1 \mathbf{x}^{(1)}(t) + \cdots + c_n \mathbf{x}^{(n)}(t), \qquad a \le t \le b .$$

So we spanned the solution $\mathbf{u}(t)$ as a linear combination of $\mathbf{x}^{(1)}(t), \ldots, \mathbf{x}^{(n)}(t)$. This is true for every solution $\mathbf{u}(t)$, so $\mathbf{x}^{(1)}(t), \ldots, \mathbf{x}^{(n)}(t)$ span the entire solution space.

We saw the importance of solutions whose Wronskian is non-zero at one point. Does there really exist a set of solutions whose Wronskian differs from 0 at a given point? The answer to this is positive. We define n solutions of the system (5.5) by n

5.4 Abel's Formula for the System

initial value vectors given by the standard basis,

$$\mathbf{x}^{(1)}(t_0) = \begin{pmatrix} 1 \\ 0 \\ \vdots \\ 0 \end{pmatrix}, \quad \mathbf{x}^{(2)}(t_0) = \begin{pmatrix} 0 \\ 1 \\ \vdots \\ 0 \end{pmatrix}, \quad \ldots, \quad \mathbf{x}^{(n)}(t_0) = \begin{pmatrix} 0 \\ 0 \\ \vdots \\ 1 \end{pmatrix}.$$

For these n solutions, the Wronskian's value at the point t_0 is 1; therefore, they satisfy the assumptions of the last two theorems. Obviously, our choice of initial conditions as the standard basis is arbitrary, and many other initial values could have been chosen. This situation is natural because every linear space has many bases. We summarize:

Theorem 5.7 *If the existence and uniqueness theorem holds for the system (5.5) of linear, homogeneous differential equations in the interval $[a, b]$ (i.e., all the coefficients are continuous in this interval), then the collection of solutions of the system in the interval is a n-dimensional linear space.*

n solutions $\mathbf{x}^{(1)}(t), \ldots, \mathbf{x}^{(n)}(t)$ form a basis for the solution space if and only if the Wronskian $W(\mathbf{x}^{(1)}(t), \ldots, \mathbf{x}^{(n)}(t))$ is non-zero at one point t_0 of the interval $[a, b]$ (and equivalently, different from 0 at all points of the interval).

5.4 Abel's Formula for the System

Theorem 5.8 *Given a system of differential, linear, homogeneous equations*

$$\mathbf{x}'(t) = P(t)\mathbf{x}(t)$$

with continuous coefficients $p_{ij}(t)$, $i, j = 1, \ldots, n$, in the interval $[a, b]$ (and therefore fulfills the conditions of the existence and uniqueness theorem) and a point t_0 in the interval. For any n solutions $\mathbf{x}^{(1)}(t), \ldots, \mathbf{x}^{(n)}(t)$ of the system,

$$W(\mathbf{x}^{(1)}, \ldots, \mathbf{x}^{(n)})(t) = W(\mathbf{x}^{(1)}, \ldots, \mathbf{x}^{(n)})(t_0) \exp\left(\int_{t_0}^{t} (p_{11} + \cdots + p_{nn})\, dt\right). \tag{5.7}$$

Here $p_{11}(t) + \cdots + p_{nn}(t) = \text{trace}(P(t))$, the trace of the matrix $P(t)$.

Proof We prove the formula only for $n = 2$, since for $n > 2$, the proof is based on the same principles, but the calculations are longer. Let

$$x_1' = p_{11}(t)x_1 + p_{12}(t)x_2,$$
$$x_2' = p_{21}(t)x_1 + p_{22}(t)x_2,$$

and its two vector solutions $\mathbf{x}^{(1)} = \begin{pmatrix} x_1^{(1)}(t) \\ x_2^{(1)}(t) \end{pmatrix}$, $\mathbf{x}^{(2)} = \begin{pmatrix} x_1^{(2)}(t) \\ x_2^{(2)}(t) \end{pmatrix}$. Their Wronskian is

$$W(t) = W\left(\mathbf{x}^{(1)}(t), \mathbf{x}^{(2)}(t)\right) = \begin{vmatrix} x_1^{(1)} & x_1^{(2)} \\ x_2^{(1)} & x_2^{(2)} \end{vmatrix} = x_1^{(1)} x_2^{(2)} - x_1^{(2)} x_2^{(1)}.$$

Let's differentiate the Wronskian and rearrange its parts:

$$\begin{aligned} W'(t) &= \left(x_1^{(1)} x_2^{(2)} - x_1^{(2)} x_2^{(1)}\right)' \\ &= \left(x_1^{(1)\prime} x_2^{(2)} + x_1^{(1)} x_2^{(2)\prime}\right) - \left(x_1^{(2)\prime} x_2^{(1)} + x_1^{(2)} x_2^{(1)\prime}\right) \\ &= \begin{vmatrix} x_1^{(1)\prime} & x_1^{(2)\prime} \\ x_2^{(1)} & x_2^{(2)} \end{vmatrix} + \begin{vmatrix} x_1^{(1)} & x_1^{(2)} \\ x_2^{(1)\prime} & x_2^{(2)\prime} \end{vmatrix}. \end{aligned}$$

But $x_1^{(1)}, x_2^{(1)}$ are the components of the solution $\mathbf{x}^{(1)}$, and $x_1^{(2)}, x_2^{(2)}$ are the components of the solution $\mathbf{x}^{(2)}$; therefore,

$$\begin{aligned} x_1^{(1)\prime} &= p_{11} x_1^{(1)} + p_{12} x_2^{(1)}, & x_1^{(2)\prime} &= p_{11} x_1^{(2)} + p_{12} x_2^{(2)}, \\ x_2^{(1)\prime} &= p_{21} x_1^{(1)} + p_{22} x_2^{(1)}, & x_2^{(2)\prime} &= p_{21} x_1^{(2)} + p_{22} x_2^{(2)}. \end{aligned}$$

We substitute this in the two determinants of which $W'(t)$ consists

$$= \begin{vmatrix} p_{11} x_1^{(1)} + p_{12} x_2^{(1)} & p_{11} x_1^{(2)} + p_{12} x_2^{(2)} \\ x_2^{(1)} & x_2^{(2)} \end{vmatrix} + \begin{vmatrix} x_1^{(1)} & x_1^{(2)} \\ p_{21} x_1^{(1)} + p_{22} x_2^{(1)} & p_{21} x_1^{(2)} + p_{22} x_2^{(2)} \end{vmatrix}$$

and simplify by subtracting rows from each other. There remains

$$= \begin{vmatrix} p_{11} x_1^{(1)} & p_{11} x_1^{(2)} \\ x_2^{(1)} & x_2^{(2)} \end{vmatrix} + \begin{vmatrix} x_1^{(1)} & x_1^{(2)} \\ p_{22} x_2^{(1)} & p_{22} x_2^{(2)} \end{vmatrix}$$

$$= p_{11} W(t) + p_{22} W(t) = (p_{11} + p_{22}) W(t).$$

The resulting linear differential equation,

$$W'(t) = (p_{11} + p_{22}) W(t) = \text{trace}(P(t)) W(t)$$

can be solved by integration of $W'(t)/W(t) = \text{trace}(P(t))$ from t_0 to t, yielding Abel's formula (5.7) for $n = 2$.

Thanks to the continuity of $p_{ij}(t)$, the integral $\int_{t_0}^{t}(p_{11} + \cdots + p_{nn}) \, dt$ exists, and $\exp\left(\int_{t_0}^{t}(p_{11} + \cdots + p_{nn}) \, dt\right)$ is always non-zero. This yields another proof of Theorem 5.5:

Theorem 5.9 *Under the assumptions above, $W(t_0) = 0$ for some t_0 if and only if $W(t) \equiv 0$ to all t in the interval $[a, b]$.*

Example 5.2 The system $\mathbf{x}'(t) = \begin{pmatrix} 0 & 1 \\ -1 & 0 \end{pmatrix} \mathbf{x}(t)$ and its solutions (Example 5.1) have the Wronskian

$$W\left(\mathbf{x}^{(1)}, \mathbf{x}^{(2)}\right) = \begin{vmatrix} \cos t & \sin t \\ -\sin t & \cos t \end{vmatrix} \equiv 1$$

for all t. It fits Abel's formula, since $\text{trace}(P(t)) \equiv 0$.

5.5 Fundamental Matrix

Every solution of a system of homogeneous equations is spanned by a basis of solutions as $\mathbf{x}(t) = c_1 \mathbf{x}^{(1)}(t) + \cdots + c_n \mathbf{x}^{(n)}(t)$ or by components as

$$\mathbf{x}(t) = c_1 \begin{pmatrix} x_1^{(1)}(t) \\ \vdots \\ x_n^{(1)}(t) \end{pmatrix} + c_2 \begin{pmatrix} x_1^{(2)}(t) \\ \vdots \\ x_n^{(2)}(t) \end{pmatrix} + \ldots + c_n \begin{pmatrix} x_1^{(n)}(t) \\ \vdots \\ x_n^{(n)}(t) \end{pmatrix}.$$

This expression can be written as the multiplication of a matrix by a vector

$$\mathbf{x}(t) = \begin{pmatrix} x_1^{(1)}(t) & x_1^{(2)}(t) & \cdots & x_1^{(n)}(t) \\ \vdots & & & \\ x_n^{(1)}(t) & x_n^{(2)}(t) & \cdots & x_n^{(n)}(t) \end{pmatrix} \begin{pmatrix} c_1 \\ \vdots \\ c_n \end{pmatrix}. \tag{5.8}$$

Note that the matrix above is constructed by placing side-by-side n independent solution vectors,

$$\left(\mathbf{x}^{(1)}(t) \,\vdots\, \mathbf{x}^{(2)}(t) \,\vdots\, \ldots \,\vdots\, \mathbf{x}^{(n)}(t)\right).$$

This matrix is called a *fundamental matrix* of the homogeneous system and is usually denoted by $\Phi(t)$.[3] The determinant of $\Phi(t)$ is precisely the Wronskian of $\mathbf{x}^{(1)}(t), \ldots, \mathbf{x}^{(n)}(t)$, and so it is non-zero for all t. Therefore, a fundamental matrix is invertible for all t. Using this matrix notation, the general solution (5.8) of a homogeneous system is written as

$$\mathbf{x}(t) = \Phi(t)\mathbf{c}, \tag{5.9}$$

where $\mathbf{c}$ is a vector of n arbitrary constants $c_1, \ldots, c_n$.

The vector-matrix notation (5.9) is effective for solving initial value problems. Given the initial value

$$\mathbf{x}(t_0) = \begin{pmatrix} \alpha_1 \\ \vdots \\ \alpha_n \end{pmatrix},$$

we mark the vector on the right-hand side by $\boldsymbol{\alpha}$. So the initial value condition is

$$\mathbf{x}(t_0) = \Phi(t_0)\mathbf{c} = \boldsymbol{\alpha}$$

and it determines $\mathbf{c} = \Phi^{-1}(t_0)\boldsymbol{\alpha}$, where Φ^{-1} denotes the inverse of matrix Φ. As $\mathbf{c}$ is determined, it is substituted back in the general solution, and we get the solution of the initial value problem as

$$\mathbf{x}(t) = \Phi(t)\Phi^{-1}(t_0)\boldsymbol{\alpha}.$$

This formula is theoretically important because it expresses a direct relationship between the initial value and the corresponding solution. Practically it is more efficient to calculate $c_1, \ldots, c_n$ by solving the system of n equations

$$c_1 \begin{pmatrix} x_1^{(1)}(t_0) \\ \vdots \\ x_n^{(1)}(t_0) \end{pmatrix} + c_2 \begin{pmatrix} x_1^{(2)}(t_0) \\ \vdots \\ x_n^{(2)}(t_0) \end{pmatrix} + \ldots + c_n \begin{pmatrix} x_1^{(n)}(t_0) \\ \vdots \\ x_n^{(n)}(t_0) \end{pmatrix} = \begin{pmatrix} \alpha_1 \\ \vdots \\ \alpha_n \end{pmatrix}$$

than to calculate the inverse matrix $\Phi^{-1}(t_0)$.

We shall encounter utilization of fundamental matrices in Sect. 5.7.

---

[3] Φ, φ - the Greek letter phi.

5.6 Systems with Constant Coefficients

So far we have not yet solved explicitly any system of differential equations. One type that can be explicitly solved and is frequently used is the family of systems with constant coefficients of the form

$$\mathbf{x}'(t) = A\mathbf{x}(t), \qquad (5.10)$$

where $A = (a_{ij})_{i,j=1}^n$ is a $n \times n$ matrix of constants.

For a linear equation (4.15) with constant coefficients and of any order, we found solutions of the form $e^{\lambda t}$, where λ is an unknown constant number, and the form of a general solution is usually $\sum c_i e^{\lambda_i t}$. For the system (5.10), we will make a similar experiment, but this time with constants that are vectors. We will look for a solution of the form

$$\mathbf{x}(t) = \mathbf{v} e^{\lambda t}$$

where λ and the vector $\mathbf{v}$ are yet unknown and $\mathbf{v} \neq \mathbf{0}$. We substitute its derivative $\mathbf{x}'(t) = \mathbf{v}\lambda e^{\lambda t}$ in (5.10) and receive

$$\mathbf{v}\lambda e^{\lambda t} = A\mathbf{v} e^{\lambda t}.$$

Dividing by $e^{\lambda t} \neq 0$ leads to equality $A\mathbf{v} = \lambda \mathbf{v}$ which is equivalent to

$$(A - \lambda I)\mathbf{v} = \mathbf{0},$$

where I is the $n \times n$ identity matrix. This is exactly the definition of an eigenvalue λ, an eigenvector $\mathbf{v}$ and the determinant $|A - \lambda I|$ is the characteristic polynomial of matrix A. Therefore, $\mathbf{x}(t) = \mathbf{v} e^{\lambda t}$ is a solution when λ is an eigenvalue of the matrix A, and $\mathbf{v}$ is the corresponding eigenvector. Remember that every multiple of an eigenvector is also an eigenvector.

Example 5.3 Let us solve the initial value problem

$$\mathbf{x}'(t) = \begin{pmatrix} 1 & 1 \\ 4 & 1 \end{pmatrix} \mathbf{x}(t), \qquad \mathbf{x}(0) = \begin{pmatrix} 5 \\ 6 \end{pmatrix}.$$

The characteristic polynomial of the matrix is

$$|A - \lambda I| = \begin{vmatrix} 1-\lambda & 1 \\ 4 & 1-\lambda \end{vmatrix} = \lambda^2 - 2\lambda - 3 = 0$$

and the eigenvalues are $\lambda_1 = 3$, $\lambda_2 = -1$. An eigenvector corresponding to $\lambda_1 = 3$ is obtained from the solution of the system

$$(A - \lambda_1 I)\mathbf{v} = \begin{pmatrix} 1-3 & 1 \\ 4 & 1-3 \end{pmatrix} \begin{pmatrix} \alpha \\ \beta \end{pmatrix} = \mathbf{0},$$

namely

$$-2\alpha + \beta = 0,$$
$$4\alpha - 2\beta = 0.$$

Note that the resulting equations are always linearly dependent because the definition of eigenvalues requires that the corresponding determinant is 0. Therefore, the equations have a non-trivial solution, and every multiple of a solution is also a solution. Take, for example, $\alpha = 1$ and $\beta = 2$. The corresponding solution of the differential system is $\mathbf{x}^{(1)}(t) = \begin{pmatrix} 1 \\ 2 \end{pmatrix} e^{3t}$. Similarly, we find for $\lambda_2 = -1$ the solution $\mathbf{x}^{(2)}(t) = \begin{pmatrix} 1 \\ -2 \end{pmatrix} e^{-t}$. The general solution is

$$\mathbf{x}(t) = c_1 \begin{pmatrix} 1 \\ 2 \end{pmatrix} e^{3t} + c_2 \begin{pmatrix} 1 \\ -2 \end{pmatrix} e^{-t}$$

and a fundamental matrix is

$$\Phi(t) = \begin{pmatrix} e^{3t} & e^{-t} \\ 2e^{3t} & -2e^{-t} \end{pmatrix}.$$

The initial conditions require that

$$\mathbf{x}(0) = c_1 \begin{pmatrix} 1 \\ 2 \end{pmatrix} e^0 + c_2 \begin{pmatrix} 1 \\ -2 \end{pmatrix} e^0 = \begin{pmatrix} 5 \\ 6 \end{pmatrix},$$

that is,

$$c_1 + c_2 = 5,$$
$$2c_1 - 2c_2 = 6.$$

That gives $c_1 = 4$, $c_2 = 1$, and the solution of the initial value problem is

$$\mathbf{x}(t) = 4 \begin{pmatrix} 1 \\ 2 \end{pmatrix} e^{3t} + \begin{pmatrix} 1 \\ -2 \end{pmatrix} e^{-t} = \begin{pmatrix} 4e^{3t} + e^{-t} \\ 8e^{3t} - 2e^{-t} \end{pmatrix}$$

5.6 Systems with Constant Coefficients

or by components,

$$x_1(t) = 4e^{3t} + e^{-t}, \quad x_2(t) = 8e^{3t} - 2e^{-t}. \tag{5.11}$$

Let's return to the general problem. For the system (5.10), we are looking for a basis of n independent vector solutions. Do there exist such n solutions, all of the form $\mathbf{v}_1 e^{\lambda_1 t}, \ldots, \mathbf{v}_n e^{\lambda_n t}$?

A matrix A has n eigenvalues, which are the roots of the characteristic polynomial $|A - \lambda I|$. These eigenvalues can be different or equal, real or complex valued. To any two different eigenvalues, there correspond independent eigenvectors. However, when an eigenvalue is of multiplicity greater than one (the algebraic multiplicity), the number of independent eigenvectors corresponding to it (the geometric multiplicity) may be smaller than the algebraic multiplicity. Therefore, it is possible that to the n eigenvalues (counted according to their multiplicities) will correspond less than n independent eigenvectors. Thus, in the search for n solutions for a differential system, we must distinguish between two cases, when the matrix A has exactly n independent eigenvectors and when it has less than n independent eigenvectors.

5.6.1 *Systems with a Basis of Eigenvectors*

In this section, we discuss the solutions of a differential system when the constant matrix A has n linearly independent eigenvectors. The following theorem of linear algebra characterizes when a matrix has n independent eigenvectors:

Theorem 5.10 *Let A be a $n \times n$ matrix. The following conditions are equivalent to each other:*

(i) The matrix A has n linearly independent eigenvectors.
(ii) The algebraic and the geometric multiplicities of each eigenvalue are equal.
(iii) The matrix A is diagonalizable, $T^{-1} A T = \mathrm{diag}\{\lambda_1, \ldots, \lambda_n\}$.

Note that the columns of the diagonalizing matrix T are exactly the n independent eigenvectors of A.

Two useful particular cases that always have n independent eigenvectors are:

(1) When all n eigenvalues are different from each other, there are of course n independent eigenvectors.
(2) A symmetric matrix always has n independent eigenvectors. The same is true for an antisymmetric matrix.

Suppose that to the n eigenvalues $\lambda_1, \ldots, \lambda_n$ of A, different or equal, there correspond n independent eigenvectors $\mathbf{v}_1, \ldots, \mathbf{v}_n$. Then the differential system has n vector solutions

$$\mathbf{x}^{(1)}(t) = \mathbf{v}_1 e^{\lambda_1 t}, \quad \ldots, \quad \mathbf{x}^{(n)}(t) = \mathbf{v}_n e^{\lambda_n t}. \tag{5.12}$$

Are they linearly independent? For this purpose, we consider their Wronskian

$$W(\mathbf{x}^{(1)}(t), \ldots, \mathbf{x}^{(n)}(t)) = \det \left(e^{\lambda_1 t} \mathbf{v}_1 \ \vdots \ e^{\lambda_2 t} \mathbf{v}_2 \ \vdots \ \cdots \ \vdots \ e^{\lambda_n t} \mathbf{v}_n \right),$$

where each column is one of the eigenvectors. We remove from each of the columns the common factors $e^{\lambda_1 t}, \ldots, e^{\lambda_n t}$, respectively, and get

$$= e^{(\lambda_1 + \cdots + \lambda_n)t} \det \left(\mathbf{v}_1 \ \vdots \ \mathbf{v}_2 \ \vdots \ \cdots \ \vdots \ \mathbf{v}_n \right).$$

Recall that the sum of the eigenvalues $\lambda_1 + \cdots + \lambda_n$ is equal to trace(A), the trace of the matrix, which is a constant number. Therefore, the last equality is just a special case of Abel's formula,

$$W(t) = W(0) \exp \left(\int_0^t \text{trace}(A) \, dt \right).$$

Obviously, $e^{(\lambda_1 + \cdots + \lambda_n)t} \neq 0$. So $W(\mathbf{x}^{(1)}, \ldots, \mathbf{x}^{(n)}) \neq 0$ precisely when the determinant consisting of the n eigenvectors is non-zero, that is, when the n eigenvectors are independent. Let's summarize this in the following theorem:

Theorem 5.11 *A differential system (5.10) with constant coefficients has a basis of n solutions of the form $\mathbf{v}_1 e^{\lambda_1 t}, \ldots, \mathbf{v}_n e^{\lambda_n t}$ if and only if the matrix A has n independent eigenvectors $\mathbf{v}_1, \ldots, \mathbf{v}_n$.*

Example 5.4 Solve the system

$$\mathbf{x}'(t) = \begin{pmatrix} 0 & 1 & 1 \\ 1 & 0 & 1 \\ 1 & 1 & 0 \end{pmatrix} \mathbf{x}(t).$$

We start with the characteristic equation

$$|A - \lambda I| = \begin{vmatrix} -\lambda & 1 & 1 \\ 1 & -\lambda & 1 \\ 1 & 1 & -\lambda \end{vmatrix} = -\lambda^3 + 3\lambda + 2 = -(\lambda - 2)(\lambda + 1)^2 = 0.$$

5.6 Systems with Constant Coefficients

We need the eigenvectors corresponding to the eigenvalues $\lambda_1 = \lambda_2 = -1$ (algebraic multiplicity of two): the system of equations

$$(A - \lambda_1 I)\mathbf{v} = \begin{pmatrix} 1 & 1 & 1 \\ 1 & 1 & 1 \\ 1 & 1 & 1 \end{pmatrix} \begin{pmatrix} \alpha \\ \beta \\ \gamma \end{pmatrix} = \begin{pmatrix} 0 \\ 0 \\ 0 \end{pmatrix}$$

reduces to a single equation $\alpha + \beta + \gamma = 0$, which has a two-parametric solution. Therefore, there are two independent eigenvectors, and the geometric multiplicity is also two,

$$\mathbf{v} = \begin{pmatrix} \alpha \\ \beta \\ \gamma \end{pmatrix} = \begin{pmatrix} \alpha \\ \beta \\ -\alpha - \beta \end{pmatrix} = \alpha \begin{pmatrix} 1 \\ 0 \\ -1 \end{pmatrix} + \beta \begin{pmatrix} 0 \\ 1 \\ -1 \end{pmatrix}.$$

Of course, this two-dimensional subspace can also be written in other forms.

For the third eigenvalue $\lambda_3 = 2$, we find similarly that the corresponding eigenvector is $\begin{pmatrix} 1 \\ 1 \\ 1 \end{pmatrix}$ (check), and the general solution can be written as

$$\mathbf{x}(t) = c_1 \begin{pmatrix} 1 \\ 0 \\ -1 \end{pmatrix} e^{-t} + c_2 \begin{pmatrix} 0 \\ 1 \\ -1 \end{pmatrix} e^{-t} + c_3 \begin{pmatrix} 1 \\ 1 \\ 1 \end{pmatrix} e^{2t}.$$

Another Proof of Theorem 5.11 At the beginning of this section, we presented the search for solutions of the form $\mathbf{x}(t) = \mathbf{v}e^{\lambda t}$ as a "motivated guess." Now we show another approach to the same question, which leads to the same solutions systematically, without any guessing.

In the differential system $\mathbf{x}' = A\mathbf{x}$, we try to replace the unknown $\mathbf{x}(t)$ by another unknown $\mathbf{y}(t)$, defined by a linear transformation

$$\mathbf{x}(t) = T\mathbf{y}(t),$$

where T is a matrix of constants that has yet to be determined. We substitute the derivative $\mathbf{x}' = (T\mathbf{y})' = T\mathbf{y}'$ in the differential system and get

$$T\mathbf{y}' = A(T\mathbf{y}),$$

namely

$$\mathbf{y}' = (T^{-1}AT)\mathbf{y}.$$

Our goal is to choose T so that $T^{-1}AT$ will be as simple as possible. It is known from linear algebra (Theorem 5.10) that when the matrix A has n linearly independent eigenvectors, it may be diagonalized, and the diagonalizing matrix T consists of the eigenvectors, placed side by side, and the resulting diagonal matrix is

$$T^{-1}AT = \text{diag}\{\lambda_1, \ldots, \lambda_n\}.$$

The differential system becomes by such substitution

$$\begin{pmatrix} y_1 \\ \vdots \\ y_n \end{pmatrix}' = \begin{pmatrix} \lambda_1 & & \\ & \ddots & \\ & & \lambda_n \end{pmatrix} \begin{pmatrix} y_1 \\ \vdots \\ y_n \end{pmatrix},$$

and its components decompose into n separate equations

$$y_1' = \lambda_1 y_1,$$
$$\vdots$$
$$y_n' = \lambda_n y_n.$$

Each equation is solved separately, and we get

$$y_1(t) = c_1 e^{\lambda_1 t}, \ldots, y_n(t) = c_n e^{\lambda_n t}.$$

From these components, we rebuild the solution vector:

$$\mathbf{y}(t) = \begin{pmatrix} y_1 \\ \vdots \\ y_n \end{pmatrix} = \begin{pmatrix} c_1 e^{\lambda_1 t} \\ \vdots \\ c_n e^{\lambda_n t} \end{pmatrix} = c_1 \begin{pmatrix} 1 \\ 0 \\ \vdots \\ 0 \end{pmatrix} e^{\lambda_1 t} + \ldots + c_n \begin{pmatrix} 0 \\ 0 \\ \vdots \\ 1 \end{pmatrix} e^{\lambda_n t}.$$

We return from $\mathbf{y}(t)$ to the original unknown $\mathbf{x}(t)$ through multiplying it by T on the left-hand side:

$$\mathbf{x}(t) = T\mathbf{y}(t) = c_1 T \begin{pmatrix} 1 \\ \vdots \\ 0 \end{pmatrix} e^{\lambda_1 t} + \ldots + c_n T \begin{pmatrix} 0 \\ \vdots \\ 1 \end{pmatrix} e^{\lambda_n t}.$$

5.6 Systems with Constant Coefficients

But the vector $T\begin{pmatrix}1\\ \vdots\\ 0\end{pmatrix}$ is exactly the first column of the matrix T, $T\begin{pmatrix}0\\ 1\\ \vdots\end{pmatrix}$ is its second column, and so on. As mentioned above, the columns of the diagonal matrix T are the eigenvectors $\mathbf{v}_1, \ldots, \mathbf{v}_n$. Thus, the general solution is

$$\mathbf{x}(t) = c_1 \mathbf{v}_1 e^{\lambda_1 t} + \ldots + c_n \mathbf{v}_1 e^{\lambda_n t},$$

as claimed in Theorem 5.11.

This example shows the importance of the change of variables and the similarity of matrices. Some problems and equations that seem complicated in their original form, look simple in another coordinate system. The simple system form is sometimes called a "*canonical form.*"

When the matrix A has complex valued eigenvalues, the situation is not fundamentally different even if the technical calculation is more complicated. Suppose that the matrix A has a complex eigenvalue λ and there corresponds to it a complex eigenvector $\mathbf{v}$,

$$A\mathbf{v} = \lambda \mathbf{v}.$$

Since the matrix A is real, the complex conjugate of this expression is

$$A\overline{\mathbf{v}} = \overline{\lambda}\,\overline{\mathbf{v}}.$$

(Here ¯ marks a complex conjugate). Therefore, also $\overline{\lambda}$ is an eigenvalue, and $\overline{\mathbf{v}}$ is the corresponding complex eigenvector. Hence, the differential system has two complex conjugate vector solutions

$$e^{\lambda t}\mathbf{v}, \quad e^{\lambda t}\overline{\mathbf{v}}$$

and from their combinations, we get two real-valued solutions

$$\mathbf{x}^{(1)} = \frac{1}{2}\left(e^{\lambda t}\mathbf{v} + e^{\overline{\lambda} t}\overline{\mathbf{v}}\right) = \operatorname{Re}\left\{e^{\lambda t}\mathbf{v}\right\}, \quad \mathbf{x}^{(1)} = \frac{1}{2i}\left(e^{\lambda t}\mathbf{v} - e^{\overline{\lambda} t}\overline{\mathbf{v}}\right) = \operatorname{Im}\left\{e^{\lambda t}\mathbf{v}\right\}.$$

Compare with the scalar solutions (4.21), (4.22).

Exercise 5.1 Prove that these two real-valued solutions are linearly independent.

Example 5.5 The system

$$\mathbf{x}'(t) = \begin{pmatrix} 3 & -2 \\ 4 & -1 \end{pmatrix} \mathbf{x}(t)$$

has the characteristic polynomial $\lambda^2 - 2\lambda + 5 = 0$ and eigenvalues $1 \pm 2i$. We will look for an eigenvector corresponding to $\lambda_1 = 1 + 2i$:

$$(A - \lambda_1 I)\mathbf{v} = \begin{pmatrix} 3 - (1+2i) & -2 \\ 4 & -1 - (1+2i) \end{pmatrix} \begin{pmatrix} \alpha \\ \beta \end{pmatrix} = \begin{pmatrix} 0 \\ 0 \end{pmatrix},$$

$$(2 - 2i)\alpha - 2\beta = 0,$$
$$4\alpha - (2 + 2i)\beta = 0.$$

According to the definition of eigenvalues, it is clear that these equations are dependent, although it is not obvious at first glance. (The second equation is obtained from the first one by multiplying it by $(1+i)$). Selecting $\alpha = 1$, $\beta = 1-i$, we get the complex valued solution

$$e^{(1+2i)t} \begin{pmatrix} 1 \\ 1-i \end{pmatrix} = e^t (\cos 2t + i \sin 2t) \begin{pmatrix} 1 \\ 1-i \end{pmatrix}$$

$$= e^t \left[\begin{pmatrix} \cos 2t \\ \cos 2t + \sin 2t \end{pmatrix} + i \begin{pmatrix} \sin 2t \\ -\cos 2t + \sin 2t \end{pmatrix} \right].$$

The complex conjugate of this solution is also another solution of the same differential system. By addition and subtraction, we get two real-valued solutions of the differential system,

$$\mathbf{x}^{(1)} = e^t \begin{pmatrix} \cos 2t \\ \cos 2t + \sin 2t \end{pmatrix}, \quad \mathbf{x}^{(2)} = e^t \begin{pmatrix} \sin 2t \\ \sin 2t - \cos 2t \end{pmatrix}.$$

Another choice of the parameters, say $\alpha = 1 + i$, $\beta = 2$, brings us to another pair of solutions, such that the connection between them and the solutions above is not clear at first glance. (Try it!) In fact, the different pairs of solutions are linear combinations of each other.

5.6.2 Systems That Have No Basis of Eigenvectors

When the geometric multiplicity of an eigenvalue is less than its algebraic multiplicity, the matrix has less than n independent eigenvectors, and there are no n independent solutions of the differential system of the form

$$\mathbf{v}_1 e^{\lambda_1 t}, \ldots, \mathbf{v}_n e^{\lambda_n t}.$$

5.6 Systems with Constant Coefficients

But from the general theory, it is known that there exists a basis of n independent solutions. What is then their form? The following steps will guide us constructively to a basis of solutions.

In the previous section, the differential system $\mathbf{x}' = A\mathbf{x}$ was transformed by the substitution $\mathbf{x} = T\mathbf{y}$ into a new system,

$$\mathbf{y}' = (T^{-1}AT)\mathbf{y}. \tag{5.13}$$

When the matrix A has n linearly independent eigenvectors, the similar matrix $T^{-1}AT$ can be selected to be diagonal. But when the matrix A has less than n eigenvectors, it cannot be diagonalized. The matrix A can be brought only to the *Jordan canonical form*[4] which is built of blocks,

$$T^{-1}AT = \begin{pmatrix} \boxed{J_1} & & & \\ & \boxed{J_2} & & \\ & & \ddots & \\ & & & \boxed{J_k} \end{pmatrix},$$

where each block is of the form

$$J = \begin{pmatrix} \lambda & 1 & & & \\ & \lambda & 1 & & \\ & & \ddots & \ddots & \\ & & & \lambda & 1 \\ & & & & \lambda \end{pmatrix}$$

and all elements that are not specified are 0.

Jordan form is extensively used in linear algebra. We mention without proof a few of its properties. Each block has a single eigenvector (calculate it), and the number of diagonal blocks is equal to the number of independent eigenvectors (the sum of the geometric multiplicities). The dimension of each block is related to the concept of *elementary divisors*, which is studied in linear algebra and will not be elaborated here. We only note that an eigenvalue can appear in more than one block and the sum of the dimensions of these blocks is its algebraic multiplicity. After a suitable

---

[4] Camille Jordan, 1838–1922.

choice of transformation T, the reduced system (5.13) is

$$\begin{pmatrix} y_1 \\ y_2 \\ \vdots \\ y_{l-1} \\ y_l \\ \vdots \\ y_n \end{pmatrix}' = \begin{pmatrix} \boxed{\begin{matrix} \lambda & 1 & & & \\ & \lambda & 1 & & \\ & & \ddots & \ddots & \\ & & & \lambda & 1 \\ & & & & \lambda \end{matrix}} & & \\ & \boxed{J_2} & \\ & & \ddots \end{pmatrix} \begin{pmatrix} y_1 \\ y_2 \\ \vdots \\ y_{l-1} \\ y_l \\ \vdots \\ y_n \end{pmatrix}.$$

For convenience, we look first only at one Jordan block of dimension $l \times l$ and consider the l components of the solution vector **y**, which it meets:

$$\begin{pmatrix} y_1 \\ y_2 \\ \vdots \\ y_{l-1} \\ y_l \end{pmatrix}' = \begin{pmatrix} \lambda & 1 & & & \\ & \lambda & 1 & & \\ & & \ddots & \ddots & \\ & & & \lambda & 1 \\ & & & & \lambda \end{pmatrix} \begin{pmatrix} y_1 \\ y_2 \\ \vdots \\ y_{l-1} \\ y_l \end{pmatrix}.$$

Component-wise, it is

$$y_1' = \lambda y_1 + y_2,$$
$$y_2' = \lambda y_2 + y_3,$$
$$\vdots$$
$$y_{l-2}' = \lambda y_{l-2} + y_{l-1},$$
$$y_{l-1}' = \lambda y_{l-1} + y_l,$$
$$y_l' = \lambda y_l.$$

This system of equations is solved recursively from the end to the beginning. The solution of the last equation, $y_l' = \lambda y_l$, is $y_l = c_l e^{\lambda t}$. It is substituted into the second equation from the end, which becomes a non-homogeneous linear equation

$$y_{l-1}' - \lambda y_{l-1} = c_l e^{\lambda t}.$$

This equation is multiplied by the integrating factor $\mu(t) = e^{-\lambda t}$,

$$\left(y_{l-1} e^{-\lambda t}\right)' = y_{l-1}' e^{-\lambda t} - \lambda e^{-\lambda t} y_{l-1} = c_l.$$

5.6 Systems with Constant Coefficients

By integration, we have that $y_{l-1}e^{-\lambda t} = c_l t + c_{l-1}$, that is,

$$y_{l-1} = (c_l t + c_{l-1})e^{\lambda t}.$$

We take another step backward and substitute y_{l-1} into equation number $l-2$:

$$y'_{l-2} - \lambda y_{l-2} = (c_l t + c_{l-1})e^{\lambda t}.$$

As above,

$$\left(y_{l-2}e^{-\lambda t}\right)' = y'_{l-2}e^{-\lambda t} - \lambda e^{-\lambda t} y_{l-2} = c_l t + c_{l-1}$$

and after one more integration

$$y_{l-2} = \left(c_l \frac{t^2}{2} + c_{l-1}t + c_{l-2}\right)e^{\lambda t}.$$

After l steps, we arrive similarly to the first component

$$y_1 = \left(c_l \frac{t^{l-1}}{(l-1)!} + c_{l-1}\frac{t^{l-2}}{(l-2)!} + \cdots + c_2 t + c_1\right)e^{\lambda t}.$$

In vector notation, it is

$$\begin{pmatrix} y_1 \\ \vdots \\ y_{l-2} \\ y_{l-1} \\ y_l \end{pmatrix} = \begin{pmatrix} c_l t^{l-1}/(l-1)! + c_{l-1}t^{l-2}/(l-2)! + \cdots + c_2 t + c_1 \\ \vdots \\ c_l t^2/2 + c_{l-1}t + c_{l-2} \\ c_l t + c_{l-1} \\ c_l \end{pmatrix} e^{\lambda t}$$

and arranged according to the parameters $c_1, \ldots, c_l$,

$$\begin{pmatrix} y_1 \\ y_2 \\ \vdots \\ y_l \end{pmatrix} = \left[c_1 \begin{pmatrix} 1 \\ 0 \\ \vdots \\ 0 \end{pmatrix} + c_2 \begin{pmatrix} t \\ 1 \\ \vdots \\ 0 \end{pmatrix} + c_3 \begin{pmatrix} t^2/2 \\ t \\ 1 \\ \vdots \end{pmatrix} + \cdots + c_l \begin{pmatrix} t^{l-1}/(l-1)! \\ t^{l-2}/(l-2)! \\ \vdots \\ 1 \end{pmatrix}\right] e^{\lambda t}$$

$$= c_1 \mathbf{e}_1 e^{\lambda t} + c_2 (t\mathbf{e}_1 + \mathbf{e}_2) e^{\lambda t} + c_3 \left(\frac{t^2}{2}\mathbf{e}_1 + t\mathbf{e}_2 + \mathbf{e}_3\right) e^{\lambda t} + \cdots$$

$$+ c_l \left(\frac{t^{l-1}}{(l-1)!}\mathbf{e}_1 + \frac{t^{l-2}}{(l-2)!}\mathbf{e}_2 + \cdots t\mathbf{e}_{l-1} + \mathbf{e}_l\right) e^{\lambda t},$$

where $\mathbf{e}_j$ are the standard basis vectors. Remember that these are the components that correspond to one block of dimension $l \times l$ of the Jordan matrix. To get the corresponding n dimensional solution vectors of a multi-block system, we complete the l dimensional vectors (belonging to one block) with $n - l$ zeros in all suitable coordinates. Finally, we return to the original unknown $\mathbf{x}(t)$ through multiplying from the left by the matrix T and obtain a solution $c_1 \mathbf{x}^{(1)}(t) + \cdots + c_l \mathbf{x}^{(l)}(t)$ where

$$
\begin{aligned}
\mathbf{x}^{(1)}(t) &= \mathbf{u}_1 e^{\lambda t}, \\
\mathbf{x}^{(2)}(t) &= (t\mathbf{u}_1 + \mathbf{u}_2) e^{\lambda t}, \\
\mathbf{x}^{(3)}(t) &= \left(\frac{t^2}{2} \mathbf{u}_1 + t\mathbf{u}_2 + \mathbf{u}_3\right) e^{\lambda t}, \ldots
\end{aligned}
\tag{5.14}
$$

and $\mathbf{u}_i = T\mathbf{e}_i$. These solutions are obviously independent. The treatment of the other blocks is similar, and in principle, we obtain n solutions.

In computational problems, we do not carry out the transformation to the Jordan form, because finding the matrix T, which leads to the Jordan form, is a considerable work. Instead, we will settle for finding the vectors $\mathbf{u}_1, \mathbf{u}_2, \ldots$ by the method of undetermined coefficients. In the first solution $\mathbf{x}^{(1)}(t) = \mathbf{u}_1 e^{\lambda t}$, the vector $\mathbf{u}_1$ must be an eigenvector corresponding to the eigenvalue λ, and it is determined by the equation

$$(A - \lambda I)\mathbf{u}_1 = \mathbf{0} .$$

If the algebraic multiplicity of λ exceeds the geometric multiplicity, we look for the next solution $\mathbf{x}^{(2)}(t) = \mathbf{u}_1 t e^{\lambda t} + \mathbf{u}_2 e^{\lambda t}$ by placing it into the differential system $\mathbf{x}' = A\mathbf{x}$:

$$\mathbf{u}_1(e^{\lambda t} + t\lambda e^{\lambda t}) + \mathbf{u}_2 \lambda e^{\lambda t} = A\left(\mathbf{u}_1 t e^{\lambda t} + \mathbf{u}_2 e^{\lambda t}\right) .$$

The comparison of coefficients is

$$
\begin{array}{rcl}
t e^{\lambda t} & : & \lambda \mathbf{u}_1 = A \mathbf{u}_1 , \\
e^{\lambda t} & : & \mathbf{u}_1 + \lambda \mathbf{u}_2 = A \mathbf{u}_2 ,
\end{array}
$$

which is equivalent to

$$
\begin{aligned}
(A - \lambda I)\mathbf{u}_1 &= \mathbf{0}, \\
(A - \lambda I)\mathbf{u}_2 &= \mathbf{u}_1 .
\end{aligned}
\tag{5.15}
$$

The vector $\mathbf{u}_1$ is determined by the first equation and is, of course, an eigenvector corresponding to the eigenvalue λ. After $\mathbf{u}_1$ is already known, $\mathbf{u}_2$ is determined from the second vector equation, but it is not an eigenvector!

5.6 Systems with Constant Coefficients

Note that it is not self-evident in advance that the non-homogeneous system $(A - \lambda I)\mathbf{u}_2 = \mathbf{u}_1$ has a solution, since $\det(A - \lambda I) = 0$. In such a case, it is known from linear algebra that the non-homogeneous system has a solution if and only if the system matrix and the extended matrix have equal ranks,

$$\text{rank}(A - \lambda I) = \text{rank}(A - \lambda I \vdots \mathbf{u}_1).$$

In our problem, the existence of a solution results from the constructive method by which (5.14) was established. The solution for $\mathbf{u}_2$ is not unique; it is determined only up to the addition of any solution of the corresponding homogeneous system.

Suppose, for example, that the algebraic multiplicity of λ is three and its geometric multiplicity is only one. In this case, we look for a third solution of the form

$$\mathbf{x}^{(3)}(t) = \mathbf{u}_1 \frac{t^2}{2} e^{\lambda t} + \mathbf{u}_2 t e^{\lambda t} + \mathbf{u}_3 e^{\lambda t}.$$

by substituting it into the differential system $\mathbf{x}' = A\mathbf{x}$:

$$\mathbf{u}_1 \left(t e^{\lambda t} + \frac{t^2}{2} \lambda e^{\lambda t} \right) + \mathbf{u}_2 (e^{\lambda t} + t \lambda e^{\lambda t}) + \mathbf{u}_3 \lambda e^{\lambda t} = A \left(\mathbf{u}_1 \frac{t^2}{2} e^{\lambda t} + \mathbf{u}_2 t e^{\lambda t} + \mathbf{u}_3 e^{\lambda t} \right)$$

and comparison of coefficients is

$$\frac{t^2}{2} e^{\lambda t} \quad : \quad \lambda \mathbf{u}_1 = A \mathbf{u}_1,$$
$$t e^{\lambda t} \quad : \quad \mathbf{u}_1 + \lambda \mathbf{u}_2 = A \mathbf{u}_2,$$
$$e^{\lambda t} \quad : \quad \mathbf{u}_2 + \lambda \mathbf{u}_3 = A \mathbf{u}_3,$$

which is equivalent to

$$(A - \lambda I)\mathbf{u}_1 = \mathbf{0},$$
$$(A - \lambda I)\mathbf{u}_2 = \mathbf{u}_1, \tag{5.16}$$
$$(A - \lambda I)\mathbf{u}_3 = \mathbf{u}_2.$$

Hence, the recursive process is clear: $\mathbf{u}_1, \mathbf{u}_2$ had been already determined during the calculation of $\mathbf{x}^{(1)}(t)$ and $\mathbf{x}^{(2)}(t)$, and the vector $\mathbf{u}_3$ will be determined by the third system $(A - \lambda I)\mathbf{u}_3 = \mathbf{u}_2$. A series of equations of the type (5.15) or (5.16) is called a *Jordan chain*. It should be remembered that $\mathbf{u}_1$ is an eigenvector but $\mathbf{u}_2, \mathbf{u}_3$ are not eigenvectors. They are called *generalized eigenvectors*, and they satisfy (why?)

$$(A - \lambda I)^2 \mathbf{u}_2 = \mathbf{0},$$
$$(A - \lambda I)^3 \mathbf{u}_3 = \mathbf{0}.$$

The next examples demonstrate some typical situations.

Example 5.6 $\mathbf{x}'(t) = \begin{pmatrix} -3 & 4 \\ -1 & 1 \end{pmatrix} \mathbf{x}(t).$

Here $|A - \lambda I| = \begin{vmatrix} -3-\lambda & 4 \\ -1 & 1-\lambda \end{vmatrix} = (\lambda+1)^2 = 0.$ We look for an eigenvector corresponding to the double eigenvalue $\lambda_1 = \lambda_2 = -1$:

$$(A - \lambda_1 I)\mathbf{v} = \begin{pmatrix} -3-(-1) & 4 \\ -1 & 1-(-1) \end{pmatrix} \begin{pmatrix} \alpha \\ \beta \end{pmatrix} = \begin{pmatrix} 0 \\ 0 \end{pmatrix},$$

$$-2\alpha + 4\beta = 0,$$
$$-\alpha + 2\beta = 0.$$

These equations are, of course, dependent and have a unique solution up to a multiplicative constant, for example, $\alpha = 2$, $\beta = 1$. One solution of the differential system is

$$\mathbf{x}^{(1)}(t) = \begin{pmatrix} 2 \\ 1 \end{pmatrix} e^{-t}.$$

Since the geometric multiplicity is one, less than the algebraic multiplicity that is two, we look for a second solution of the form $\mathbf{x}^{(2)}(t) = \mathbf{v} t e^{-t} + \mathbf{u} e^{-t}$. As we know, $\mathbf{v}$ is an eigenvector determined by $(A - \lambda_1 I)\mathbf{v} = (A+I)\mathbf{v} = \mathbf{0}$ and was calculated above, and the vector $\mathbf{u}$ is determined by $(A+I)\mathbf{u} = \mathbf{v}$, namely,

$$\begin{pmatrix} -3+1 & 4 \\ -1 & 1+1 \end{pmatrix} \begin{pmatrix} \gamma \\ \delta \end{pmatrix} = \begin{pmatrix} 2 \\ 1 \end{pmatrix}.$$

This system consists of two dependent equations

$$-2\gamma + 4\delta = 2,$$
$$-\gamma + 2\delta = 1,$$

which have a one-parametric family of solutions

$$\mathbf{u} = \begin{pmatrix} \gamma \\ \delta \end{pmatrix} = \begin{pmatrix} -1+2\delta \\ \delta \end{pmatrix} = \begin{pmatrix} -1 \\ 0 \end{pmatrix} + \delta \begin{pmatrix} 2 \\ 1 \end{pmatrix}$$

for all δ. The corresponding solution of the differential system is

$$\mathbf{x}^{(2)}(t) = \mathbf{v} t e^{-t} + \mathbf{u} e^{-t} = \begin{pmatrix} 2 \\ 1 \end{pmatrix} t e^{-t} + \left[\begin{pmatrix} -1 \\ 0 \end{pmatrix} + \delta \begin{pmatrix} 2 \\ 1 \end{pmatrix} \right] e^{-t}.$$

5.6 Systems with Constant Coefficients

The last term, $\delta \begin{pmatrix} 2 \\ 1 \end{pmatrix} e^{-t}$, is a multiple of the first solution $\mathbf{x}^{(1)}(t)$, and it adds no new information. Therefore, we choose $\delta = 0$ and obtain a second solution

$$\mathbf{x}^{(2)}(t) = \begin{pmatrix} 2 \\ 1 \end{pmatrix} t e^{-t} + \begin{pmatrix} -1 \\ 0 \end{pmatrix} e^{-t} = \begin{pmatrix} 2t - 1 \\ t \end{pmatrix} e^{-t}.$$

Remark From the form of the second solution, $\mathbf{x}^{(2)}(t)$, we see that it contains a vector whose terms are polynomials of the first degree. It is, therefore, possible to search from the beginning a solution of the form

$$\begin{pmatrix} \alpha t + \gamma \\ \beta t + \delta \end{pmatrix} e^{-t}.$$

This approach is correct but less convenient because in this way, we encounter all four unknown $\alpha, \beta, \gamma, \delta$ at once, while it is possible to treat first the pair α, β and then separately γ, δ.

Finally, we solve a problem where an eigenvalue has an algebraic multiplicity three while its geometric multiplicity is two.

Example 5.7 $\mathbf{x}'(t) = \begin{pmatrix} 1 & 0 & 2 \\ 0 & 1 & 1 \\ 0 & 0 & 1 \end{pmatrix} \mathbf{x}(t)$. Here $|A - \lambda I| = (\lambda - 1)^3 = 0$, and its roots are $\lambda_1 = \lambda_2 = \lambda_3 = 1$. First, we look for eigenvectors. The system of equations $(A - \lambda I)\mathbf{v} = \mathbf{0}$ becomes for $\lambda = 1$

$$\begin{pmatrix} 0 & 0 & 2 \\ 0 & 0 & 1 \\ 0 & 0 & 0 \end{pmatrix} \begin{pmatrix} \alpha \\ \beta \\ \gamma \end{pmatrix} = \begin{pmatrix} 0 \\ 0 \\ 0 \end{pmatrix} = \mathbf{0}$$

whose solutions are $\gamma = 0$ and arbitrary α, β. So

$$\mathbf{v} = \begin{pmatrix} \alpha \\ \beta \\ 0 \end{pmatrix} = \alpha \begin{pmatrix} 1 \\ 0 \\ 0 \end{pmatrix} + \beta \begin{pmatrix} 0 \\ 1 \\ 0 \end{pmatrix}.$$

The geometric multiplicity is two, and we found two solutions for the differential system

$$\mathbf{x}^{(1)}(t) = \begin{pmatrix} 1 \\ 0 \\ 0 \end{pmatrix} e^t, \quad \mathbf{x}^{(2)}(t) = \begin{pmatrix} 0 \\ 1 \\ 0 \end{pmatrix} e^t.$$

A third solution of the form $\mathbf{v}\,te^t + \mathbf{u}e^t$ is obtained from

$$(A - I)\mathbf{v} = \mathbf{0},$$
$$(A - I)\mathbf{u} = \mathbf{v}.$$

The vector $\mathbf{v}$ was already calculated above, and it depends on two parameters α, β; therefore, the equations for $\mathbf{u}$ are

$$\begin{pmatrix} 0 & 0 & 2 \\ 0 & 0 & 1 \\ 0 & 0 & 0 \end{pmatrix} \begin{pmatrix} u_1 \\ u_2 \\ u_3 \end{pmatrix} = \begin{pmatrix} \alpha \\ \beta \\ 0 \end{pmatrix},$$

that is,

$$2u_3 = \alpha, \qquad u_3 = \beta.$$

These equations are compatible only if $\alpha = 2\beta$ and then $\mathbf{u}$ is

$$\mathbf{u} = \begin{pmatrix} u_1 \\ u_2 \\ u_3 \end{pmatrix} = \begin{pmatrix} c_1 \\ c_2 \\ \beta \end{pmatrix},$$

where c_1, c_2 are arbitrary. To sum up, a third solution $\mathbf{v}\,te^t + \mathbf{u}e^t$ is

$$\mathbf{x}(t) = \begin{pmatrix} 2\beta \\ \beta \\ 0 \end{pmatrix} te^t + \begin{pmatrix} c_1 \\ c_2 \\ \beta \end{pmatrix} e^t = \beta \left[\begin{pmatrix} 2 \\ 1 \\ 0 \end{pmatrix} te^t + \begin{pmatrix} 0 \\ 0 \\ 1 \end{pmatrix} e^t \right] + \begin{pmatrix} c_1 \\ c_2 \\ 0 \end{pmatrix} e^t.$$

The last term is just the linear combination $c_1\mathbf{x}^{(1)}(t) + c_2\mathbf{x}^{(2)}(t)$ of the first two solutions already known to us. Therefore, as a third solution, we can take

$$\mathbf{x}^{(3)}(t) = \begin{pmatrix} 2 \\ 1 \\ 0 \end{pmatrix} te^t + \begin{pmatrix} 0 \\ 0 \\ 1 \end{pmatrix} e^t = \begin{pmatrix} 2t \\ t \\ 1 \end{pmatrix} e^t.$$

During the calculation of this solution, we had to make ad hoc decisions regarding the implementation of the Jordan chain of Eqs. (5.16) and selecting parameters. These improvisations may be avoided, and the solutions can be calculated directly if one knows more about the structure of the Jordan form.

From the general forms (5.12) or (5.14) of solutions of systems with constant coefficients, it is easy to determine their stability:

Theorem 5.12 *The trivial solution* $\mathbf{x}(t) \equiv \mathbf{0}$ *of the differential system* $\mathbf{x}'(t) = A\mathbf{x}(t)$ *is asymptotically stable if and only if all the eigenvalues of the matrix* A *are negative or have a negative real part. Furthermore, in this case, all solutions of the system are asymptotically stable.*

The trivial solution is stable if and only if all the eigenvalues (or their real parts) are non-positive and those with zero real parts have equal algebraic and geometric multiplicities.

Indeed, if $\text{Re}\,\lambda < 0$, then $t^k e^{\lambda t} \to 0$ as $t \to +\infty$ and the vector solutions of the system tend to the zero vector.

If $\text{Re}\,\lambda = 0$, $\lambda = \beta i$, then the equality of the algebraic and geometric multiplicities ensures that the solution consists only of $\cos bt$, $\sin bt$ with no powers of t.

5.7 Non-Homogeneous Systems

So far, we have dealt with homogeneous differential systems of the form (5.5),

$$\mathbf{x}'(t) = P(t)\mathbf{x}(t) \,.$$

Now we turn to non-homogeneous systems whose form is

$$\mathbf{x}'(t) = P(t)\mathbf{x}(t) + \mathbf{q}(t) \,, \tag{5.17}$$

or in component notation

$$x_i'(t) = p_{i1}(t)x_1 + p_{i2}(t)x_2 + \ldots + p_{in}(t)x_n + q_i(t), \quad i = 1, 2, \ldots, n \,.$$

As for non-homogeneous linear equations of any order, also for non-homogeneous linear systems, the general solution consists of one particular solution plus all solutions of the corresponding homogeneous system. The goal of this chapter is to suggest a method to find one particular solution.

For non-homogeneous linear systems, we use the idea of the variation of parameters, as we did for non-homogeneous equations. The general solution of a homogeneous system is of the form

$$\mathbf{x}_H(t) = c_1 \mathbf{x}^{(1)}(t) + \cdots + c_n \mathbf{x}^{(n)}(t)$$

and according to (5.9), it can be written as

$$\mathbf{x}(t) = \Phi(t)\mathbf{c} \,,$$

where $\Phi(t)$ is the fundamental matrix whose columns are the solutions $\mathbf{x}^{(1)}(t)$, ..., $\mathbf{x}^{(n)}(t)$ and $\mathbf{c}$ is a vector of n arbitrary components $c_1, \ldots, c_n$. For the non-homogeneous system (5.17), we look for a solution of the form

$$\mathbf{x}_{\text{NH}}(t) = c_1(t)\mathbf{x}^{(1)}(t) + \cdots + c_n(t)\mathbf{x}^{(n)}(t),$$

namely,

$$\mathbf{x}_{\text{NH}}(t) = \Phi(t)\mathbf{c}(t),$$

where $\Phi(t)$ is a fundamental matrix constructed from the solutions of the corresponding homogeneous system and $\mathbf{c}(t)$ is a vector function whose components are n unknown functions $c_1(t), \ldots, c_n(t)$. We substitute the derivative of the supposed solution,

$$\bigl(\Phi(t)\mathbf{c}(t)\bigr)' = \Phi'(t)\mathbf{c}(t) + \Phi(t)\mathbf{c}'(t)$$

into the system (5.17). (The derivative of a matrix-vector product is calculated as the derivative of a product of two scalar functions, but the order of multiplication must be preserved. That is because the product of a matrix by a vector consists of sums of products of scalar functions). The non-homogeneous system (5.17) becomes

$$\Phi'(t)\mathbf{c}(t) + \Phi(t)\mathbf{c}'(t) = P(t)\bigl(\Phi(t)\mathbf{c}(t)\bigr) + \mathbf{q}(t). \tag{5.18}$$

What is $\Phi'(t)$? We differentiate each of its columns separately:

$$\Phi'(t) = \left(\mathbf{x}^{(1)\prime} \vdots \mathbf{x}^{(2)\prime} \vdots \ldots \vdots \mathbf{x}^{(n)\prime}\right)$$

$$= \left(P\mathbf{x}^{(1)} \vdots P\mathbf{x}^{(2)} \vdots \ldots \vdots P\mathbf{x}^{(n)}\right)$$

$$= P(t)\left(\mathbf{x}^{(1)} \vdots \mathbf{x}^{(2)} \vdots \ldots \vdots \mathbf{x}^{(n)}\right) = P(t)\Phi(t).$$

The substitution $\Phi'(t) = P(t)\Phi(t)$ into (5.18) results with

$$P(t)\Phi(t)\mathbf{c}(t) + \Phi(t)\mathbf{c}'(t) = P(t)\Phi(t)\mathbf{c}(t) + \mathbf{q}(t).$$

After multiplication by $\Phi^{-1}(t)$, there remains only

$$\mathbf{c}'(t) = \Phi^{-1}(t)\mathbf{q}(t).$$

The vector function $\mathbf{c}(t)$ is obtained by integration. (Integration of a vector function is performed on each component separately, like differentiation). In order to distinguish between the variable t outside the integral and the variable of integration,

5.7 Non-Homogeneous Systems

we denote the last one by some different letter, say s:

$$\mathbf{c}(t) = \int^t \Phi^{-1}(s)\mathbf{q}(s)\,ds + \mathbf{K},$$

where the vector $\mathbf{K}$ is a vector of constants of integration. To summarize, the solution of the non-homogeneous system is

$$\mathbf{x}_{NH}(t) = \Phi(t)\mathbf{c}(t) = \Phi(t)\int^t \Phi^{-1}(s)\mathbf{q}(s)\,ds + \Phi(t)\mathbf{K}. \tag{5.19}$$

The last term, $\Phi(t)\mathbf{K}$, is a general solution of the corresponding homogeneous system (5.5) with an arbitrary vector of constants of integration $\mathbf{K}$. The first term is a particular solution of the non-homogeneous system (5.17),

$$\mathbf{x}_p(t) = \Phi(t)\int^t \Phi^{-1}(s)\mathbf{q}(s)\,ds.$$

The factor $\Phi(t)$ may be inserted into the integral, since as a function of t it is constant with respect the integration according to s. The result is

$$\mathbf{x}_p(t) = \int^t \Phi(t)\Phi^{-1}(s)\mathbf{q}(s)\,ds.$$

It is mandatory to keep the order of the multiplications because matrix multiplication is not commutative. This formula is analogous to the particular solution (4.37) of linear equations of any order, and the kernel of the integral $\Phi(t)\Phi^{-1}(s)$ is the analog of the Cauchy kernel $K(x,t)$. Also, here, the kernel $\Phi(t)\Phi^{-1}(s)$ depends only on the corresponding homogeneous system, while the factor $\mathbf{q}(s)$ originates on the right-hand side of the non-homogeneous system.

Let us add to the non-homogeneous system (5.17) an initial value condition $\mathbf{x}(t_0) = \boldsymbol{\alpha}$. To solve the initial value problem, it is convenient to choose for the integral in (5.19) the lower limit t_0:

$$\mathbf{x}(t) = \Phi(t)\int_{t_0}^t \Phi^{-1}(s)\mathbf{q}(s)\,ds + \Phi(t)\mathbf{K}.$$

With $t = t_0$ and the initial value condition $\mathbf{x}(t_0) = \boldsymbol{\alpha}$, we get

$$\boldsymbol{\alpha} = \mathbf{x}(t_0) = \mathbf{0} + \Phi(t_0)\mathbf{K}$$

which determines $\mathbf{K}$ to be

$$\mathbf{K} = \Phi^{-1}(t_0)\boldsymbol{\alpha}.$$

Finally, the solution of the non-homogeneous initial value problem is

$$\mathbf{x}(t) = \Phi(t) \int_{t_0}^{t} \Phi^{-1}(s)\mathbf{q}(s)\,ds + \Phi(t)\Phi^{-1}(t_0)\boldsymbol{\alpha} \ . \tag{5.20}$$

The first term is the solution of the non-homogeneous system that equals $\mathbf{0}$ at t_0, and the second term is the solution of the corresponding homogeneous system that has at the point t_0 the value $\boldsymbol{\alpha}$.

Example 5.8

$$\mathbf{x}'(t) = \begin{pmatrix} 1 & 1 \\ 4 & 1 \end{pmatrix}\mathbf{x}(t) + \begin{pmatrix} e^{2t} \\ e^{t} \end{pmatrix}, \qquad \mathbf{x}(0) = \begin{pmatrix} 5 \\ 6 \end{pmatrix}.$$

We saw in Example 4.5 that the homogeneous system $\mathbf{x}'(t) = \begin{pmatrix} 1 & 1 \\ 4 & 1 \end{pmatrix}\mathbf{x}(t)$ has a fundamental matrix

$$\Phi(t) = \begin{pmatrix} e^{3t} & e^{-t} \\ 2e^{3t} & -2e^{-t} \end{pmatrix}.$$

The inversion formula for 2×2 matrices,

$$\begin{pmatrix} a & b \\ c & d \end{pmatrix}^{-1} = \frac{1}{ad - bc}\begin{pmatrix} d & -b \\ -c & a \end{pmatrix},$$

(verify by direct multiplication) is in our case

$$\Phi^{-1}(t) = \frac{1}{-4e^{2t}}\begin{pmatrix} -2e^{-t} & -e^{-t} \\ -2e^{3t} & e^{3t} \end{pmatrix} = \frac{1}{4}\begin{pmatrix} 2e^{-3t} & e^{-3t} \\ 2e^{t} & -e^{t} \end{pmatrix}.$$

According to (5.20), the solution of the non-homogeneous initial value problem is

$$\mathbf{x}(t) = \frac{1}{4}\begin{pmatrix} e^{3t} & e^{-t} \\ 2e^{3t} & -2e^{-t} \end{pmatrix}\int_{0}^{t}\begin{pmatrix} 2e^{-3s} & e^{-3s} \\ 2e^{s} & -e^{s} \end{pmatrix}\begin{pmatrix} e^{2s} \\ e^{s} \end{pmatrix}ds$$

$$+ \begin{pmatrix} e^{3t} & e^{-t} \\ 2e^{3t} & -2e^{-t} \end{pmatrix}\frac{1}{4}\begin{pmatrix} 2 & 1 \\ 2 & -1 \end{pmatrix}\begin{pmatrix} 5 \\ 6 \end{pmatrix}$$

$$= \frac{1}{4}\begin{pmatrix} e^{3t} & e^{-t} \\ 2e^{3t} & -2e^{-t} \end{pmatrix}\int_{0}^{t}\begin{pmatrix} 2e^{-s} + e^{-2s} \\ 2e^{3s} - e^{2s} \end{pmatrix}ds + \begin{pmatrix} 4e^{3t} + e^{-t} \\ 8e^{3t} - 2e^{-t} \end{pmatrix}.$$

Complete the calculations!

5.7 Non-Homogeneous Systems

Problems

5.1 Given the two vector functions $\mathbf{x}^{(1)}(t) = \begin{pmatrix} t \\ 1 \end{pmatrix}$, $\mathbf{x}^{(2)}(t) = \begin{pmatrix} t^2 \\ 2t \end{pmatrix}$.

(a) Calculate $W(\mathbf{x}^{(1)}(t), \mathbf{x}^{(2)}(t))$.
(b) On which t intervals the vector functions $\mathbf{x}^{(1)}(t)$, $\mathbf{x}^{(2)}(t)$ are linearly independent?
(c) Find a differential system $\mathbf{x}'(t) = A(t)\mathbf{x}(t)$ that $\mathbf{x}^{(1)}(t)$, $\mathbf{x}^{(2)}(t)$ are its solutions.
(d) Where is the differential system defined and where not?

5.2 Convert the equation $y''' + a_1 y'' + a_2 y' + a_3 y = 0$, a_1, a_2, a_3 constants, to a system $\mathbf{x}'(t) = A\mathbf{x}(t)$ by substituting

$$x_1(t) = y(t), \quad x_2(t) = y'(t), \quad x_3(t) = y''(t),$$

as done in Eq. (5.3). Show that the characteristic polynomial of the third-order equation and the characteristic polynomial of the 3×3 matrix A are the same.

5.3 Prove that if $A(t)$ is an $n \times n$ antisymmetric matrix of continuous real functions, then the Wronskian of the system $\mathbf{x}'(t) = A(t)\mathbf{x}(t)$ is constant.

Solve the following systems of equations, from the easiest to the hardest:

5.4 $\mathbf{x}' = \begin{pmatrix} -2 & 1 \\ 1 & -2 \end{pmatrix} \mathbf{x}$.

5.5 $\mathbf{x}'(t) = \begin{pmatrix} 1 & 1 & 1 \\ 2 & 1 & -1 \\ -8 & -5 & -3 \end{pmatrix} \mathbf{x}$.

5.6 $\mathbf{x}'(t) = \begin{pmatrix} 3 & 1 & 1 \\ 1 & 0 & 2 \\ 1 & 2 & 0 \end{pmatrix} \mathbf{x}$.

5.7 $\mathbf{x}'(t) = \begin{pmatrix} 1789 & 1870 \\ 2014 & 2039 \end{pmatrix} \mathbf{x}, \quad \mathbf{x}(0) = \begin{pmatrix} 0 \\ 0 \end{pmatrix}$.

5.8 $\mathbf{x}'(t) = \begin{pmatrix} 3 & -4 \\ 1 & -1 \end{pmatrix} \mathbf{x}, \quad \mathbf{x}(0) = \begin{pmatrix} 3 \\ 5 \end{pmatrix}$.

5.9 $\mathbf{x}'(t) = \begin{pmatrix} 0 & 1 & 1 \\ 1 & 0 & 1 \\ 1 & 1 & 0 \end{pmatrix} \mathbf{x}$.

5.10 $\mathbf{x}'(t) = \begin{pmatrix} 1 & 1 & 1 \\ 1 & 1 & 1 \\ 1 & 1 & 1 \end{pmatrix} \mathbf{x}$.

5.11 $\mathbf{x}'(t) = \begin{pmatrix} 3 & 2 \\ -5 & 1 \end{pmatrix} \mathbf{x}$.

5.12 $\mathbf{x}'(t) = \begin{pmatrix} 1 & 2 \\ -5 & -1 \end{pmatrix} \mathbf{x}$.

5.13 $\mathbf{x}'(t) = \begin{pmatrix} 0 & 8 & 0 \\ 0 & 0 & -2 \\ 2 & 8 & -2 \end{pmatrix} \mathbf{x}$.

5.14 $\mathbf{x}'(t) = \begin{pmatrix} 0 & 1 & 0 \\ 0 & 0 & 1 \\ 4 & -4 & 1 \end{pmatrix} \mathbf{x}, \quad \mathbf{x}(0) = \begin{pmatrix} 2 \\ -2 \\ -8 \end{pmatrix}$.

5.15 $\mathbf{x}'(t) = \begin{pmatrix} 0 & 1 & 0 \\ 0 & 0 & 1 \\ -1 & 5 & 5 \end{pmatrix} \mathbf{x}$. Note that according to (5.3), this system is equivalent

to Problem 4.27(a) in Chap. 4.

5.16 $\mathbf{x}'(t) = \begin{pmatrix} 1 & 3 \\ 0 & 1 \end{pmatrix} \mathbf{x}$.

5.17 $\mathbf{x}'(t) = \begin{pmatrix} 1 & 1 & 1 \\ 2 & 1 & -1 \\ -3 & 2 & 4 \end{pmatrix} \mathbf{x}$.

5.18 $\mathbf{x}'(t) = \begin{pmatrix} 1 & 0 & 2 \\ 1 & 1 & 0 \\ 0 & 0 & 1 \end{pmatrix} \mathbf{x}$.

5.19 Given the system of differential equations $\mathbf{x}'(t) = \begin{pmatrix} 1 & 1 & 1 \\ 2 & 2 & 2 \\ 3 & 3 & 3 \end{pmatrix} \mathbf{x}$.

(a) Calculate the eigenvalues and the corresponding eigenvectors of the coefficient matrix. It is convenient to solve directly the equations

$$\begin{pmatrix} 1 & 1 & 1 \\ 2 & 2 & 2 \\ 3 & 3 & 3 \end{pmatrix} \begin{pmatrix} \alpha \\ \beta \\ \gamma \end{pmatrix} = \lambda \begin{pmatrix} \alpha \\ \beta \\ \gamma \end{pmatrix}.$$

What are the corresponding algebraic and geometric multiplicities?

5.7 Non-Homogeneous Systems

(b) Solve the given system of the differential equations.

(d) Solve the differential system $\mathbf{x}'(t) = \begin{pmatrix} 1 & 1 & 1 & 1 & 1 \\ 2 & 2 & 2 & 2 & 2 \\ 3 & 3 & 3 & 3 & 3 \\ 4 & 4 & 4 & 4 & 4 \\ 5 & 5 & 5 & 5 & 5 \end{pmatrix} \mathbf{x}$.

5.20 Find a basis of real solutions for the system $\mathbf{x}'(t) = \begin{pmatrix} 1 & 0 & -2 \\ 0 & 1 & 0 \\ 1 & -1 & -1 \end{pmatrix} \mathbf{x}$.

Determine the initial conditions $\mathbf{x}(0) = \mathbf{v}_0$ so that the corresponding solutions will be periodic.

5.21 The system $\mathbf{x}' = \begin{pmatrix} 5 & -3 & -2 \\ 8 & -5 & -4 \\ -4 & 3 & 3 \end{pmatrix} \mathbf{x}$ has the eigenvalues $\lambda_1 = \lambda_2 = \lambda_3 = 1$,

and corresponding to them, there are two independent eigenvectors. Find a third solution of the form $\mathbf{x}^{(3)}(t) = \mathbf{v}\, t e^t + \mathbf{u}\, e^t$ so that

(1) $(A - I)\mathbf{v} = \mathbf{0}$, (2) $(A - I)\mathbf{u} = \mathbf{v}$.

Hint: what are all the solutions $\mathbf{v}$ of (1), and on how many parameters do they depend on? For which vectors $\mathbf{v}$, there exist solutions $\mathbf{u}$ for (2)?

5.22 Solve the system of equations $\mathbf{x}' = \mathbf{x}$.

5.23 Given that $\mathbf{x}^{(1)}(t) = \mathbf{w} e^t$, $\mathbf{x}^{(2)}(t) = \mathbf{v}\, t e^t + \mathbf{u}\, e^t$ ($\mathbf{u}, \mathbf{v}, \mathbf{w}$ are constant vectors), are independent solutions of a system $\mathbf{x}' = A\mathbf{x}$, where A is a 2×2 constant matrix.

(a) What are the eigenvalues of the matrix A?
(b) What are the eigenvectors of A?
(c) Is the matrix A diagonalizable?
(d) What are $\det(A)$, $\operatorname{trace}(A)$?
(e) What are the connections between $\mathbf{u}, \mathbf{v}$ and $\mathbf{w}$?
(d) Describe a process for finding the matrix A with the aid of $\mathbf{u}, \mathbf{v}, \mathbf{w}$.

5.24 Show that for the differential system $\mathbf{x}'(t) = A\, \mathbf{x}(t)$ of any order with constant coefficients, it is impossible to have a non-trivial solution of the form $\mathbf{x}(t) = \mathbf{v}\, t e^{at}$, where $\mathbf{v}$ is a constant vector.

5.25 What is the minimal dimension $n \times n$ of a matrix A such that the system $\mathbf{x}' = A\mathbf{x}$ has a non-trivial solution of the form $\mathbf{v}_1 \sin t + \mathbf{v}_2 \cos^2 t$, with two constant vectors $\mathbf{v}_1, \mathbf{v}_2$.

5.26 Let A be a matrix of dimension $n \times n$. Find the minimal order n of the system $\mathbf{x}'(t) = A\mathbf{x}(t)$ if it is known that it has four non-zero solutions such that one of them is bounded on the whole t axis, the second is bounded on $(0, \infty)$ but unbounded on

$(-\infty, 0)$, the third is bounded on $(-\infty, 0)$ but unbounded on $(0, \infty)$, and finally, the fourth is unbounded both on $(0, \infty)$ and on $(-\infty, 0)$.

Give an example of a matrix A that fulfills all the conditions above, including the minimality of order, so that all its elements are only 0 or $+1$.

5.27 Given a system $\mathbf{x}'(t) = B\,\mathbf{x}(t)$ where B is a constant real matrix of dimension 4×4. It is also given that:

(a) trace$(B) = 0$, det$(B) = 1$.
(b) The system has solutions that are bounded on the whole t axis and some solutions that are unbounded there.

Find the form of the general solution of this differential system. Explain what is determined in the solution unequivocally by the data above and what remains subject to free choice.

5.28 Find a 2×2 system $\mathbf{x}' = A\mathbf{x}$ so that

$$\mathbf{x}^{(1)} = \begin{pmatrix} 1 \\ 1 \end{pmatrix} e^{-2t}, \quad \mathbf{x}^{(2)} = \begin{pmatrix} t \\ t-1 \end{pmatrix} e^{-2t},$$

are its solutions. Is the matrix A diagonalizable?

5.29 Prove that all solutions $\mathbf{x}'(t) = \begin{pmatrix} a & b \\ c & d \end{pmatrix} \mathbf{x}$ tend to $\mathbf{0}$ when $t \to +\infty$ if and only if $ad - bc > 0$, $a + d < 0$.

5.30 Given an initial value problem $\mathbf{x}' = A\mathbf{x}$, $\mathbf{x}(0) = \boldsymbol{\alpha}$, where A is a constant matrix. The successive iterations

$$y_{m+1}(x) = y_0 + \int_{x_0}^{x} f(x, y_m(x))\, dx$$

become for this problem

$$\mathbf{x}^{(m+1)}(t) = \boldsymbol{\alpha} + \int_{0}^{t} A\mathbf{x}^{(m)}(t)\, dt,$$

where $\mathbf{x}^{(m)}(t)$ indicates the m-th iteration. Prove the following claims:

(a) If we start the iterations with $\mathbf{x}^{(0)}(t) \equiv \boldsymbol{\alpha}$, then

$$\mathbf{x}^{(1)} = \boldsymbol{\alpha} + \int_0^t A\boldsymbol{\alpha}\, dt = (I + At)\boldsymbol{\alpha}.$$

(Remember that the integration of a vector is an integration of each of its component separately!)

5.7 Non-Homogeneous Systems

(b) Repeat the iteration, and calculate $\mathbf{x}^{(2)}, \mathbf{x}^{(3)}$. Show that after m iterations we get

$$\mathbf{x}^{(m)} = \left(I + At + A^2 \frac{t^2}{2!} + \cdots + A^n \frac{t^m}{m!} \right) \alpha .$$

(c) As $m \to \infty$, an infinite series of matrices is obtained. (The sum of an infinite series of matrices is a matrix in which each (i, j)-th element is the sum of an infinite series). What does the above series remind you of?

(d) Calculate explicitly the successive iterations when $A = \begin{pmatrix} 0 & 1 \\ -1 & 0 \end{pmatrix}$. Identify explicitly the four infinite series obtained in this way as the four elements of the sum matrix.

What is the solution of the initial value problem

$$\mathbf{x}'(t) = \begin{pmatrix} 0 & 1 \\ -1 & 0 \end{pmatrix} \mathbf{x}, \quad \mathbf{x}(0) = \alpha \, ?$$

Solve the following non-homogeneous systems using the variation of parameters method. The corresponding homogeneous systems appeared in previous problems in this chapter.

5.31 $\mathbf{x}' = \begin{pmatrix} 1 & 3 \\ 0 & 1 \end{pmatrix} \mathbf{x} + \begin{pmatrix} e^{-t} \\ 2e^{-t} \end{pmatrix}$.

5.32 $\mathbf{x}' = \begin{pmatrix} 1 & 3 \\ 0 & 1 \end{pmatrix} \mathbf{x} + \begin{pmatrix} t \\ 1 \end{pmatrix}$.

5.33 $\mathbf{x}' = \begin{pmatrix} -2 & 1 \\ 1 & -2 \end{pmatrix} \mathbf{x} + \begin{pmatrix} 2e^{-t} \\ 3t \end{pmatrix}$.

5.34 $\mathbf{x}' = \begin{pmatrix} 1 & 1 & 1 \\ 2 & 1 & -1 \\ -8 & -5 & -3 \end{pmatrix} \mathbf{x} + \begin{pmatrix} e^t \\ e^{2t} \\ 0 \end{pmatrix}$.

5.35 $\mathbf{x}' = \begin{pmatrix} 3 & 1 & 1 \\ 1 & 0 & 2 \\ 1 & 2 & 0 \end{pmatrix} \mathbf{x} + \begin{pmatrix} 3 \\ 5 \\ 7 \end{pmatrix}$.

5.36 $\mathbf{x}' = \begin{pmatrix} 1 & 1 & 1 \\ 1 & 1 & 1 \\ 1 & 1 & 1 \end{pmatrix} \mathbf{x} - \begin{pmatrix} 2 \\ 2 \\ 2 \end{pmatrix}$. Verify that the three vectors $\begin{pmatrix} 2 \\ 0 \\ 0 \end{pmatrix}$, $\begin{pmatrix} 0 \\ 2 \\ 0 \end{pmatrix}$ and $\begin{pmatrix} 0 \\ 0 \\ 2 \end{pmatrix}$ are solutions of this non-homogeneous system. What is the general solution of the system?

Chapter 6
The Qualitative Theory and the Phase Plane

6.1 Trajectories and Critical Points in the Phase Plane

We have emphasized several times that we do not know how to solve most differential equations and systems explicitly. The purpose of this chapter is to draw qualitative conclusions about the solutions of a system of differential equations utilizing geometric considerations without solving the equations explicitly. It turns out that even in cases where the explicit solution is known, such as for linear systems with constant coefficients, it may be more convenient and useful to look at a geometric picture rather than at an explicit but complicated formula. It is especially so for the following family of systems.

Definition 6.1 A system of differential equations is called *autonomous* if the independent variable t does not appear explicitly on its right-hand side. An autonomous system with two unknown functions is

$$\begin{aligned} x_1'(t) &= F_1(x_1, x_2), \\ x_2'(t) &= F_2(x_1, x_2). \end{aligned} \quad (6.1)$$

For the convenience of printing, we will write a solution as a row vector $\mathbf{x}(t) = (x_1(t), x_2(t))$ instead of a column vector.

In particular, a linear, homogeneous system with constant coefficients

$$\begin{aligned} x_1'(t) &= ax_1 + bx_2, \\ x_2'(t) &= cx_1 + dx_2, \end{aligned} \quad (6.2)$$

is an autonomous system. The autonomous equations $y' = f(y)$ and $y'' = f(y)$ had already been encountered in the examples of Sect. 1.2, in (3.14) and in Sect. 2.6.

Our discussion in the next section will only be about homogeneous linear equations with constant coefficients (6.2). But the following discussion is suitable for any autonomous system, whether it is linear or not. We will formulate the ideas for a general autonomous system (6.1), since there is no benefit in limiting the discussion to linear systems only.

The foundation of the geometric discussion is the observation that every solution of the system (6.1) is also a parametric description of a curve in the (x_1, x_2)-plane:

$$x_1 = x_1(t),$$
$$x_2 = x_2(t),$$

where $a \leq t \leq b$ or $-\infty < t < \infty$.

Definition 6.2 The (x_1, x_2)-plane is called the *phase plane*.

A curve whose parametric description is given by a solution of the autonomous system (6.1) is called the *trajectory* or the *orbit* of the solution. Each trajectory $(x_1(t), x_2(t))$ has a *direction*, which is determined by the motion of the points on it as t grows. The set of the trajectories corresponding to all solutions of the differential system is called the *phase portrait* of the system.

Example 6.1 One of the solutions of the system (Example 5.1)

$$\begin{aligned} x_1' &= x_2, \\ x_2' &= -x_1, \end{aligned} \qquad (6.3)$$

is $\mathbf{x}(t) = (x_1(t), x_2(t)) = (R \sin t, R \cos t)$. The equations

$$x_1(t) = R \sin t, \quad x_2(t) = R \cos t, \quad -\infty < t < \infty,$$

are a parametric description of its trajectory, in this case, a circle.

A basic property of autonomous systems is that their solutions are preserved under a time shift $t \to t + c$. More precisely,

Theorem 6.1 *If $(x_1(t), x_2(t))$ is a solution of the autonomous system (6.1), then for every constant c also $(x_1(t + c), x_2(t + c))$ is a solution of the same system.*

Proof This property is called *time invariance*, and it follows from the fact that the independent variable t does not appear explicitly on the right-hand side of (6.1). Indeed, $(y_1(t), y_2(t)) = (x_1(t + c), x_2(t + c))$ satisfies the same system,

$$\frac{dy_j(t)}{dt} = \frac{dx_j(t+c)}{dt} = F_j(x_1(t+c), x_2(t+c)) = F_j(y_1(t), y_2(t)), \quad j = 1, 2,$$

therefore, also $(x_1(t + c), x_2(t + c))$ is a solution of (6.1).

6.1 Trajectories and Critical Points in the Phase Plane

What is the connection and what is the difference between the concepts "solution" and "trajectory"? First, a solution is a function (a vector function, to be more precise), while a trajectory is a geometric being—a curve. To each solution $(x_1(t), x_2(t))$, there corresponds a certain trajectory. But the same trajectory corresponds also to all infinitely many time-shifted solutions $(x_1(t + c), x_2(t + c))$, for every constant c. Therefore, the correspondence between "solution" and "trajectory" is not one to one. Geometrically, all these solutions pass through the same points of the plane; however, different solutions pass through the same point at different times. For example,

$$(x_1(t), x_2(t)) = (R\sin(t - \pi/3), \; R\cos(t - \pi/3))$$

is also a solution of the system (6.3) and plots the same circular path in the same direction, but it passes at every point after a lag of $\pi/3$ time units after the solution $(R\sin t, R\cos t)$.

The following example shows the essential difference between finding a solution of the system and finding its trajectory.

Example 6.2 Given the autonomous system

$$\begin{aligned} x'(t) &= x^2 - 6xy, \\ y'(t) &= 3y^2 - 2xy. \end{aligned} \qquad (6.4)$$

If we divide the system's equations by each other, we get the first-order differential equation

$$\frac{dy}{dx} = \frac{dy/dt}{dx/dt} = \frac{3y^2 - 2xy}{x^2 - 6xy},$$

which can be solved by substituting $v = y/x$. It can also be written as

$$(3y^2 - 2xy)\,dx + (-x^2 + 6xy)\,dy = 0,$$

which is an exact equation since $(3y^2 - 2xy)_y = 6y - 2x = (-x^2 + 6xy)_x$. (See Chap. 2, Exercise 2.44.) $F(x, y) = 3xy^2 - x^2 y$ is the corresponding potential function because

$$\frac{\partial}{\partial x}(3xy^2 - x^2 y) = 3y^2 - 2xy, \quad \frac{\partial}{\partial y}(3xy^2 - x^2 y) = -x^2 + 6xy.$$

Therefore, the general solution of the first-order exact differential equation is

$$3xy^2 - x^2 y = \text{const}.$$

These are all the trajectories of the system (6.4) in the (x, y)-plane, but it does not help us find the solutions $(x(t), y(t))$ of the system as functions of t.

Example 6.3 The two autonomous equations of the Volterra-Lotka system in Example 1.3, $x' = ax - bxy$ and $y' = -cy + dxy$, lead to the equation

$$\frac{dy}{dx} = \frac{y'}{x'} = \frac{ax - bxy}{-cy + dxy}$$

for $y(x)$, which can be solved by separation of variables as $\int \frac{y\,dy}{a - by} = \int \frac{x\,dx}{-c + dx}$.

Theorem 6.2 *If the system (6.1) fulfills the conditions of the existence and uniqueness theorem, then any two different trajectories never meet.*

Proof Suppose on the contrary that two different trajectories meet at a point (α, β). Then the solution that corresponds to one of the trajectories, say $(x_1(t), x_2(t))$, passes through this point at a certain time $t = t_1$, while the solution $(\tilde{x}_1(t), \tilde{x}_2(t))$ that generates the second trajectory passes through the same point at $t = t_2$. In other words, these two solutions satisfy, respectively, the initial value conditions

$$\left(x_1(t_1), x_2(t_1)\right) = (\alpha, \beta) \quad \text{and} \quad \left(\tilde{x}_1(t_2), \tilde{x}_2(t_2)\right) = (\alpha, \beta).$$

In order to take advantage of the existence and uniqueness theorem, we define a third solution $\left(\tilde{\tilde{x}}_1(t), \tilde{\tilde{x}}_2(t)\right) = \left(\tilde{x}_1(t + t_2 - t_1), \tilde{x}_2(t + t_2 - t_1)\right)$. Its values at $t = t_1$ are

$$\left(\tilde{\tilde{x}}_1(t_1), \tilde{\tilde{x}}_2(t_1)\right) = (\tilde{x}_1(t_2), \tilde{x}_2(t_2)) = (\alpha, \beta),$$

the same as the values of the solution $(x_1(t), x_2(t))$ at the same time t_1. According to the existence and uniqueness theorem, it follows that they are identical,

$$\left(\tilde{x}_1(t + t_2 - t_1), \tilde{x}_2(t + t_2 - t_1)\right) \equiv (x_1(t), x_2(t)) \quad \text{for every } t.$$

This means that the solution $\left(\tilde{x}_1(t), \tilde{x}_2(t)\right)$ is obtained from $(x_1(t), x_2(t))$ by a time shift of $t_2 - t_1$ and their trajectories are identically the same, contrary to the assumption of the theorem.

So far we have considered the meeting of two different trajectories. If a trajectory meets itself after a time, e.g.,

$$(x_1(t_1), x_2(t_1)) = (x_1(t_1 + T), x_2(t_1 + T)),$$

then according to the reasoning above $(x_1(t), x_2(t)) \equiv (x_1(t + T), x_2(t + T))$ for every t. This means that the solution is periodic with period T and its trajectory is a

6.1 Trajectories and Critical Points in the Phase Plane

closed curve. For example, the solution $x_1 = R \sin t$, $x_2 = R \cos t$ of Example 6.1 returns to the same points when t increases by 2π.

The explanation above highlights another face of the difference between solutions and trajectories. A solution is a pair of functions and is determined by three parameters t_0, α, β. In contrast, a trajectory is a curve and is determined by two parameters α, β.

A solution that is identically constant for every t has a special role in the investigation of autonomous systems.

Definition 6.3 If an autonomous system has a constant solution

$$x_1(t) \equiv c_1, \quad x_2(t) \equiv c_2, \quad -\infty < t < \infty,$$

the point (c_1, c_2) is called an *equilibrium point* or a *critical point*. Any other point is called a *regular point*.

This name comes from the fact that a solution starting at this point never leaves it. Its corresponding trajectory consists of a single point. Because of the existence and uniqueness theorem, no other solution enters the equilibrium point or leaves it for any value of t.

At an equilibrium point $(x_1(t), x_2(t)) \equiv (c_1, c_2)$, we have of course, $x_1'(t) \equiv x_2'(t) \equiv 0$. Therefore, (c_1, c_2) is an equilibrium point of the autonomous system (6.1) when c_1, c_2 satisfy the equations

$$F_1(c_1, c_2) = 0,$$
$$F_2(c_1, c_2) = 0.$$

For example, the linear system (6.2) and the system (6.4) have an equilibrium point $(0, 0)$.

The equilibrium point has an additional meaning. A curve with a parametric description

$$x_1 = x_1(t), \quad x_2 = x_2(t),$$

has at every point a tangent vector $\mathbf{T} = (x_1'(t), x_2'(t))$, which points to the direction of the motion on the trajectory. For an autonomous system, a vector

$$\mathbf{T} = (x_1'(t), x_2'(t)) = (F_1(x_1, x_2), F_2(x_1, x_2))$$

has a well-defined direction as long as $\mathbf{T} \neq \mathbf{0}$, that is, at any point that is not an equilibrium point. Moreover, due to the continuity of $F_1(x_1, x_2), F_2(x_1, x_2)$, the direction of the tangent vector $\mathbf{T}$ also changes continuously. Therefore, the direction of progress of the system solutions is well-defined at non-equilibrium points. At an equilibrium point $\mathbf{T}$ is the zero vector and has no definite direction; hence, the behavior of the system in a neighborhood of such point may be complicated.

Fig. 6.1 A direction field and a critical point of a planar autonomous system

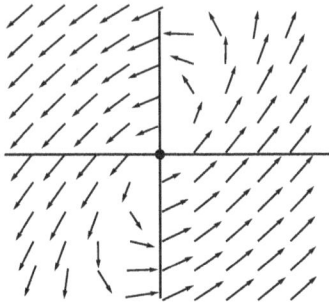

At each point (x_1, x_2), we draw an arrow aimed at the direction of the tangent vector $\mathbf{T} = \big(F_1(x_1, x_2), F_2(x_1, x_2)\big)$ and thus get the direction field corresponding to the system (6.1). Figure 6.1 is an example of a direction field with a critical point $(0, 0)$. Free and efficient programs that plot the direction field and solutions of a system are available on the Internet.

We will adjust the definition of stability for 2×2 systems:

Definition 6.4 A solution $\mathbf{x}^{(1)}(t) = \big(x_1^{(1)}(t), x_2^{(1)}(t)\big)$ of a differential system is called *stable* when $t \to \infty$ if

1. $x_1^{(1)}(t), x_2^{(1)}(t)$ are defined for all t, $t_0 \leq t < \infty$.
2. For every $\varepsilon > 0$ there exists $\delta = \delta(\varepsilon) > 0$ such that for any other solution $\mathbf{x}(t) = \big(x_1(t), x_2(t)\big)$, the inequalities

$$|x_1^{(1)}(t_0) - x_1(t_0)|, \quad |x_2^{(1)}(t_0) - x_2(t_0)| < \delta,$$

at t_0 imply

$$|x_1^{(1)}(t) - x_1(t)|, \quad |x_2^{(1)}(t) - x_2(t)| < \varepsilon$$

for all t, $t_0 \leq t < \infty$. Otherwise, the solution $\mathbf{x}^{(1)}(t) = \big(x_1^{(1)}(t), x_2^{(1)}(t)\big)$ is called *unstable*.

If, in addition, also $\lim_{t \to +\infty} |x_1^{(1)}(t) - x_1(t)| = 0$, $\lim_{t \to +\infty} |x_2^{(1)}(t) - x_2(t)| = 0$, then the solution $\mathbf{x}^{(1)}(t) = \big(x_1^{(1)}(t), x_2^{(1)}(t)\big)$ is called *asymptotically stable*.

Later on we will study the stability of the fixed solutions that correspond to equilibrium points.

6.2 Classification of Linear Systems in the Phase Plane

Our goal is to draw the different possible phase images of the system (6.2) of two linear, homogeneous differential equations with real valued constant coefficients,

$$x_1' = ax_1 + bx_2 ,$$
$$x_2' = cx_1 + dx_2 .$$

It may seem that the discussion is unnecessary because the solutions of such a system are already known from Sect. 5.6. If the eigenvalues λ_1, λ_2 of the coefficient matrix correspond to two independent eigenvectors, then the general solution of the differential system is

$$\begin{pmatrix} x_1(t) \\ x_2(t) \end{pmatrix} = c_1 e^{\lambda_1 t} \mathbf{v}_1 + c_2 e^{\lambda_2 t} \mathbf{v}_2 .$$

If $\lambda_1 = \lambda_2$ and its geometric multiplicity is 1, then

$$\begin{pmatrix} x_1(t) \\ x_2(t) \end{pmatrix} = c_1 e^{\lambda_1 t} \mathbf{v}_1 + c_2 \left[t e^{\lambda_1 t} \mathbf{v}_1 + e^{\lambda_1 t} \mathbf{u} \right] ,$$

and if the eigenvalues are complex, the solution also contains trigonometric functions. But this approach is not justified. It will be seen that even when the solution is explicitly given by familiar formulas, the behavior of the system is easier to understand through its geometric picture.

Let's start with some comments about the equilibrium points (critical points) of the linear system (6.2). Every linear system has of course the trivial solution

$$\bigl(x_1(t), x_2(t)\bigr) \equiv (0, 0)$$

and its corresponding equilibrium point $(0, 0)$. When λ_1, λ_2 are negative, all solutions tend to $(0, 0)$ when $t \to +\infty$, but due to existence and uniqueness theorem, they do not meet the trivial solution. "To tend" and "to meet" are different things, and we are dealing with separate trajectories.

Does a linear system with constant coefficients have additional equilibrium points? If $x_1(t), x_2(t)$ are constant, then $x_1' \equiv 0, x_2' \equiv 0$, namely,

$$ax_1 + bx_2 = 0 ,$$
$$cx_1 + dx_2 = 0 . \tag{6.5}$$

The trivial solution $(0, 0)$ is the unique solution of this system if and only if its determinant is non-zero. Therefore, we will assume that

$$\det(A) = ad - bc \neq 0$$

and then the differential system has a single equilibrium point (critical point) $(0, 0)$. Since the eigenvalues of a matrix satisfy $\lambda_1 \lambda_2 = \det(A)$, we essentially assume that

$$\lambda_1, \lambda_2 \neq 0 \, .$$

In the opposite situation, if $\lambda_1 \lambda_2 = ad - bc = 0$, the system has many equilibrium points. This situation will be briefly described at the end of this section.

Let's draw the different phase images that correspond to different eigenvalues.

Case A λ_1, λ_2 are real, different, of the same sign.

When $\lambda_1 \neq \lambda_2$, the solutions of the differential system (6.2) are

$$\mathbf{x}(t) = \begin{pmatrix} x_1(t) \\ x_2(t) \end{pmatrix} = c_1 e^{\lambda_1 t} \mathbf{v}_1 + c_2 e^{\lambda_2 t} \mathbf{v}_2 \, .$$

Assume, for example, that both eigenvalues are negative and $\lambda_1 < \lambda_2 < 0$. The case $\lambda_1, \lambda_2 > 0$ will be described at the end of this discussion.

If $c_1 = 0$, the trajectory that corresponds to the solutions $\mathbf{x}(t) = c_2 e^{\lambda_2 t} \mathbf{v}_2$ is a half-line from the origin either in the direction of $\mathbf{v}_2$ or of $-\mathbf{v}_2$, if $c_2 > 0$ or $c_2 < 0$. Similarly, if $c_2 = 0$, the trajectories of the solutions $\mathbf{x}(t) = c_1 e^{\lambda_1 t} \mathbf{v}_1$ are two rays from the origin in the directions of $\mathbf{v}_1$ and $-\mathbf{v}_2$, if $c_1 > 0$ or $c_1 < 0$. In all cases, when $t \to +\infty$, the solutions tend to $(0, 0)$ (since $\lambda_1, \lambda_2 < 0$), but they never meet the critical point $(0, 0)$. When $t \to -\infty$, the solutions arrive from infinity along the half-lines. Thus, we found four different trajectories, the rays in the directions of the eigenvectors $\mathbf{v}_1, -\mathbf{v}_1, \mathbf{v}_2, -\mathbf{v}_2$. See Fig. 6.2.

Now let $c_1, c_2 \neq 0$. Because of $\lambda_1 < \lambda_2 < 0$, we rewrite the solutions as

$$\mathbf{x}(t) = e^{\lambda_2 t} \left[c_1 e^{(\lambda_1 - \lambda_2) t} \mathbf{v}_1 + c_2 \mathbf{v}_2 \right] \, .$$

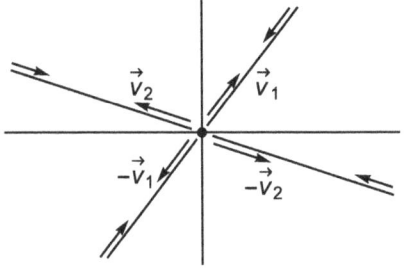

Fig. 6.2 Eigenvectors and five trajectories (where is the fifth one?)

6.2 Classification of Linear Systems in the Phase Plane

Fig. 6.3 Phase portrait of a node

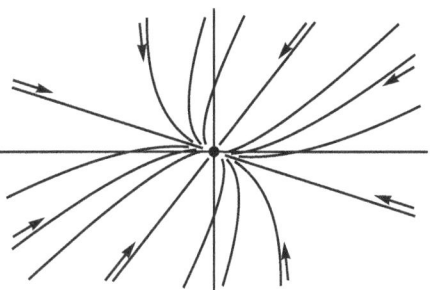

When $t \to +\infty$, the term $c_1 e^{(\lambda_1 - \lambda_2)t} \mathbf{v}_1$ tends to $\mathbf{0}$ and is negligible compared to $c_2 \mathbf{v}_2 \neq \mathbf{0}$. Thus, the direction of the trajectory is dominated by the direction of the vector $c_2 \mathbf{v}_2$, and the scalar factor $e^{\lambda_2 t}$ determines the rate of approach of the solution to the origin. Hence, as $t \to +\infty$, every trajectory of a solution with $c_2 \neq 0$ approaches $(0, 0)$ either in the direction of $\mathbf{v}_2$ or that of $-\mathbf{v}_2$. Recall that if $c_2 = 0$, then the solutions $c_1 e^{\lambda_1 t} \mathbf{v}_1$ tend to $(0, 0)$ along the directions of $\mathbf{v}_1$ or $-\mathbf{v}_1$. See Fig. 6.3. In this case, the critical point is called a *node*. The trivial solution $\mathbf{x}(t) \equiv (0, 0)$ (or the critical point $(0, 0)$) is asymptotically stable because every solution tends to $(0, 0)$ when $t \to +\infty$. We remind that the definition of stability requires this only for solutions starting close to $(0, 0)$, but here all solutions satisfy it.

When $t \to -\infty$, we write the general solution as

$$\mathbf{x}(t) = e^{\lambda_1 t} \left[c_1 \mathbf{v}_1 + c_2 e^{(\lambda_2 - \lambda_1)t} \mathbf{v}_2 \right].$$

This time $c_2 e^{(\lambda_2 - \lambda_1)t} \mathbf{v}_2$ is negligible relative to $c_1 \mathbf{v}_1 \neq \mathbf{0}$, and the factor $e^{\lambda_1 t}$ grows to infinity. Therefore, when $t \to -\infty$, each trajectory diverges to infinity so that its direction becomes closer to the direction of $\mathbf{v}_1$ or $-\mathbf{v}_1$.

The case when λ_1, λ_2 are real, different but both positive, is treated by replacing t with $(-t)$. Their phase portraits are similar, except one difference —the direction of movement on each trajectory is reversed. In this case $(0, 0)$ is an unstable solution.

Case B λ_1, λ_2 are real, different, of opposite signs.

Suppose, for example, that $\lambda_2 < 0 < \lambda_1$. The solution is still

$$\mathbf{x}(t) = c_1 e^{\lambda_1 t} \mathbf{v}_1 + c_2 e^{\lambda_2 t} \mathbf{v}_2$$

but despite the similarity in the formulas, the geometric picture is completely different from the previous one.

When $c_1 = 0$ (and since $\lambda_2 < 0$), any solution $\mathbf{x}(t) = c_2 e^{\lambda_2 t} \mathbf{v}_2$ tends, as $t \to +\infty$, to $(0, 0)$ along the half-line in the direction of $\mathbf{v}_2$ or $-\mathbf{v}_2$. Conversely, in the case $c_2 = 0$, when $t \to +\infty$, the solution $\mathbf{x}(t) = c_1 e^{\lambda_1 t} \mathbf{v}_1$ moves away to infinity along half-lines in the direction $\mathbf{v}_1$ or $-\mathbf{v}_1$. That happens because $\lambda_1 > 0$.

Fig. 6.4 Phase portrait of a saddle point

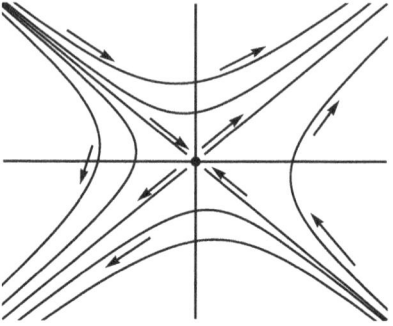

When $c_1, c_2 \neq 0$, we rewrite the solutions in the form

$$\mathbf{x}(t) = e^{\lambda_1 t}\left[c_1 \mathbf{v}_1 + c_2 e^{(\lambda_2 - \lambda_1)t} \mathbf{v}_2\right].$$

When $t \to +\infty$, the term $c_2 e^{(\lambda_2 - \lambda_1)t} \mathbf{v}_2$ is small relative to $c_1 \mathbf{v}_1$, which determines the eventual direction of the trajectory. So the trajectory runs away to infinity approximately the direction of $\mathbf{v}_1$ or $-\mathbf{v}_1$.

When $t \to -\infty$ it is better to write the solutions as

$$\mathbf{x}(t) = e^{\lambda_2 t}\left[c_1 e^{(\lambda_1 - \lambda_2)t} \mathbf{v}_1 + c_2 \mathbf{v}_2\right].$$

The factor $e^{\lambda_2 t}$ tends to zero when $t \to -\infty$, and $c_2 \mathbf{v}_2$ is the dominant term. So the solution arrives from infinity approximately from the direction of $\mathbf{v}_2$ or $-\mathbf{v}_2$. See Fig. 6.4.

The trivial solution $\mathbf{x}(t) \equiv 0$ is unstable because almost all the trajectories (except for two) that pass near it move away in one of the two possible directions. Due to the shape of the phase portrait, the critical point $(0, 0)$ is called a *saddle*.

In Fig. 6.4, we see that infinitely many different trajectories approach the same asymptote, and so each one gets closer to the others as well. Does it mean that all corresponding solutions are asymptotically stable? This time, the geometric intuition misleads, and the answer is negative. Although the trajectories are close to each other, the corresponding solutions are not so. Take two solutions

$$\mathbf{x}(t) = c_1 e^{\lambda_1 t} \mathbf{v}_1 + c_2 e^{\lambda_2 t} \mathbf{v}_2, \qquad \mathbf{y}(t) = \tilde{c}_1 e^{\lambda_1 t} \mathbf{v}_1 + \tilde{c}_2 e^{\lambda_2 t} \mathbf{v}_2.$$

Due to $\lambda_2 < 0 < \lambda_1$, their difference is, when $t \to +\infty$,

$$\|\mathbf{x}(t) - \mathbf{y}(t)\| = \|(c_1 - \tilde{c}_1)e^{\lambda_1 t}\mathbf{v}_1 + (c_2 - \tilde{c}_2)e^{\lambda_2 t}\mathbf{v}_2\| \approx |c_1 - \tilde{c}_1| \cdot \|\mathbf{v}_1\| e^{\lambda_1 t} \to \infty.$$

Hence, these solutions indeed move along similar trajectories; however, they pass near the same point at different times, and the distance between them increases with

6.2 Classification of Linear Systems in the Phase Plane

t. That is the case even for two solutions $\mathbf{x}(t)$ and $\mathbf{x}(t+c)$, which have the same trajectory.

Exercise 6.1 Two armies fight each other. $x_1(t), x_2(t)$ denote the sizes of the armies as functions of the time, and k_1, k_2 are two positive constants representing their respective efficiency. The strength of each army erodes proportionally to the size of the opposing army and its effectiveness; therefore,

$$x_1'(t) = -k_2 x_2 ,$$
$$x_2'(t) = -k_1 x_1 .$$

What can be concluded from this model about the outcome of the campaign for different initial value conditions?

There is no need to take this model too seriously because life is more complicated than a two-dimensional linear model with two parameters. Despite this, it is surprising that reasonable conclusions can be drawn even from the behavior of such a simple model. We will summarize this in the following guiding questions:

1. Calculate the eigenvalues and the eigenvectors of the system

$$\mathbf{x}' = \begin{pmatrix} 0 & -k_2 \\ -k_1 & 0 \end{pmatrix} \mathbf{x}$$

   and sketch the corresponding phase portrait schematically. With the help of the signs of x_1', x_2', mark the direction of motion on the trajectories. Refer mainly to the positive quarter of the plane that corresponds to the story of the model.

2. Dividing the equations by each other,

$$\frac{dx_1}{dx_2} = \frac{x_1'(t)}{x_2'(t)} = \frac{k_2 x_2}{k_1 x_1},$$

   show that the equations of the trajectories are $k_1 x_1^2 - k_2 x_2^2 = $ const. What are these curves?

3. Which trajectory corresponds to the case where the initial value conditions satisfy $k_1 x_1^2(0) > k_2 x_2^2(0)$? Which point on the phase plane represents the end of the war, and what will be its outcome? What happens when $k_1 x_1^2(0) < k_2 x_2^2(0)$?

4. Which trajectory fits the initial value conditions $k_1 x_1^2(0) = k_2 x_2^2(0)$? What will be the result of the war? When will the war end?

Case C λ_1, λ_2 are real, equal with geometric multiplicity two.

If $\lambda_1 = \lambda_2$ and there correspond two independent eigenvectors $\mathbf{v}_1, \mathbf{v}_2$, then the general solution of the differential system is

$$\mathbf{x}(t) = c_1 e^{\lambda_1 t} \mathbf{v}_1 + c_2 e^{\lambda_1 t} \mathbf{v}_2 = e^{\lambda_1 t} [c_1 \mathbf{v}_1 + c_2 \mathbf{v}_2] .$$

Fig. 6.5 Phase portrait of a star, $\lambda_1 = \lambda_2 < 0$

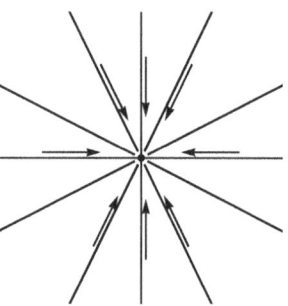

The linear combinations $c_1\mathbf{v}_1 + c_2\mathbf{v}_2$ span the whole plane, and in fact, every vector is an eigenvector. This situation actually occurs only for the matrices $A = \lambda_1 I$. Each half-line from the origin in every direction is a trajectory, and the direction of motion along it depends on the sign of $\lambda_1 = \lambda_2$. This phase portrait is called a *star*. See Fig. 6.5.

Case D λ_1, λ_2 are real, equal and with geometric multiplicity one.

The general solution is $\mathbf{x}(t) = c_1 e^{\lambda_1 t} \mathbf{v}_1 + c_2 \left[t e^{\lambda_1 t} \mathbf{v}_1 + e^{\lambda_1 t} \mathbf{u} \right]$ where $\mathbf{v}_1$ is an eigenvector and the vector $\mathbf{u}$ is determined by $(A - \lambda_1 I)\mathbf{u} = \mathbf{v}_1$. It is convenient to write the general solution in the form

$$\mathbf{x}(t) = e^{\lambda_1 t} \left[(c_1 \mathbf{v}_1 + c_2 \mathbf{u}) + c_2 t \mathbf{v}_1 \right] . \tag{6.6}$$

For $c_2 = 0$ the trajectories of the solutions $\mathbf{x}(t) = c_1 e^{\lambda_1 t} \mathbf{v}_1$ are two disjoint half-lines from the origin in the directions $\pm \mathbf{v}_1$. These two trajectories separate the phase plane into two halves.

Consider a solution $\mathbf{x}(t)$ with $c_2 \neq 0$. The vector part

$$\mathbf{y}(t) = (c_1 \mathbf{v}_1 + c_2 \mathbf{u}) + c_2 t \mathbf{v}_1 \tag{6.7}$$

determines the direction of $\mathbf{x}(t)$ in the plane. (6.7) is a parametric equation of a straight line that passes through the point $c_1 \mathbf{v}_1 + c_2 \mathbf{u}$ and is parallel to the vector $\mathbf{v}_1$, and it lies in one of the half-planes that were mentioned above. The direction of motion on the line depends on the sign of c_2. The trajectory is shaped as a point moves on the line $\mathbf{y}(t)$ and is multiplied by the exponential factor $e^{\lambda_1 t}$, which determines the exponential growth and decay of $\mathbf{x}(t)$ as $t \to \pm\infty$. See Fig. 6.6.

Suppose that $\lambda_1 = \lambda_2 < 0$. Since $\lambda_1 < 0$, the solution $\mathbf{x}(t)$ tends to $(0, 0)$ when $t \to +\infty$, and it runs away to infinity when $t \to -\infty$. When $t \to \pm\infty$, the size of the vector $c_1 \mathbf{v}_1 + c_2 \mathbf{u}$ is negligible compared to the size of $c_2 t \mathbf{v}_1$. Therefore, as $t \to +\infty$, the trajectory approaches the origin in the direction of the vector $c_2 \mathbf{v}_1$, and when $t \to -\infty$, the trajectory runs away to infinity, and its direction comes close to that of $(-c_2 \mathbf{v}_1)$. Thus, during its movement between the origin and infinity, the trajectory makes a U-turn between the opposite directions $\pm c_2 \mathbf{v}_1$. See Fig. 6.6.

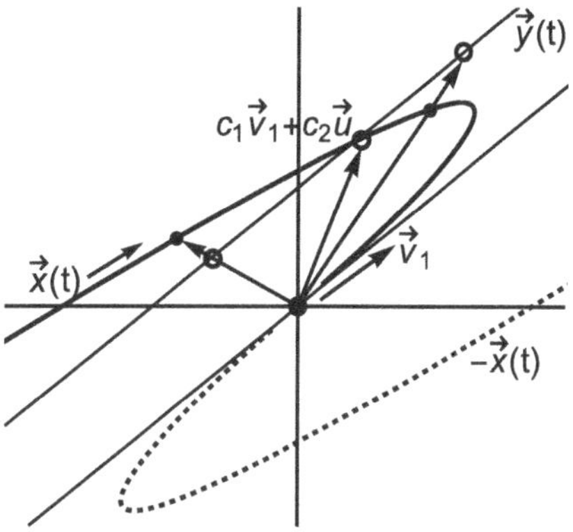

6.2 Classification of Linear Systems in the Phase Plane

Fig. 6.6 Reference diagram: ○, a point on the straight line $\mathbf{y}(t)$; •, a point on the trajectory $\mathbf{x}(t)$. The trajectories of $\mathbf{x}(t)$ and $-\mathbf{x}(t)$ are shown

Fig. 6.7 Improper node

Fig. 6.8 Improper node in opposite direction

If $\mathbf{x}(t)$ is a solution, then $-\mathbf{x}(t)$, which corresponds to the parameters $-c_1, -c_2$, is so as well. This helps us complete the phase portrait of the system. This phase portrait that corresponds to a single eigenvector is named an *improper node*. If $\lambda_1 = \lambda_2 < 0$, the trivial solution (the critical point) $(0, 0)$ is asymptotically stable since all solutions tend to it. If $\lambda_1 = \lambda_2 > 0$, the trivial solution is unstable.

The shape of the U-turn depends on the relative locations of the vectors $\mathbf{v}_1, \mathbf{u}$. See Figs. 6.7 and 6.8.

We can decide which of the Figs. 6.7 and 6.8 corresponds to a given system, by looking at the directions of the tangents to some trajectory. For example, let's look at the system

$$\mathbf{x}' = \begin{pmatrix} -3 & 4 \\ -1 & 1 \end{pmatrix} \mathbf{x},$$

which was solved in Example 5.6. Its eigenvalues are $\lambda_1 = \lambda_2 = -1$, and it has a single eigenvector, $\begin{pmatrix} 2 \\ 1 \end{pmatrix}$; therefore, its phase portrait is an improper node. The tangent to the trajectory through the point $(x_1, x_2) = (1, 0)$ is

$$\mathbf{T} = \begin{pmatrix} x_1'(t) \\ x_2'(t) \end{pmatrix} = \begin{pmatrix} -3 & 4 \\ -1 & 1 \end{pmatrix} \begin{pmatrix} 1 \\ 0 \end{pmatrix} = \begin{pmatrix} -3 \\ -1 \end{pmatrix}.$$

When we place the tangent vector $\mathbf{T} = (-3, -1)$ at the point $(1, 0)$, it turns out that the phase portrait of this system is of the form of Fig. 6.7.

Case E λ_1, λ_2 are pure imaginary.

Suppose that the system has two pure imaginary eigenvalues. Since the system is real valued, its eigenvalues are conjugate, say $\lambda_1 = i\beta$, $\lambda_2 = -i\beta$. Without loss of generality, we take $\beta > 0$. One real-valued solution is

$$\mathbf{x}^{(1)}(t) = \operatorname{Re}\left\{e^{i\beta t}\mathbf{v}\right\} = \operatorname{Re}\left\{(\cos \beta t + i \sin \beta t) \begin{pmatrix} p + iq \\ r + is \end{pmatrix}\right\}$$

$$= \begin{pmatrix} p \cos \beta t - q \sin \beta t \\ r \cos \beta t - s \sin \beta t \end{pmatrix}.$$
(6.8)

Its trajectory has the parametric description

$$x_1(t) = p \cos \beta t - q \sin \beta t,$$
$$x_2(t) = r \cos \beta t - s \sin \beta t, \quad -\infty < t < +\infty.$$

These are periodic functions with period $2\pi/\beta$, so the curve is closed. It is also symmetric with respect to the origin because the replacement of t by $t + \pi/\beta$ transforms $(x_1(t), x_2(t))$ into $(-x_1(t), -x_2(t))$. This curve does not pass through the point $(0, 0)$ (Why?). Thus we have a closed, symmetrical curve that surrounds the origin. See Fig. 6.9.

Fig. 6.9 A center

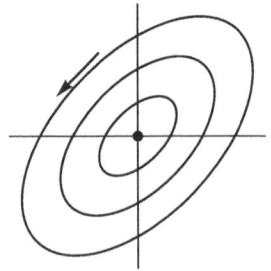

6.2 Classification of Linear Systems in the Phase Plane

Another real solution is

$$\mathbf{x}^{(2)}(t) = \operatorname{Im}\left\{e^{i\beta t}\mathbf{v}\right\} = \begin{pmatrix} p\sin\beta t + q\cos\beta t \\ r\sin\beta t + s\cos\beta t \end{pmatrix}.$$

Note that $\mathbf{x}^{(2)}(t) \equiv \mathbf{x}^{(1)}\left(t - \dfrac{\pi}{2\beta}\right)$, so $\mathbf{x}^{(1)}(t)$, $\mathbf{x}^{(2)}(t)$ are two linearly independent solutions that outline the same trajectory.

For obvious reasons, the phase image of Case E is called a *center*. The trivial solution $(0, 0)$ is stable because any solution that starts close to it always stays close to it. However, it is not asymptotically stable because no solution tends to $(0, 0)$ when $t \to +\infty$.

Exercise 6.2 The trajectories of Case E are ellipses.

To avoid indices, we write the differential system (6.2) as

$$x' = ax + by,$$
$$y' = cx + dy.$$

We use two properties of eigenvalues, namely, $\lambda_1 + \lambda_2 = \operatorname{trace}(A)$, $\lambda_1\lambda_2 = \det(A)$. In our case, $\lambda_1 = i\beta$, $\lambda_2 = -i\beta$, so

$$a + d = \operatorname{trace}(A) = \lambda_1 + \lambda_2 = i\beta + (-i\beta) = 0,$$
$$ad - bc = \det(A) = \lambda_1\lambda_2 = \beta^2 > 0.$$

Divide the two differential equations by each other and replace d with $(-a)$:

$$\frac{dy}{dx} = \frac{dy/dt}{dx/dt} = \frac{cx + dy}{ax + by} = \frac{cx - ay}{ax + by}.$$

This equation is equivalent to $(-cx + ay)\,dx + (ax + by)\,dy = 0$, which is an exact equation. (See Problem 2.37a in Chap. 2). After a few calculations, we get the general solution $-cx^2 + 2axy + by^2 = K$, that is,

$$-c\left[\left(x - \frac{a}{c}y\right)^2 + \frac{-a^2 - bc}{c^2}y^2\right] = K.$$

But $a = -d$; therefore, $-a^2 - bc = ad - bc = \det(A) = \lambda_1\lambda_2 = \beta^2$, so our trajectory is

$$\left(x - \frac{a}{c}y\right)^2 + \frac{\beta^2}{c^2}y^2 = K'.$$

This is a second-order curve that is bounded in the real plane; hence, it is an ellipse.

Case F λ_1, λ_2 are complex conjugates.

Suppose that the system has two complex conjugate eigenvalues $\lambda_1 = \alpha + i\beta$, $\lambda_2 = \alpha - i\beta$, which are not pure imaginary. In this case, $\alpha \neq 0$, and without loss of generality, we take $\beta > 0$. Two real-valued solutions are of the form

$$\mathbf{x}^{(1)}(t) = \text{Re}\left\{e^{\lambda_1 t} \mathbf{v}\right\}, \qquad \mathbf{x}^{(2)}(t) = \text{Im}\left\{e^{\lambda_1 t} \mathbf{v}\right\}.$$

We consider the first one in detail:

$$\begin{aligned}
\mathbf{x}^{(1)}(t) = \text{Re}\left\{e^{\lambda_1 t} \mathbf{v}\right\} &= \text{Re}\left\{e^{(\alpha+i\beta)t} \begin{pmatrix} p + iq \\ r + is \end{pmatrix}\right\} \\
&= e^{\alpha t} \text{Re}\left\{(\cos \beta t + i \sin \beta t) \begin{pmatrix} p + iq \\ r + is \end{pmatrix}\right\} \quad (6.9) \\
&= e^{\alpha t} \begin{pmatrix} p \cos \beta t - q \sin \beta t \\ r \cos \beta t - s \sin \beta t \end{pmatrix}.
\end{aligned}$$

As we have seen in Case E, the curve with the parametric description

$$\begin{aligned}
x_1(t) &= p \cos \beta t - q \sin \beta t, \\
x_2(t) &= r \cos \beta t - s \sin \beta t, \qquad -\infty < t < +\infty,
\end{aligned}$$

is a closed, symmetrical curve that surrounds the origin. Now we add the factor $e^{\alpha t}$, $\alpha \neq 0$, and draw a trajectory, which corresponds to (6.9). When t increases by $2\pi/\beta$, the values of the trigonometric functions repeat themselves, but due to the factor $e^{\alpha t} = e^{2\pi \alpha/\beta}$, the distance of the curve points from the origin increases or decreases depending on whether $\alpha > 0$ or $\alpha < 0$. That's why all trajectories are *spirals*. The second solution $\mathbf{x}^{(2)}(t)$ is discussed similarly. The phase portrait is naturally called a *spiral*.

The trivial solution is asymptotically stable when $\alpha = \text{Re}\{\lambda\} < 0$ and unstable for positive α.

Figures 6.10 and 6.11 are two spirals moving away from the origin; one rotates clockwise and the other anti-clockwise, depending on the sign of β.

Fig. 6.10 Counterclockwise spiral

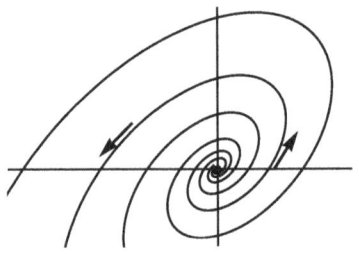

6.2 Classification of Linear Systems in the Phase Plane

Fig. 6.11 Clockwise spiral

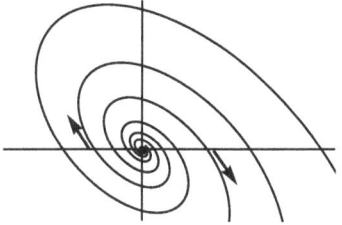

We can decide which of Figs. 6.10 and 6.11 corresponds to a given system, by looking at the directions of the tangents to some trajectory. Consider again the system

$$\begin{aligned} x_1' &= 3x_1 - 2x_2 \,, \\ x_2' &= 4x_1 - x_2 \end{aligned} \qquad (6.10)$$

(Example 5.5), whose eigenvalues are $\lambda_{1,2} = 1 \pm 2i$. Since $\mathrm{Re}\,\lambda_1 > 0$, the trajectories move away from the origin as t increases. At the point $(x_1, x_2) = (1, 0)$, the tangent to the trajectory is

$$\mathbf{T} = \begin{pmatrix} x_1{}'(t) \\ x_2{}'(t) \end{pmatrix}' = \begin{pmatrix} 3 & -2 \\ 4 & -1 \end{pmatrix} \begin{pmatrix} 1 \\ 0 \end{pmatrix} = \begin{pmatrix} 3 \\ 4 \end{pmatrix} \,.$$

Therefore, the movement on the spiral is counterclockwise as in Fig. 6.10.

Case G An eigenvalue is 0.

So far, we have studied the phase portrait of the differential system (6.2),

$$\begin{aligned} x_1' &= ax_1 + bx_2 \,, \\ x_2' &= cx_1 + dx_2 \,. \end{aligned}$$

when λ_1, λ_2 are both non-zero. When at least one of the eigenvalues is zero, then $\det(A) = ad - bc = 0$; the algebraic equations (6.5),

$$\begin{aligned} ax_1 + bx_2 &= 0 \,, \\ cx_1 + dx_2 &= 0 \,, \end{aligned}$$

are linearly dependent and have infinitely many non-trivial solutions. Therefore, all points on the line $ax_1 + bx_2 = 0$ are equilibrium points (critical points) of the system (6.2). These equilibrium points are, of course, non-isolated and form a continuum.

Fig. 6.12 A continuum of critical points and trajectories

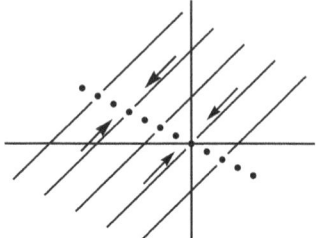

Example 6.4 The system of equations

$$\mathbf{x}' = \begin{pmatrix} -1 & -2 \\ -1 & -2 \end{pmatrix} \mathbf{x}$$

has eigenvalues $\lambda_1 = 0$, $\lambda_2 = -3$ and a general solution

$$\mathbf{x}(t) = \begin{pmatrix} x_1(t) \\ x_2(t) \end{pmatrix} = c_1 \begin{pmatrix} 2 \\ -1 \end{pmatrix} + c_2 \begin{pmatrix} 1 \\ 1 \end{pmatrix} e^{-3t} = \begin{pmatrix} 2c_1 + c_2 e^{-3t} \\ -c_1 + c_2 e^{-3t} \end{pmatrix}.$$

Therefore, for $c_2 = 0$, every point $(2c_1, -c_1)$ is a critical point and a complete trajectory of the solution. For all $c_2 \neq 0$, the solution $(x_1(t), x_2(t))$ draws a straight line $x_1(t) - x_2(t) = 3c_1$. For $c_2 > 0$ and for $c_2 < 0$, the plotted trajectories are, respectively, two rays approaching the equilibrium point $(2c_1, -c_1)$ when $t \to +\infty$. Each equilibrium point is stable but not asymptotically stable because a solution that is given by another equilibrium point does not come nearer. See Fig. 6.12.

Example 6.5 The system of equations

$$\mathbf{x}' = \begin{pmatrix} 1 & 1 \\ -1 & -1 \end{pmatrix} \mathbf{x}$$

has eigenvalues $\lambda_1 = \lambda_2 = 0$ with geometric multiplicity 1 and a general solution

$$\mathbf{x}(t) = \begin{pmatrix} x_1(t) \\ x_2(t) \end{pmatrix} = c_1 \begin{pmatrix} 1 \\ -1 \end{pmatrix} + c_2 \left[t \begin{pmatrix} 1 \\ -1 \end{pmatrix} + \begin{pmatrix} 1 \\ 0 \end{pmatrix} \right] = \begin{pmatrix} c_1 + c_2(t+1) \\ -c_1 - c_2 t \end{pmatrix}.$$

For $c_2 = 0$, every point $(c_1, -c_1)$ is an equilibrium point lying on the line $x_1 + x_2 = 0$. Every other trajectory is a parallel straight line $x_1 + x_2 = c_2$. See Fig. 6.13. It is clear that every equilibrium point is unstable.

Exercise 6.3 Find all 2×2 systems with eigenvalues $\lambda_1 = \lambda_2 = 0$ and geometric multiplicity 2.

In cases A–F, we described the different phase portraits corresponding to different eigenvalues of a linear, homogeneous system with constant coefficients.

6.2 Classification of Linear Systems in the Phase Plane

Fig. 6.13 Eigenvalue zero of multiplicity 2

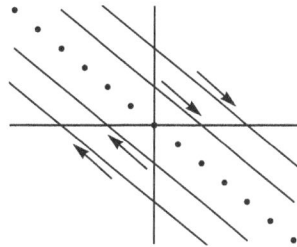

We summarize the comments regarding stability and instability, which are scattered throughout the section:

- If $\lambda_1, \lambda_2 < 0$ or if $\text{Re}\{\lambda_1\}, \text{Re}\{\lambda_2\} < 0$, then the critical point $(0, 0)$ is asymptotically stable because every solution tends to $(0, 0)$ when $t \to +\infty$. In fact, in this case, every solution is asymptotically stable when $t \to +\infty$ because each solution approaches every other one.
- If at least one eigenvalue is positive or $\text{Re}\{\lambda\} > 0$, then the critical point $(0, 0)$ is unstable because there are solutions that move away from it to infinity.
- If $\text{Re}\{\lambda_1\} = \text{Re}\{\lambda_2\} = 0$, that is, $\lambda_1 = \overline{\lambda}_2 = i\beta$, then the critical point $(0, 0)$ is stable but not asymptotically stable, because the trajectories are closed and surround the origin, but they do not tend to the origin nor move away from it.

Example 6.6 Analyze the phase portraits of the system of equations

$$\mathbf{x}' = \begin{pmatrix} -2 & 1-\alpha \\ -1 & -4 \end{pmatrix} \mathbf{x}(t)$$

for all values of α, and check the stability of the equilibrium point $(0, 0)$.

The characteristic polynomial of the matrix is $\lambda^2 + 6\lambda + 9 - \alpha = 0$; its eigenvalues are $\lambda_1 = -3 + \sqrt{\alpha}$, $\lambda_2 = -3 - \sqrt{\alpha}$. The possible phase portraits are:

- $\alpha < 0$: λ_1, λ_2 are complex valued, the phase portrait is a spiral, and all the solutions are asymptotically stable.
- $\alpha = 0$: $\lambda_1 = \lambda_2 = -3$, and the matrix has a single eigenvector (check). This is a degenerate node, and it is asymptotically stable.
- $0 < \alpha < 9$: $\lambda_1 \neq \lambda_2$, and both are negative. The phase portrait is a node, and it is asymptotically stable.
- $\alpha = 9$: $\lambda_1 = 0, \lambda_2 = -6$. The situation is similar to Example 6.4 and Fig. 6.12. There is a continuum of equilibrium points. Each equilibrium point is stable but not asymptotically stable, because a solution starting at different equilibrium points do not approach one another nor move far away.
- $\alpha > 9$: $\lambda_1 < 0 < \lambda_2$, The phase portrait is a saddle, and the trivial solution $(0, 0)$ is obviously unstable.

To conclude the analysis of the phase portraits, we will return and emphasize the role of the critical point $(0, 0)$ of a linear system with constant coefficients.

Fig. 6.14 A view through the microscope at an improper node and at a regular point

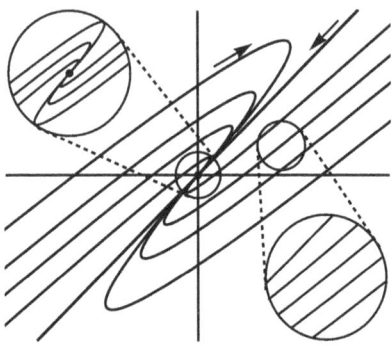

If $(x_1(t), x_2(t))$ is a solution, then thanks to the homogeneity of the system of equations, for every constant k, also $(kx_1(t), kx_2(t))$ is a solution. But multiplying by a constant is simply changing the scale of the axes. Therefore, solution behavior in a square millimeter around $(0, 0)$ is similar to its behavior in a square kilometer around the same point. The conclusion is surprising: It means that each small neighborhood of the critical point $(0, 0)$ contains all the information about the linear differential system in the whole plane.

Figure 6.14 demonstrates what happens when we look through the microscope at different points of the phase plane. Each small neighborhood of the origin contains the shape of the entire phase plane, but neighborhoods of other regular points have no such role.

6.3 Small Linear Perturbations of Linear Systems

One of the interesting considerations in the investigation of linear systems is how the phase portrait of a linear system changes when the coefficient matrix is slightly changed. A small perturbation of the coefficients of the system (6.2),

$$x_1' = ax_1 + bx_2 ,$$
$$x_2' = cx_1 + dx_2 ,$$

turns it into another linear system with other constant coefficients,

$$x_1' = \tilde{a}x_1 + \tilde{b}x_2 ,$$
$$x_2' = \tilde{c}x_1 + \tilde{d}x_2 ,$$

where $|a - \tilde{a}|$, $|b - \tilde{b}|$, $|c - \tilde{c}|$, $|d - \tilde{d}|$ are small. (Note that we only discuss small changes that maintain linearity. A general perturbation can add nonlinear terms that make the system nonlinear.) When the elements of a matrix change a

6.3 Small Linear Perturbations of Linear Systems

little, its eigenvalues also change a little, and in some situations, it can change the nature of the phase portrait and its stability.

Since we discuss only real-valued matrices and their real-valued perturbations, the location of the eigenvalues in the complex plane $\mathbb{C}$ is limited: either both are real or both are complex conjugates. After a perturbation, the eigenvalues must obey these constraints as they move in the complex plane.

What are the possible changes following a small perturbation? When the eigenvalues are real and different from each other, then after a small perturbation, they may move only on the real axis. If both eigenvalues are non-zero, so will be the perturbed ones, and their signs will be preserved. Therefore, in this case, there will be no substantial change in the phase portrait (a node or a saddle), and the stability of the equilibrium point stays unchanged.

Substantial changes can occur in the phase portrait following a small perturbation of a real eigenvalue of algebraic multiplicity two. $\lambda_1 = \lambda_2 = \alpha$ may split into two different real eigenvalues $\tilde{\alpha}, \tilde{\tilde{\alpha}}$ or into complex conjugate eigenvalues $\tilde{\alpha}+i\tilde{\beta}, \tilde{\alpha}-i\tilde{\beta}$. So an improper node can change into a node or a spiral.

If the eigenvalues are complex conjugates, $\lambda_1 = \alpha+i\beta, \lambda_1 = \alpha-i\beta$, so will they be after a small perturbation. If in addition $\alpha = \text{Re}\{\lambda_i\} \neq 0$, then even this feature will be preserved. Therefore, a spiral phase portrait remains of the same type after a small perturbation, and the stability of its critical point is also preserved.

On the other hand, if the eigenvalues are pure imaginary, $\lambda_1 = i\beta, \lambda_2 = -i\beta$, then after a small perturbation, they may turn into $\tilde{\alpha} + i\tilde{\beta}, \tilde{\alpha} - i\tilde{\beta}$, where $\tilde{\alpha}$ may be positive, negative, or zero. Thus, a phase portrait of a center may remain a center or become a stable or an unstable spiral.

Example 6.7 What are the phase portraits of the system

$$\mathbf{x}' - \begin{pmatrix} 2+h & -5 \\ 1 & -2 \end{pmatrix} \mathbf{x}$$

for h close to zero?

The characteristic polynomial of the matrix is

$$\begin{vmatrix} 2+h-\lambda & -5 \\ 1 & -2-\lambda \end{vmatrix} = \lambda^2 - h\lambda + 1 - 2h = 0,$$

and the eigenvalues are $\lambda_{1,2} = h/2 \pm i\sqrt{1 - 2h - h^2/4}$. For $h = 0$, the eigenvalues are $\pm i$, the phase portrait is a center, and the critical point is stable but not asymptotically stable. For all h positive and small, the phase portrait consists of spirals escaping from the origin when $t \to +\infty$. In contrast, when h is negative and small, the trajectories are spirals tending to the origin, which is asymptotically stable.

6.4 Notes on Nonlinear Systems in the Plane

After the description of the behavior of linear autonomous systems (i.e., linear systems with constant coefficients) in great detail, we will make a few comments about arbitrary autonomous systems,

$$\begin{aligned} x'(t) &= F(x, y) \,, \\ y'(t) &= G(x, y) \,. \end{aligned} \quad (6.11)$$

It would be naive to believe that we can study nonlinear systems to the same extent as we did with linear systems with constant coefficients. Linear systems with constant coefficients have a uniform structure, and we know how to solve them explicitly. Nonlinear autonomous systems, on the other hand, have nothing in common except that they are complicated. We do not know how to solve most of them explicitly and may only approximate the solutions numerically. We suggest a modest goal—to describe certain systems around their equilibrium points.

Recall that a point (x_0, y_0) is an equilibrium point (critical point) of the system (6.11) if $(x(t), y(t)) \equiv (x_0, y_0)$ is an identically constant solution. The condition for this is

$$\begin{aligned} F(x_0, y_0) &= 0 \,, \\ G(x_0, y_0) &= 0 \,. \end{aligned}$$

According to definition of stability, the equilibrium solution (x_0, y_0) is said to be asymptotically stable if every other solution that starts near (x_0, y_0) satisfies

$$\lim_{t \to \infty} x(t) = x_0, \quad \lim_{t \to \infty} y(t) = y_0.$$

To study a nonlinear system near an equilibrium point (x_0, y_0), we assume that the functions $F(x, y)$, $G(x, y)$ are differentiable at least twice. The corresponding Taylor's formula for $F(x, y)$ around (x_0, y_0) is

$$F(x, y) = F(x_0, y_0) + \left[F_x(x_0, y_0)(x - x_0) + F_y(x_0, y_0)(y - y_0) \right] + \cdots$$

where $(\cdots)$ indicates terms of the order of magnitude of $(x - x_0)^2 + (y - y_0)^2$. $G(x, y)$ is treated similarly. As $F(x_0, y_0) = G(x_0, y_0) = 0$, the first terms of the two Taylor's formulas are 0. Therefore, using vector notation, we get

$$\begin{pmatrix} x'(t) \\ y'(t) \end{pmatrix} = \begin{pmatrix} F(x, y) \\ G(x, y) \end{pmatrix} = \begin{pmatrix} F_x(x_0, y_0) & F_y(x_0, y_0) \\ G_x(x_0, y_0) & G_y(x_0, y_0) \end{pmatrix} \begin{pmatrix} x - x_0 \\ y - y_0 \end{pmatrix} + \cdots . \quad (6.12)$$

6.4 Notes on Nonlinear Systems in the Plane

Here also (...) indicates the remainder terms of Taylor formula of the order of magnitude $(x - x_0)^2 + (y - y_0)^2$ that are, in a small neighborhood of (x_0, y_0), negligible compared to the linear terms.

For simplicity, we denote $u = x - x_0$, $v = y - y_0$, that is, we place a new set of coordinates u, v at the equilibrium point (x_0, y_0). This shift does not affect the derivatives, $u'(t) = x'(t)$, $v'(t) = y'(t)$, and the system (6.11) of differential equations becomes

$$\begin{pmatrix} u \\ v \end{pmatrix}' = \begin{pmatrix} F_x(x_0, y_0) & F_y(x_0, y_0) \\ G_x(x_0, y_0) & G_y(x_0, y_0) \end{pmatrix} \begin{pmatrix} u \\ v \end{pmatrix} + \cdots .$$

The remainder terms that are not explicitly written are of the order of magnitude of $u^2 + v^2$. Except for these remainder terms, we obtained a linear system with constant coefficients

$$\begin{pmatrix} u \\ v \end{pmatrix}' = A \begin{pmatrix} u \\ v \end{pmatrix}, \quad \text{where } A = \begin{pmatrix} F_x(x_0, y_0) & F_y(x_0, y_0) \\ G_x(x_0, y_0) & G_y(x_0, y_0) \end{pmatrix}. \quad (6.13)$$

This process is called *linearization* of the system of differential equations near a critical point (x_0, y_0). Note that the determinant of the matrix A is the Jacobian $\dfrac{\partial(F, G)}{\partial(x, y)}$ at (x_0, y_0).

The following theorem, quoted without proof, attempts to describe the stability of a system of differential equations near a critical point.

Theorem 6.3 *Given the system of autonomous equations (6.11),*

$$x'(t) = F(x, y),$$
$$y'(t) = G(x, y)$$

and (x_0, y_0) is its isolated critical point. Suppose that (6.13) is the linearization of this system around the critical point (x_0, y_0) and that the two eigenvalues λ_1, λ_2 of the matrix A are either real and nonzero or complex conjugate such that $\text{Re}\{\lambda_1\}, \text{Re}\{\lambda_2\} \neq 0$. Then the equilibrium solution $(x(t), y(t)) \equiv (x_0, y_0)$ is an asymptotically stable or unstable solution of the nonlinear system (6.11) if and only if the trivial solution $(u(t), v(t)) \equiv (0, 0)$ is, respectively, an asymptotically stable or unstable solution of the corresponding linearized system (6.13).

If $\lambda_1 = 0$ or $\text{Re}\{\lambda_1\} = \text{Re}\{\lambda_2\} = 0$, the nature of the stability of the given system and its linearization are not necessarily equal.

If the linearized system (6.13) has a phase portrait of type node, saddle, or spiral, so is also the phase portrait of the original system (6.11) in a small neighborhood of the critical point (x_0, y_0). A degenerate node, star, and center are not necessarily preserved.

The formulation of the last claim should be understood only intuitively because we have not defined the different types of critical points for nonlinear systems.

Example 6.8 The critical points of the system

$$x' = y + x^2,$$
$$y' = -x + y^2,$$

are determined by the two equations $y + x^2 = 0$ and $-x + y^2 = 0$. By a simple calculation, one finds two critical points: $(x, y) = (1, -1)$, $(0, 0)$.

Near the critical point $(x, y) = (0, 0)$, we write our system in the form

$$\begin{pmatrix} x \\ y \end{pmatrix}' = \begin{pmatrix} 0 & 1 \\ -1 & 0 \end{pmatrix} \begin{pmatrix} x \\ y \end{pmatrix} + \begin{pmatrix} x^2 \\ y^2 \end{pmatrix}.$$

x^2, y^2 are the small remainder terms of Taylor's formula. The eigenvalues of the linear part are $i, -i$, and we cannot deduce anything about the behavior of the nonlinear system by Theorem 6.3. Other means are needed.

In order to linearize the system near the critical point $(1, -1)$, we define new variables $u = x - 1$, $v = y - (-1)$ and rewrite the system as

$$u' = (v - 1) + (u + 1)^2 = 2u + v + u^2,$$
$$v' = -(u + 1) + (v - 1)^2 = -u + 2v + v^2,$$

i.e.,

$$\begin{pmatrix} u \\ v \end{pmatrix}' = \begin{pmatrix} 2 & 1 \\ -1 & -2 \end{pmatrix} \begin{pmatrix} u \\ v \end{pmatrix} + \begin{pmatrix} u^2 \\ v^2 \end{pmatrix}.$$

The linear part of this system has the eigenvalues $\lambda_1 = \sqrt{3}$, $\lambda_2 = -\sqrt{3}$, so its phase portrait is a saddle and the critical point $(u, v) = (0, 0)$ is unstable. Therefore, $(x(t), y(t)) \equiv (1, -1)$ is an unstable equilibrium solution of the given nonlinear system.

Theorem 6.3 about nonlinear systems is local, i.e., an attempt to describe the behavior of the differential system in a small neighborhood of an equilibrium point. However, this is only a preliminary step. The fundamental problem is the global behavior of the system: How does the whole phase plane look like? What are its critical points, how does a trajectory develop, where does it "begin," and where does it "end?" Which trajectories are closed, which run away to infinity, and which of them link between different equilibrium points? Which elements in the phase plane are stable and which are not? These questions are the introduction to dynamical systems theory.

6.4 Notes on Nonlinear Systems in the Plane

We conclude this chapter with a nonlinear example where the phase portrait can be completely analyzed. In this example, we will apply phase plane methods to the nonlinear pendulum equation introduced in Example 1.8.

Example 6.9 For the nonlinear pendulum equation $\theta'' = -\frac{g}{L} \sin \theta$, we define two new functions $x(t)$, $y(t)$,

$$x(t) = \theta(t), \quad y(t) = \theta'(t).$$

Then, of course, $x'(t) = y(t)$, and $y'(t) = \theta''(t) = -\frac{g}{\ell} \sin \theta = -\frac{g}{\ell} \sin x$. In other words, we are dealing with the autonomous nonlinear system

$$\begin{aligned} x' &= y, \\ y' &= -\frac{g}{\ell} \sin x. \end{aligned} \qquad (6.14)$$

Our goal is to plot the phase plane with its trajectories and equilibrium points.

The equilibrium points are the zero points of the right-hand side of (6.14), $y = 0$, $\sin x = 0$, namely, the points $(x, y) = (n\pi, 0)$, $n = 0, \pm 1, \pm 2, \ldots$. The matrix of the linearization at each equilibrium point is

$$A = \begin{pmatrix} F_x & F_y \\ G_x & G_y \end{pmatrix} = \begin{pmatrix} 0 & 1 \\ -\frac{g}{\ell} \cos x & 0 \end{pmatrix} \bigg|_{(x,y)=(n\pi,0)}. \qquad (6.15)$$

For odd integers n, the linearization at the critical points $(\pm \pi, 0)$, $(\pm 3\pi, 0)$, $\ldots$, is $A = \begin{pmatrix} 0 & 1 \\ g/\ell & 0 \end{pmatrix}$, and its eigenvalues are $\lambda_1 = +\sqrt{g/\ell}$, $\lambda_2 = -\sqrt{g/\ell}$. By Theorem 6.3, the equilibrium points $(\pm \pi, 0)$, $(\pm 3\pi, 0), \ldots$ are unstable saddle points and are marked by S in Fig. 6.15.

For even integers n, at the equilibrium points $(0, 0)$, $(\pm 2\pi, 0)$, $(\pm 4\pi, 0)$, $\ldots$, the linearization matrix (6.15) is $A = \begin{pmatrix} 0 & 1 \\ -g/\ell & 0 \end{pmatrix}$, its characteristic polynomial is $\lambda^2 + g/\ell = 0$, and its eigenvalues are pure imaginary. In this case, Theorem 6.3 does not provide information about the nature of the equilibrium points of the nonlinear (6.14). Later, we will determine the nature of these points by other considerations.

Next we obtain the trajectories (but not the solutions $(x(t), y(t))$!) of the system (6.14). Dividing the two equations of (6.14) by each other, we get

$$\frac{dy}{dx} = \frac{dy/dt}{dx/dt} = \frac{(g/\ell) \sin x}{y}.$$

Let's separate the variables x, y:

$$\int y\,dy = -\frac{g}{\ell}\int \sin x\,dx ,$$

namely, $\frac{1}{2}y^2 - \frac{g}{\ell}\cos x = \text{const}$. We add g/ℓ to both sides to get

$$\frac{1}{2}y^2 + \frac{g}{\ell}(1-\cos x) = K . \tag{6.16}$$

The two terms $y^2/2$ and $(g/\ell)(1-\cos x)$ of (6.16) are never negative and have a physical interpretation: we multiply the two sides of (6.16) by $m\ell^2$, where m denotes the mass of the pendulum and ℓ its length and substitute $x=\theta$, $y=\theta'$. The result is

$$\frac{1}{2}m(\ell\theta')^2 + mg\ell(1-\cos\theta) = \text{const} .$$

This is the energy conservation law, with the first term indicating the kinetic energy of the rotating pendulum and the second term the potential energy measured from the lowest point reached by the pendulum.

It is clear that the collection of trajectories (6.16) maintains symmetry in the phase plane when (x, y) is replaced by $(\pm x, \pm y)$. We want to draw the trajectories

$$y = \pm\sqrt{2\left(K - \frac{g}{\ell} + \frac{g}{\ell}\cos x\right)} \tag{6.17}$$

for different values of K.

1. $K < 0$ is impossible since $y^2 \geq 0$, $1-\cos x \geq 0$.
2. $K = 0$ implies $y = 0$, $1-\cos x = 0$, that is, the equilibrium points $(0,0)$, $(\pm 2\pi, 0)$, $(\pm 4\pi, 0)$, ... These are the equilibrium points whose nature is not determined by Theorem 6.3, and we will return to them later.
3. For the critical points $(\pm\pi, 0)$, $(\pm 3\pi, 0)$, ... (saddles), Eq. (6.16) is $\frac{1}{2}0^2 + \frac{g}{\ell}(1-(-1)) = K$, i.e., $K = 2g/\ell$. For this value of K, Eq. (6.16) becomes

$$\frac{1}{2}y^2 + \frac{g}{\ell}(1-\cos x) = \frac{2g}{\ell},$$

that is,

$$y = \pm\sqrt{\frac{2g}{\ell}(1+\cos x)} . \tag{6.18}$$

6.4 Notes on Nonlinear Systems in the Plane

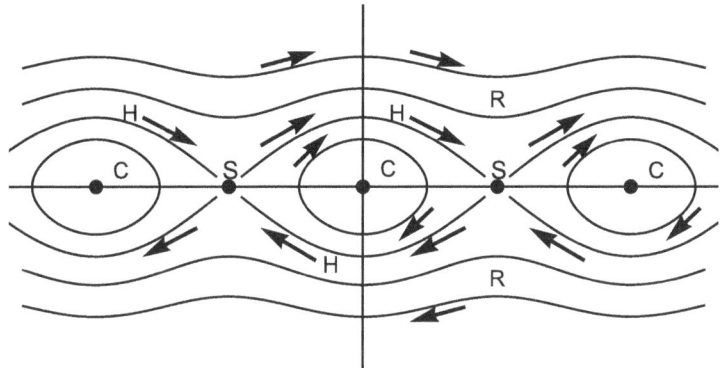

Fig. 6.15 Phase portrait of the pendulum equation

In the strip $-\pi \leq x \leq \pi$, (6.18) represents two disjoint trajectories that extend between the saddle points $(\pm\pi, 0)$. The directions of the movement on the trajectories will be determined later. Of course, horizontal shifts by multiples of 2π yield additional trajectories. The trajectories that spread out between two saddle points are marked by H in Fig. 6.15.

It remains to discuss the cases $K > 2g/\ell$ and $0 < K < 2g/\ell$.

4. If $K > 2g/\ell$, then $K - \dfrac{g}{\ell} + \dfrac{g}{\ell}\cos x > \dfrac{g}{\ell}(1 + \cos x) \geq 0$. Hence, for each such K, the two branches in (6.17),

$$y = \pm\sqrt{2\left(K - \dfrac{g}{\ell} + \dfrac{g}{\ell}\cos x\right)}$$

are defined for all $-\infty < x < \infty$ and represent two trajectories, which are marked in Fig. 6.15 by R.

5. Finally, let's look at $0 < K < 2g/\ell$. This time, we write Eq. (6.17) as

$$y = \pm\sqrt{\dfrac{2g}{\ell}\left(\cos x - \left(1 - K\dfrac{\ell}{g}\right)\right)}.$$

Since $-1 < 1 - K\dfrac{\ell}{g} < 1$, it is convenient to denote it as $1 - K\dfrac{\ell}{g} = \cos\alpha$, $-\pi < \alpha < \pi$ and

$$y = \pm\sqrt{\dfrac{2g}{\ell}\left(\cos x - \cos\alpha\right)}.$$

The two branches are defined for $-\alpha \leq x \leq \alpha$, and together, they form a single closed curve, which is one trajectory of the system. It surrounds the critical point

(0, 0); therefore, the phase portrait around (0, 0) is a center, and the critical point is stable but not asymptotically stable. The shifts by multiples of 2π generate the same portrait around the critical points $(\pm 2\pi n, 0)$ and are denoted in Fig. 6.15 by C.

It remains to determine the directions of movement on the trajectories in Fig. 6.15. Since $x' = y$, the $x(t)$ component proceeds to the right for $y > 0$ and moves to the left in the lower half-plane. These directions demonstrate that equilibrium points $(\pm \pi, 0)$, $(\pm 3\pi, 0)$, ... are indeed unstable saddle points.

The stable equilibrium point (0, 0) describes a pendulum that hangs without motion, and the closed trajectories C around it represent periodic oscillation around the equilibrium point. The saddle points $(\pm \pi, 0)$, denoted by S, correspond to a pendulum that stands upside down. This is a very unstable situation. The trajectories H describe a pendulum that starts falling from the uppermost point $x = \theta = -\pi$ and "almost" completes a whole rotation to $x = \theta = \pi$ or vice versa. The last two cases are practically difficult to happen. The trajectories R plot complete rotations of a pendulum due to high kinetic energy. As $y = \theta'$, the upper half-plane represents clockwise motion and the lower half-plane —the opposite direction.

Problems

6.1 The system $\mathbf{x}' = A\mathbf{x}$ has a general solution $\mathbf{x} = c_1\mathbf{u}(t) + c_1\mathbf{v}(t)$. What are the solutions of $\mathbf{y}' = 2A\mathbf{y}$? What is the relationship between the phase portraits of the two systems?

6.2 It is given that the system $\mathbf{x}' = A\mathbf{x}$ has a phase portrait of spirals moving away from the origin. What is the phase portrait of the system $\mathbf{x}' = (A + I)\mathbf{x}$?

6.3

(a) The matrix A has eigenvalues λ_1, λ_2 and eigenvectors $\mathbf{v}_1$, $\mathbf{v}_2$. What are the eigenvalues and eigenvectors of the matrix A^2?
(b) The system $\mathbf{x}' = A\mathbf{x}$ has a phase portrait of "center" type. What is the type of the phase portrait of system $\mathbf{x}' = A^2\mathbf{x}$?

6.4 Explore the phase portrait of the system $\mathbf{x}' = \begin{pmatrix} \alpha & 1 \\ -1 & -1 \end{pmatrix} \mathbf{x}$ for all values of α.
Draft schematically the shape of the phase plane when the eigenvalues are non-zero.

6.5 Given the system of differential equations $\mathbf{x}' = \begin{pmatrix} -2 & 1-\alpha \\ -1 & -4 \end{pmatrix} \mathbf{x}$.

(a) For each α, $-\infty < \alpha < \infty$, find the type of the phase portrait of the system, and determine whether the trivial solution $\mathbf{x}(t) \equiv (0, 0)$ is stable, asymptotically stable, or unstable.

6.4 Notes on Nonlinear Systems in the Plane

(b) For $\alpha = -1$, draw the phase portrait schematically. Mark the direction of progress on the trajectories, and explain how did you determine it.

6.6 Given the system of equations $\mathbf{x}'(t) = \begin{pmatrix} \alpha & -2 \\ 2 & \alpha \end{pmatrix} \mathbf{x}(t)$.

(a) Find all values α for which every solution of the system is asymptotically stable, and determine for them the type of critical point of the system.
(b) For which values of α is every solution of the system is stable but not asymptotically stable?
(c) For $\alpha = -6$, draw schematically the phase portrait of the system, and mark on it the directions of the motion on the trajectories.

6.7

(a) For each value of β, find the type of the phase portrait of the system

$$\mathbf{x}'(t) = \begin{pmatrix} -2 & 1 \\ \beta & -2 \end{pmatrix} \mathbf{x}(t).$$ Draw schematically the phase portrait, and determine whether the equilibrium point $(0, 0)$ is stable, asymptotically stable, or unstable.

(b) For $\beta = 1$, give an example of a trajectory that satisfies $x_1(t) > 0$, $x_2(t) > 0$ for all t, and mark it on the phase portrait.

6.8 Explore the system of equations $\mathbf{x}'(t) = \begin{pmatrix} c-1 & 1 \\ -4 & 1 \end{pmatrix} \mathbf{x}(t)$ for all values of c, and determine the type of the equilibrium point and its stability.

6.9 Given the system of equations $\begin{pmatrix} x \\ y \end{pmatrix}' = \begin{pmatrix} -1 & 8 \\ 1 & 1 \end{pmatrix} \begin{pmatrix} x \\ y \end{pmatrix}$.

(a) Sketch its phase portrait and mark the directions of motion on the trajectories.
(b) Given the initial value conditions $x(0) = \alpha$, $y(0) = \beta$. Determine what the conditions on α, β guarantee that the first component $x(t)$ will satisfy $x(t) \geq 0$ for all $t \geq 0$.

6.10

(a) Draw the phase portrait of the system $\begin{pmatrix} x \\ y \end{pmatrix}' = \begin{pmatrix} -2 & 3 \\ 0 & -2 \end{pmatrix} \begin{pmatrix} x \\ y \end{pmatrix}$, and mark the directions of movement on the trajectories.
(b) Find and draw all system trajectories that do not meet the line $x = y$.

6.11 The purpose of this exercise is to show that when $\lambda_1 = \lambda_2$, then a small change in the system can change the type of critical point.

(a) Determine the type of the critical point $(0, 0)$ for $\mathbf{x}' = \begin{pmatrix} -1 & 1 \\ 0 & -1 \end{pmatrix} \mathbf{x}$.

(b) Determine the type of the critical point $(0, 0)$ for $\mathbf{x}' = \begin{pmatrix} -1 & 1 \\ \alpha & -1 \end{pmatrix} \mathbf{x}$ where α is positive, arbitrarily small. What is the difference between the two systems?

6.12

(a) Determine the type of the critical point $(0, 0)$ for $\mathbf{x}' = \begin{pmatrix} 1 & -4 \\ 4 & -7 \end{pmatrix} \mathbf{x}$ and check its stability.

(b) Now we will slightly change the system above, and consider the system

$$\mathbf{x}' = \begin{pmatrix} 1 - \hbar & -4 \\ 4 & -7 + \hbar \end{pmatrix} \mathbf{x}$$

where $\hbar = 6.62 \times 10^{-34}$ is Planck's constant. Determine the nature of the critical point in the new system.

6.13

(a) Determine the type of the critical point $(0, 0)$ for $\mathbf{x}' = \begin{pmatrix} 0 & 1 \\ -1 & 0 \end{pmatrix} \mathbf{x}$ and check its stability.

(b) Do as above for the system $\mathbf{x}' = \begin{pmatrix} \alpha & 1 \\ -1 & \alpha \end{pmatrix} \mathbf{x}$ where α is an arbitrarily small, positive number.

(c) The same question when α is an arbitrarily small, negative number.

(d) What is the conclusion about a possible change in stability following a small perturbation of the system? For which eigenvalues the critical point is sensitive to changes, and when is it insensitive?

6.14

(a) Show that when the eigenvalues of the matrix A are $\lambda_1 = i\beta$, $\lambda_2 = -i\beta$, then the real-valued solution

$$\mathbf{x}^{(1)}(t) = \operatorname{Re}\left\{e^{\lambda_1 t} \mathbf{v}\right\} = \begin{pmatrix} p \cos \beta t - q \sin \beta t \\ r \cos \beta t - s \sin \beta t \end{pmatrix}$$

of $\mathbf{x}' = A\mathbf{x}$ can be written as

$$\mathbf{x}^{(1)}(t) = \begin{pmatrix} \sqrt{p^2 + q^2} \sin(\beta t + \varphi_1) \\ \sqrt{r^2 + s^2} \sin(\beta t + \varphi_2) \end{pmatrix},$$

for $\varphi_1 = \arctan(q/p)$, $\varphi_2 = \arctan(s/r)$. How can one write the second real-valued solution, $\mathbf{x}^{(2)}(t) = \operatorname{Im}\left\{e^{\lambda_1 t} \mathbf{v}\right\}$? What is its trajectory?

6.4 Notes on Nonlinear Systems in the Plane

(c) Prove that every linear combination $c_1 \mathbf{x}^{(2)}(t) + c_2 \mathbf{x}^{(2)}(t)$ can be written as

$$\sqrt{c_1^2 + c_2^2} \begin{pmatrix} \sqrt{p^2 + q^2}\, \sin(\beta t + \varphi) \\ \sqrt{r^2 + s^2}\, \sin(\beta t + \psi) \end{pmatrix}$$

for suitable φ, ψ.

6.15

(a) It is given that the system $\mathbf{x}' = \begin{pmatrix} a & b \\ c & d \end{pmatrix} \mathbf{x}$ has complex eigenvalues, and its phase portrait consists of spirals or ellipses.

(b) By checking the direction of the tangent to a trajectory at the point $(x_1, x_2) = (1, 0)$, prove that the trajectory surrounds the origin counterclockwise if $c > 0$.

(c) By checking the direction of the tangent to a trajectory at the point $(x_1, x_2) = (0, 1)$, prove that the trajectory surrounds the origin counterclockwise if $b < 0$.

(d) It seems that if both b, c have the same sign, there is a contradiction. Explain this seeming contradiction.

6.16

(a) Show that the system $\mathbf{x}'(t) = \begin{pmatrix} 1 & 2 \\ -5 & -1 \end{pmatrix} \mathbf{x}$ has a real-valued general solution

$$\mathbf{x}(t) = c_1 \begin{pmatrix} 2\cos 3t \\ \cos 3t + 3\sin 3t \end{pmatrix} + c_2 \begin{pmatrix} 2\sin 3t \\ \sin 3t - 3\cos 3t \end{pmatrix}.$$

(See Problem 5.12, Chap. 5).

(b) Show that the parametric description $x = 2\cos 3t$, $y = \cos 3t + 3\sin 3t$ is equivalent to the curve $(3x)^2 + (2y - x)^2 = 36$. Explain why is it an ellipse.

(c) Draw the trajectories corresponding to the family of solutions

$$\mathbf{x}(t) = c_1 \begin{pmatrix} 2\cos 3t \\ \cos 3t + 3\sin 3t \end{pmatrix}$$

for all values of c_1. Show that through every point of the plane, except the origin, there passes such a trajectory.

(d) Now draw the trajectories of the solutions

$$\mathbf{x}(t) = c_2 \begin{pmatrix} 2\sin 3t \\ \sin 3t - 3\cos 3t \end{pmatrix}$$

for all values of c_2. Explain how is it possible that these two different families of trajectories both fill the entire phase plane.

6.17

(a) Solve the system

$$x' = x - y,$$
$$y' = 5x - y.$$

(b) In order to plot y as a function of x (independent of t!), divide the system equations by each other,

$$\frac{dy}{dx} = \frac{dy/dt}{dx/dt} = \frac{5x - y}{x - y},$$

and solve this differential equation.

(c) Show that the family of ellipses $5x^2 - 2xy + y^2 = C$, $C > 0$ is a general solution of the equation in (b).
Determine the directions of the two axes of ellipses from this family by solving the two extremum problems $\min(x^2 + y^2)$ and $\max(x^2 + y^2)$ subject to constrain $5x^2 - 2xy + y^2 = 1$.

6.18

(a) Find the system $\mathbf{x}'(t) = A\mathbf{x}(t)$ whose general solution is

$$\mathbf{x}(t) = c_1 \begin{pmatrix} 1 \\ 1 \end{pmatrix} e^t + c_2 \begin{pmatrix} 1 \\ -1 \end{pmatrix} e^{-t}.$$

(b) Look at the trajectories that correspond to the two solutions

$$\mathbf{x}^{(1)}(t) = \begin{pmatrix} 1 \\ 1 \end{pmatrix} e^t + \begin{pmatrix} 1 \\ -1 \end{pmatrix} e^{-t}, \quad \mathbf{x}^{(2)}(t) = \frac{1}{2} \begin{pmatrix} 1 \\ 1 \end{pmatrix} e^t + 2 \begin{pmatrix} 1 \\ -1 \end{pmatrix} e^{-t}.$$

Do these trajectories have common points? If not, explain why there are not; if there are, explain how is it possible.

6.19

(a) Show that the trajectories of $\mathbf{x}'(t) = \begin{pmatrix} a & b \\ c & d \end{pmatrix} \mathbf{x}$ and of $\mathbf{x}'(t) = \begin{pmatrix} -c & -d \\ a & b \end{pmatrix} \mathbf{x}$ are orthogonal to each other at any point of intersection of two trajectories.

(b) Calculate and draw the trajectories of $\mathbf{x}'(t) = \begin{pmatrix} 1 & -1 \\ 5 & -1 \end{pmatrix} \mathbf{x}$ and the trajectories of the orthogonal system. See Problem 6.17.

6.4 Notes on Nonlinear Systems in the Plane

6.20

(a) Find all the critical points of the system

$$x' = 1 - xy,$$
$$y' = x - y^3.$$

(b) One of the critical points is $(x, y) = (1, 1)$. It is moved to the origin by the transformation $v = y - 1$, $u = x - 1$. Find the appropriate system for (u, v).
(c) Suppose that for small values of u, v, high powers and products of u, v are negligible compared to the linear expressions. What is the linear system that remains after this approximation?
(d) What is the type of critical point $(u, v) = (0, 0)$ for the linearized system?
(e) Repeat this process for the second critical point of the original system.

6.21 The same questions for the system

$$x' = x + y^2,$$
$$y' = y + x^2.$$

6.22 Find the equilibrium points of the system

$$x' = x^2 + y^2 - 4,$$
$$y' = y,$$

and describe the behavior of the phase portrait in their neighborhoods.

Chapter 7
Solution of Differential Equations by Power Series

We have already mentioned several times the difficulty to find explicit solutions of an ordinary differential equation. One of the possible ways to circumvent this difficulty is to expand a solution to a power series.

Solving problems with infinite series has advantages and disadvantages. A main advantage is that the power series allows approximate numerical calculation. But the rate of convergence and calculation efficiency differ from series to series and vary from point to point in the convergence domain. We will return to consider this later.

A major disadvantage of representing a solution by a series is that looking at an infinite series gives no idea of its qualitative behavior. It is difficult to see whether it represents an increasing, decreasing, or oscillating function. For example, the series $\sum_0^\infty (-1)^n x^n/n!$ describes a decreasing function (e^{-x}), while the seemingly similar series $\sum_0^\infty (-1)^n x^{2n}/(2n)!$ describes an oscillating function ($\cos x$). Can you guess the appearance of the function described by the series $\sum_0^\infty (-1)^n x^{3n}/(3n)!$? See Problem 7.10 at the end of this chapter. Despite all these shortcomings, a solution by an infinite series is sometimes the least evil.

7.1 Reminder About Power Series

Let's mention several properties of power series, which are also called *Taylor series*.[1] Each power series around $x = x_0$,

$$\sum_{n=0}^\infty a_n(x-x_0)^n = a_0 + a_1(x-x_0) + a_2(x-x_0)^2 + \cdots \tag{7.1}$$

---
[1] Brook Taylor, 1685–1731.

has some *radius of convergence* R, whose value is given by

$$\frac{1}{R} = \limsup \sqrt[n]{|a_n|}.$$

Here $0 \le R \le \infty$. If $0 < R < \infty$, the series (7.1) converges (and converges absolutely) whenever $|x - x_0| < R$, that is, in the open interval $(x_0 - R, x_0 + R)$, and diverges whenever $|x - x_0| > R$. The convergence of the series at the endpoints of the interval of convergence must be checked on a case-by-case basis for each series. In each closed interval $[a, b]$, which is contained in the open interval of convergence, the power series converges uniformly. When $R = 0$, the series converges only at the point $x = x_0$, and it is of no use. If $R = \infty$, the series converges for every x.

Within the interval of convergence, the power series represents a differentiable function, and its derivative is calculated by formal term-by-term differentiation,

$$\left(\sum_{n=0}^{\infty} a_n (x - x_0)^n\right)' = \sum_{n=1}^{\infty} a_n n (x - x_0)^{n-1} \qquad (7.2)$$
$$= a_1 + 2a_2(x - x_0) + 3a_3(x - x_0)^2 + \cdots.$$

The radius of convergence of the resulting series is the same R. Note that the power series of the derivative starts with $n = 1$ because the derivative of the constant term a_0 is zero. Since the derivative series is also a power series with the same radius of convergence R, the above process can be repeated over and over and get derivatives of any order. For example,

$$\left(\sum_{n=0}^{\infty} a_n (x - x_0)^n\right)'' = \sum_{n=2}^{\infty} a_n \cdot n(n-1)(x - x_0)^{n-2}$$
$$= 2a_2 + 3 \cdot 2a_3(x - x_0) + 4 \cdot 3a_4(x - x_0)^2 + \cdots.$$
$$(7.3)$$

Thus, a power series represents within its domain of convergence a function with derivatives of any order.

If a power series (7.1) converges to a function $f(x)$,

$$f(x) = \sum_{n=0}^{\infty} a_n (x - x_0)^n, \qquad x_0 - R < x < x_0 + R,$$

we consider the series as the *Taylor series* of the function $f(x)$. The coefficients of the Taylor series are given by

$$a_k = \frac{f^{(k)}(x_0)}{k!}, \qquad k = 0, 1, 2, \ldots. \qquad (7.4)$$

7.1 Reminder About Power Series

It should be emphasized that although our power series converges in the interval $|x - x_0| < R$ and converges uniformly in every closed subinterval of it, the rate of convergence is different for every value $x_0 - R < x < x_0 + R$. The series converges rapidly for values of x in the vicinity of x_0 because $|x - x_0|$ is small there, and it converges slowly when $|x - x_0|$ is close to R. Therefore, the series can be a good means of numerical calculation near x_0, but away from this point, its effectiveness decreases. Far from x_0, the series is sometimes practically useless, since many terms must be summed and the accumulating error resulting from the need to truncate digits is too great.

We formulated the properties of power series for a real variable x. However, the study of functions of a complex variable makes it clear that the natural area to discuss power series is the complex plane $\mathbb{C}$. It turns out that if a function $f(z)$ is differentiable according to the complex variable z (even one derivative) in a domain $D \subset \mathbb{C}$, it can be expanded into a power series around each point z_0 of D,

$$f(z) = \sum_{n=0}^{\infty} a_n (z - z_0)^n .$$

This series converges in a certain disc $|z - z_0| < R$, which is the largest disc around z_0, which is contained in the domain of differentiability of $f(z)$. This is a powerful property of functions of a complex variable. It follows automatically that a function $f(z)$, which has one derivative according to the complex variable z, has derivatives of any order. Such a function is called an *analytic function*.

Among the analytic functions in the entire complex plane $\mathbb{C}$, we name the polynomials, the exponential function e^z, $\sin z$, and the like. Any rational function (a quotient of two polynomials) is analytic in any domain where its denominator is not zero. In contrast, the functions $z^{1/p}$, $1/z^n$, and $\log(z)$ are not analytic in any neighborhood of $z = 0$.

To demonstrate the meaning of the radius of convergence in the complex plane, let's look, for example, at the function $f(z) = 1/(1 + z^2)$. According to what we know about a geometric series, it follows that the expansion of $f(z)$ into a power series around $z_0 = 0$ is

$$\frac{1}{1+z^2} = \sum_{n=0}^{\infty} (-z^2)^n = 1 - z^2 + z^4 - + \cdots .$$

and its domain of convergence is $|z| < 1$. What is the radius of convergence of the power series that corresponds to $f(z)$ around $z_0 = 2$? We can say, of course, that

$$\frac{1}{1+z^2} = \sum_{n=0}^{\infty} a_n (z - 2)^n , \tag{7.5}$$

and its coefficients are given by

$$a_n = \frac{1}{n!} \left(\frac{1}{1+z^2}\right)^{(n)} \bigg|_{z=2}, \qquad n = 0, 1, 2, \ldots,$$

but it is practically difficult to calculate more than a few of these derivatives, and it won't help us in calculating the radius of convergence. However, $f(z) = 1/(1 + z^2)$ is a rational function and is analytic for all $z \neq i, -i$, that is, in the domain $\mathbb{C} \setminus \{i, -i\}$. Therefore, it can be expanded into a power series around any point $z_0 \neq i, -i$. The radius of convergence of such series is the distance from z_0 to the nearest among the points i and $-i$. In particular, the radius of convergence around $z_0 = 2$ is the distance from $z = 2$ to $z = i$, namely, $R = |2 - i| = \sqrt{5}$, and the series (7.5) converges for $|z - 2| < \sqrt{5}$. This is quite surprising: If we discuss the same question on the real line, there is seemingly no reason why should the series expansion $f(x) = 1/(1 + x^2) = \sum_{n=0}^{\infty} a_n (x - 2)^n$ converge precisely for $2 - \sqrt{5} < x < 2 + \sqrt{5}$.

7.2 Solution of Differential Equations by Power Series

We start to study under which conditions the solutions of a second-order, normalized, linear equation may be written as power series and how to find them. The method can be applied to many other equations, but we choose specifically second-order equations and not equations of a lower or higher order for two main reasons: Second-order linear equations appear frequently in applications, and they demonstrate all the difficulties that arise for higher-order equations.

The basis of our work is the following theorem:

Theorem 7.1 *Given the functions $P(x)$, $Q(x)$ that can be expanded into power series around a point x_0 in the intervals $|x - x_0| < R_2$, $|x - x_0| < R_1$, respectively. Then the initial value problem*

$$y'' + P(x)y' + Q(x)y = 0, \tag{7.6}$$

$$y(x_0) = \alpha, \quad y'(x_0) = \beta, \tag{7.7}$$

has a unique solution, and it can be expanded into a power series around x_0 as

$$y(x) = \sum_{n=0}^{\infty} a_n (x - x_0)^n$$

at least in the interval $|x - x_0| < R = \min\{R_1, R_2\}$. x_0 is called a regular point of the differential equation (7.6).

7.2 Solution of Differential Equations by Power Series

We will not prove this theorem, but we will clarify some of its assumptions and conclusions. First, if the coefficients $P(x)$, $Q(x)$ can be written as power series, they are certainly continuous in the domains of convergence of the series. Therefore, the very existence of a solution to the initial value problem is a particular case of Theorem 4.3. The innovation of the current theorem is only in the presentation of the solution as a power series, that is, an analytic function in the complex plane. How is it analitically proved? In the proof of Theorem 3.1 in Sect. 3.3, we constructed the solution as a limit of a series of iterations. In Step 4 of that proof, we took advantage of the fact that if a series of continuous functions converges uniformly, the limit of the series is also a continuous function. In the present case, a similar conclusion of the theory of analytic functions is used: If a series of analytic functions converges uniformly, the limit function is also an analytic function.

Once we know that an equation has solutions that can be expanded into power series, we find the series by calculating the coefficients of the powers. We substitute the desired series with its unknown coefficients into the differential equation and compare the coefficients of the powers on both sides. This is exactly the method of the undetermined coefficients, but this time, it is applied to infinitely many coefficients. We will see that by this method, the coefficients are indeed determined, and we get a series that satisfies the differential equation, at least formally. This series converges according to Theorem 7.1; its term-by-term derivative series converges as well; therefore, it is really a solution of the differential equation in the suitable interval.

The next example demonstrates the method for a simple differential equation, whose solutions are well-known to us from Chap. 4.

Example 7.1 Consider the differential equation $y'' + y = 0$. According to the notation of Eq. (7.6), the coefficients of our differential equation are, respectively, $P(x) \equiv 0$, $Q(x) \equiv 1$, and they are expandable to trivial powers series around any point. Choose for simplicity $x_0 = 0$; then the power series expansion of $Q(x)$ around this point is

$$Q(x) = 1 + 0 \cdot x + 0 \cdot x^2 + \cdots, \qquad -\infty < x < \infty.$$

Consequently, all solutions can be written as power series. We look for a power series solution around $x_0 = 0$, $y(x) = \sum_{n=0}^{\infty} a_n x^n$. Its first two derivatives are

$$y'(x) = \sum_{n=1}^{\infty} a_n n x^{n-1} = a_1 + 2a_2 x + 3a_3 x^2 + \cdot$$

$$y''(x) = \sum_{n=2}^{\infty} a_n n(n-1) x^{n-2} = 2a_2 + 3 \cdot 2 \, a_3 x + 4 \cdot 3 \, a_4 x^2 + \cdots$$

which we substitute into the given differential equation:

$$y'' + y = \sum_{n=2}^{\infty} a_n n(n-1) x^{n-2} + \sum_{n=0}^{\infty} a_n x^n = 0 . \tag{7.8}$$

It will be convenient that the powers of x will be written in the same form in both series. There should be no difficulty to do this: in every power series, there appear the same powers $x^0, x^1, x^2, \ldots$, and we have to care only about a suitable notation. To do that, we present a useful technique for handling series, *shifting of the indices*.

For example, in the series $\sum_{n=2}^{\infty} a_n n(n-1) x^{n-2}$ (corresponding to y''), we replace the summation index n by $n = k + 2$: When n runs through the integers $n = 2, 3, 4, \ldots$, the new variable k has the values $k = 0, 1, 2, \ldots$, respectively. Therefore,

$$\sum_{n=2}^{\infty} a_n n(n-1) x^{n-2} \equiv \sum_{k=0}^{\infty} a_{k+2}(k+2)(k+1) x^k . \tag{7.9}$$

In fact, the terms that are written on both sides of the last identity are the same,

$$2a_2 + 3 \cdot 2\, a_3 x + 4 \cdot 3\, a_4 x^2 + \cdots . \tag{7.10}$$

The index of summation is a dummy symbol, and the value of the series itself does not depend on the name of the index n and not on that of the index k. Thus, we may replace on the right-hand side of (7.9) the symbol "k" by the symbol "n" and write it as

$$\sum_{n=2}^{\infty} a_n n(n-1) x^{n-2} \equiv \sum_{n=0}^{\infty} a_{n+2}(n+2)(n+1) x^n . \tag{7.11}$$

Practically, we replaced each appearance of the index n by $n+2$ and shifted the range of summation $2 \leq n < \infty$ in the opposite direction by 2, to $0 \leq n < \infty$.

Equation (7.8) becomes

$$y'' + y = \sum_{n=0}^{\infty} a_{n+2}(n+2)(n+1) x^n + \sum_{n=0}^{\infty} a_n x^n = 0 .$$

The two series are easily added:

$$\sum_{n=0}^{\infty} \left[a_{n+2}(n+2)(n+1) + a_n \right] x^n = 0 .$$

7.2 Solution of Differential Equations by Power Series

We consider the zero right-hand side as the power series $\sum_0^\infty 0 \cdot x^n$. Comparison of the coefficient of x^n on both sides compels

$$a_{n+2}(n+2)(n+1) + a_n = 0, \qquad n = 0, 1, 2, \ldots$$

or

$$a_{n+2} = -\frac{a_n}{(n+1)(n+2)}, \qquad n = 0, 1, 2, \ldots . \qquad (7.12)$$

This is a *recurrence formula* by which each coefficient is calculated by the previous ones. For the consecutive integers $n = 0, 1, \ldots$, (7.12) implies

$$a_2 = -\frac{a_0}{1 \cdot 2}, \qquad a_3 = -\frac{a_1}{2 \cdot 3}.$$

Later on the recurrence formula yields

$$a_4 = -\frac{a_2}{3 \cdot 4} = +\frac{a_0}{1 \cdot 2 \cdot 3 \cdot 4}, \qquad a_5 = -\frac{a_3}{4 \cdot 5} = +\frac{a_1}{2 \cdot 3 \cdot 4 \cdot 5},$$

and so on. It turns out that for every even index

$$a_{2k} = \frac{(-1)^k}{(2k)!} a_0, \qquad k = 0, 1, 2, \ldots$$

and for each odd index,

$$a_{2k+1} = \frac{(-1)^k}{(2k+1)!} a_1, \qquad k = 0, 1, 2, \ldots .$$

With these coefficients, the infinite power series solution $y(x) = \sum_{n=0}^\infty a_n x^n$ is

$$\begin{aligned}
y(x) &= a_0 + a_1 x + a_2 x^2 + a_3 x^3 + \cdots \\
&= a_0 + a_1 x - \frac{a_0}{1 \cdot 2} x^2 - \frac{a_1}{2 \cdot 3} x^3 + \frac{a_0}{4!} x^4 + \frac{a_1}{5!} x^5 - \cdots \\
&= a_0 \left[1 - \frac{x^2}{2!} + \frac{x^4}{4!} - + \cdots + \frac{(-1)^k}{(2k)!} x^{2k} + \cdots \right] \\
&\quad + a_1 \left[x - \frac{x^3}{3!} + \frac{x^5}{5!} - + \cdots + \frac{(-1)^k}{(2k+1)!} x^{2k+1} + \cdots \right].
\end{aligned}$$

The two series written here are nothing but $\cos x$ and $\sin x$, respectively. So as expected, we found the general solution $y(x) = a_0 \cos x + a_1 \sin x$.

The unique role of a_0, a_1 stands out. All coefficients are determined by a_0, a_1, but a_0, a_1 themselves are not determined from (7.12). This situation is natural. According to the formula $a_n = y^{(n)}(0)/n!$, $n = 0, 1, 2, \ldots$ for the Taylor series coefficients, we get for $n = 0, 1$,

$$a_0 = y(0), \quad a_1 = y'(0).$$

These are exactly the two natural initial value conditions at the point $x_0 = 0$. Thus, a_0, a_1 are determined by the initial value conditions, and they determine recursively all other coefficients.

It goes without saying that for a simple equation with constant coefficients, there is no need of a solution by an infinite series, since we know how to solve it more efficiently. But equations with constant coefficients are almost the only ones that we know to solve explicitly. In the following example, a solution using an infinite series is inevitable, because we do not know any other method to solve the equation.

Example 7.2 Given the equation $y'' + e^x y = 0$, its coefficients are $P(x) \equiv 0$, $Q(x) = e^x$, and they can be expanded into a power series around any point with an infinite radius. We choose for convenience $x_0 = 0$, substitute into the equation $Q(x) = e^x = \sum_0^\infty x^n/n!$, and look for a power series solution $y(x) = a_0 + a_1 x + a_2 x^2 + \cdots$. The term $e^x y(x)$ is the product of two infinite series; therefore, we limit ourselves to the comparison of a few coefficients of the lowest powers and find so the first terms of the solution series. In the same way, additional coefficients can be calculated, as far as we wish.

We substitute the series of y, y'' and e^x into the equation and get

$$y'' + e^x y = \left(2a_2 + 3 \cdot 2 a_3 x + 4 \cdot 3 a_4 x^2 + 5 \cdot 4 a_5 x^3 + \cdots\right)$$
$$+ \left(1 + x + \frac{x^2}{2} + \frac{x^3}{6} + \cdots\right)\left(a_0 + a_1 x + a_2 x^2 + a_3 x^3 + \cdots\right) = 0.$$

The coefficients of $x^0, x^1, x^2, \ldots$, are

$$x^0: \quad 2a_2 + 1 \cdot a_0 = 0,$$
$$x^1: \quad 3 \cdot 2a_3 + (1 \cdot a_1 + 1 \cdot a_0) = 0,$$
$$x^2: \quad 4 \cdot 3a_4 + (1 \cdot a_2 + 1 \cdot a_1 + \frac{1}{2} a_0) = 0,$$
$$x^3: \quad 5 \cdot 4a_5 + (1 \cdot a_3 + 1 \cdot a_2 + \frac{1}{2} a_1 + \frac{1}{6} a_0) = 0.$$

7.2 Solution of Differential Equations by Power Series

Hence, $a_2 = -\dfrac{a_0}{2}$, $a_3 = -\dfrac{a_0 + a_1}{6}$, and with the aid of recursion,

$$a_4 = -\frac{a_0/2 + a_1 + a_2}{12} = -\frac{a_0/2 + a_1 - a_0/2}{12} = -\frac{a_1}{12},$$

$$a_5 = -\frac{a_3 + a_2 + a_1/2 + a_0/6}{20} = -\frac{-(a_0 + a_1)/6 - a_0/2 + a_1/2 + a_0/6}{20}$$

$$= \frac{a_0}{40} - \frac{a_1}{60}$$

and so on. Thus,

$$y(x) = a_0 + a_1 x + \left(-\frac{a_0}{2}\right)x^2 + \left(-\frac{a_0+a_1}{6}\right)x^3 + \left(-\frac{a_1}{12}\right)x^4 + \left(\frac{a_0}{40} - \frac{a_1}{60}\right)x^5 + \cdots$$

$$= a_0 \left(1 - \frac{x^2}{2} - \frac{x^3}{6} + \frac{x^5}{40} + \cdots\right) + a_1 \left(x - \frac{x^3}{6} - \frac{x^4}{12} - \frac{x^5}{60} + \cdots\right).$$

As expected, this is a linear combination of two series, which are two solutions. Since the coefficients $P(x) \equiv 0$, $Q(x) = e^x$ of the differential equation can be expanded into power series with infinite radius, these two series also converge for all x.

Another example of a solution using series is the Legendre[2] equation.

Example 7.3 The equation

$$(1 - x^2)y'' - 2xy' + \alpha(\alpha + 1)y = 0, \tag{7.13}$$

where α is a constant called the Legendre equation of order α, and it appears in the solution of many classical problems. We will solve, for example, the equation

$$(1 - x^2)y'' - 2xy' + 6y = 0, \tag{7.14}$$

that corresponds to $\alpha = 2$. To use Theorem 7.1, we normalize Eq. (7.14) as

$$y'' - \frac{2x}{1-x^2} y' + \frac{6}{1-x^2} y = 0. \tag{7.15}$$

The coefficients

$$P(x) = -\frac{2x}{1-x^2}, \quad Q(x) = \frac{6}{1-x^2}$$

---
[2] Adrien-Marie Legendre, 1782–1853.

are rational functions and can be expanded to power series around every point except $x = 1, -1$. We will solve the equation around $x_0 = 0$. According to what is known about geometric series,

$$Q(x) = \frac{6}{1-x^2} = 6(1 + x^2 + x^4 + \cdots), \qquad |x| < 1,$$

$$P(x) = \frac{-2x}{1-x^2} = -2(x + x^3 + x^5 + \cdots), \qquad |x| < 1.$$

(7.16)

It is therefore clear in advance that the solutions of the equation can be expanded into power series around $x_0 = 0$, and they converge at least in the interval $|x| < 1$. Therefore, we look for a solution $y(x) = \sum_{n=0}^{\infty} a_n x^n$.

The normalization (7.15) is needed to justify by Theorem 7.1 the use of series and to determine its radius of convergence. However, it is more convenient to substitute the series for y, y', y'' in the original equation (7.14) rather than in the normalized equation (7.15), since it is easier to multiply the series by the factors $1 - x^2, -2x, 2$ in Eq. (7.14) than to multiply them by the infinite series (7.16).

By substituting the series for y, y', y'' in (7.14), we obtain

$$(1 - x^2)y'' - 2xy' + 6y =$$

$$= (1 - x^2) \sum_{n=2}^{\infty} a_n n(n-1) x^{n-2} - 2x \sum_{n=1}^{\infty} a_n n x^{n-1} + 6 \sum_{n=0}^{\infty} a_n x^n$$

$$= \sum_{n=2}^{\infty} a_n n(n-1) x^{n-2} - \sum_{n=2}^{\infty} a_n n(n-1) x^n - 2 \sum_{n=0}^{\infty} a_n n x^n + 6 \sum_{n=0}^{\infty} a_n x^n = 0.$$

In the first infinite sum, we shift the index n by 2 (to replace x^{n-2} by x^n). In the second sum, we do not change the indices, but it is convenient to extend the range of the summation from $2 \le n < \infty$ to $0 \le n < \infty$, since the terms that correspond to $n = 0, 1$ are zero anyway. After these changes, we have

$$\sum_{n=0}^{\infty} a_{n+2}(n+2)(n+1) x^n - \sum_{n=0}^{\infty} a_n n(n-1) x^n - 2 \sum_{n=0}^{\infty} a_n n x^n + 6 \sum_{n=0}^{\infty} a_n x^n = 0.$$

These four series are combined into one series

$$\sum_{n=0}^{\infty} \Big[a_{n+2}(n+2)(n+1) - a_n n(n-1) - 2a_n n + 6a_n \Big] x^n = 0.$$

7.2 Solution of Differential Equations by Power Series

This series is identically zero if all its coefficients are 0:

$$a_{n+2}(n+2)(n+1) - a_n[n(n-1) + 2n - 6] = 0,$$

that is,

$$a_{n+2} = \frac{(n-2)(n+3)}{(n+1)(n+2)} a_n, \quad n = 0, 1, 2, \ldots.$$

All the coefficients are determined with the aid of this recursive formula. For $n = 0, 1, 2, \ldots$, we receive

$$n = 0: \quad a_2 = -3a_0,$$
$$n = 1: \quad a_3 = -\frac{2}{3} a_1,$$
$$n = 2: \quad a_4 = 0,$$
$$n = 3: \quad a_5 = \frac{3}{10} a_3 = -\frac{1}{5} a_1.$$

Thanks to the recursive formula, $a_4 = 0$ implies $a_6 = a_8 = \cdots = 0$. So

$$y(x) = \sum_{n=0}^{\infty} a_n x^n$$

$$= a_0 + a_1 x + (-3a_0) x^2 + \left(-\frac{2a_1}{3}\right) x^3 + \left(-\frac{a_1}{5}\right) x^5 + \cdots$$

$$= a_0 \left(1 - 3x^2\right) + a_1 \left(x - \frac{2}{3} x^3 - \frac{1}{5} x^5 + \cdots\right),$$

a linear combination of two linearly independent solutions. It is interesting to note that one of them is the polynomial $y_1 = 1 - 3x^2$ and does not require an infinite series. Indeed, Theorem 7.1 guarantees that every solution converges *at least* in radius $\min\{R_1, R_2\}$. In our case, the solution $y = 1 - 3x^2$ is a "series" that converges for all x.

Exercise 7.1 Prove that if $\alpha = 2k$ is an even integer, then the Legendre equation of order $\alpha = 2k$ has a solution that is an even polynomial of order $2k$.
If $\alpha = 2k + 1$ is an odd integer, then the Legendre equation of order $\alpha = 2k + 1$ has a solution that is an odd polynomial of order $2k + 1$. These polynomials are called the *Legendre polynomials*.

7.3 Solutions Around a Regular-Singular Point

In the previous section, we found the solutions of the equation $y'' + P(x)y' + Q(x)y = 0$ near a *regular point* $x = x_0$, that is, when the functions $P(x), Q(x)$ can be expanded into power series in a neighborhood of $x = x_0$. A point where this property does not hold, i.e., at least one of the functions $P(x), Q(x)$ cannot be expanded into a power series, is known as a *singular point* of the differential equation.

There is no reason to believe that we can find a general theory describing the solutions of a differential equation around *all* kinds of singular points. After all, singular points are characterized not by a common feature but by the absence of a certain feature, and there are many reasons why a function cannot be expanded to a power series. We will therefore restrict our discussion to one subfamily of the so-called *regular-singular points*.

Definition 7.1 A point $x = x_0$ is called a *regular-singular point* of the differential equation

$$y'' + P(x)y' + Q(x)y = 0 \tag{7.17}$$

if at least one of the two coefficients cannot be expanded to a power series around x_0, but the coefficients can be written as

$$P(x) = \frac{p(x)}{x - x_0}, \qquad Q(x) = \frac{q(x)}{(x - x_0)^2}, \tag{7.18}$$

where both $p(x), q(x)$ can be expanded to power series around x_0 with certain positive radii:

$$p(x) = p_0 + p_1(x - x_0) + p_2(x - x_0)^2 + \cdots, \qquad |x - x_0| < R_1,$$
$$q(x) = q_0 + q_1(x - x_0) + q_2(x - x_0)^2 + \cdots, \qquad |x - x_0| < R_2.$$

Another way to write (7.18) is

$$P(x) = \frac{p_0}{x - x_0} + p_1 + p_2(x - x_0) + p_3(x - x_0)^2 + \cdots$$
$$Q(x) = \frac{q_0}{(x - x_0)^2} + \frac{q_1}{x - x_0} + q_2 + q_3(x - x_0) + q_4(x - x_0)^2 + \cdots. \tag{7.19}$$

Regular-singular points are studied due to several reasons:

- Equations with regular-singular points appear in many applications in physics and mathematics.

7.3 Solutions Around a Regular-Singular Point

- Sometimes, regular-singular points are the most interesting points of a differential equation, and the behavior of the solutions in their neighborhood is the most important. Even when a differential equation has regular (seemingly simple) points and regular-singular points (seemingly more complicated), sometimes, it is better to solve the equation around the "complicated" points.
- Fortunately, there is a systematic way to solve equations around a regular-singular point, and it will be described late.

Remark 7.1 ($P(x)$ **and** $Q(x)$) The roles of $P(x)$ and $Q(x)$ in (7.18) are different: the denominator of the first coefficient $P(x)$ may contain the factor $(x - x_0)$, while the denominator of the second coefficient $Q(x)$ may contain $(x - x_0)^2$, but not vice versa. For example, $x = 0$ is not a regular-singular point of the equation $y'' + \dfrac{2}{x^2} y' + \dfrac{3}{x} y = 0$.

Remark 7.2 (Missing Powers) The definition of regular-singular point permits that, in fact, part of the negative powers of $x - x_0$ will not appear. For example, for the equation $y'' + \dfrac{2}{x} y' + \dfrac{3}{x} y = 0$, the point $x = 0$ is a regular-singular point because the coefficients can be written as

$$P(x) = \frac{2}{x} + 0 + 0x + \cdots, \qquad Q(x) = \frac{0}{x^2} + \frac{3}{x} + \cdots.$$

If all negative powers are missing, namely, $p_0, q_0, q_1 = 0$, we effectively return to a regular point, and the present discussion is unnecessary.

Remark 7.3 (Complex Variables) In terms of the theory of complex functions, Eqs. (7.19) have an obvious meaning. For the complex variable z, $P(z)$ and $Q(z)$ are *meromorphic* functions in the neighborhood of the point $z = x_0$, $P(z)$ has a *simple pole*, and $Q(z)$ has a second-order pole at x_0. Equations (7.19) are the Laurent[3] series of $P(z)$ and $Q(z)$, respectively.

Remark 7.4 (n-th Order Equation) The discussion in this section can be extended to a linear equation of any order. $x = x_0$ is called a regular-singular point of the equation

$$y^{(n)} + P_1(x) y^{(n-1)} + \cdots + P_n(x) y = 0$$

if it is possible to write each of the n coefficients in the form

$$P_k(x) = \frac{p_k(x)}{(x - x_0)^k}, \qquad k = 1, \ldots, n,$$

---

[3] Pierre Alphonse Laurent, 1813–1854.

where $p_k(x)$ are analytic functions around x_0. The second-order equations that we discuss are the most familiar.

Example 7.4 In Example 7.3, we discussed a Legendre equation

$$(1 - x^2)y'' - 2xy' + \alpha(\alpha + 1)y = 0,$$

whose normalized form is

$$y'' + \frac{2x}{x^2 - 1} y' - \frac{\alpha(\alpha + 1)}{x^2 - 1} y = 0,$$

and solved it around the regular point $x = 0$. The points $x = 1, -1$ are regular-singular points for this equation. Near $x = 1$, for example, we write

$$P(x) = \frac{2x}{x^2 - 1} = \frac{2x/(x+1)}{x - 1}, \qquad Q(x) = -\alpha(\alpha + 1)\frac{(x-1)/(x+1)}{(x-1)^2}.$$

The numerators $p(x) = \frac{2x}{x+1}$, $q(x) = -\alpha(\alpha+1)\frac{x-1}{x+1}$, being rational functions whose denominators are not 0 at $x = 1$, can be written as power series around $x = 1$. For example,

$$p(x) = \frac{2x}{x+1} = 2 - \frac{2}{x+1} = 2 - \frac{1}{1 + (x-1)/2} = 2 - \sum_{n=0}^{\infty}(-1)^n \left(\frac{x-1}{2}\right)^n.$$

As a sum of a geometric series, it converges for all $\left|\frac{x-1}{2}\right| < 1$. $q(x)$ is treated similarly. Thus, $x = 1$ is a regular-singular point of the Legendre equation.

We list several differential equations that have regular-singular points.

- The Bessel equation $x^2 y'' + xy' + (x^2 - \alpha^2)y = 0$ around $x = 0$,
- The Tchebycheff equation $(1 - x^2)y'' - xy' + \alpha^2 y = 0$ around $x = 1, -1$,
- The Laguerre equation $xy'' + (1 - x)y' + \alpha y = 0$ around $x = 0$,
- The hypergeometric equation $x(1 - x)y'' + [\gamma - x(1 + \alpha + \beta)]y' - \alpha\beta y = 0$, α, β, γ constants, has two regular-singular points $x = 0, 1$.

See also the equations in Problems 4.19 and 4.20 at the end of Chap. 4, which demonstrate reduction of order. Each of them has a regular-singular point.

The method of solution around a regular-singular point is called the *Frobenius method*.[4] Without loss of generality, let $x = 0$ be a regular-singular point of the

---

[4] Ferdinand Georg Frobenius, 1849–1917.

7.3 Solutions Around a Regular-Singular Point

differential equation $y'' + P(x)y' + Q(x)y = 0$. That is, according to (7.18), the equation can be written as

$$y'' + \frac{p_0 + p_1 x + p_2 x^2 + \cdots}{x} y' + \frac{q_0 + q_1 x + q_2 x^2 + \cdots}{x^2} y = 0.$$

We multiply by x^2 and mark the resulting operator by $L[y]$:

$$L[y] \equiv x^2 y'' + x(p_0 + p_1 x + p_2 x^2 + \cdots)y' + (q_0 + q_1 x + q_2 x^2 + \cdots)y = 0. \quad (7.20)$$

If in each of the series we consider only the leading coefficient and ignore near $x = 0$ all higher powers, we get the approximate equation

$$x^2 y'' + p_0 x y' + q_0 y = 0, \quad (7.21)$$

which is the familiar Euler equation from Sect. 4.8. This equation motivates the following study. The normalized form of (7.21) is

$$y'' + \frac{p_0}{x} y' + \frac{q_0}{x^2} y = 0.$$

It is of the form of (7.19) with $x_0 = 0$, and according to our definition, $x = 0$ is a regular-singular point. Since for the Euler equation (7.21) we look for solutions of the form x^r, the Frobenius method suggests to search for solutions of the original Eq. (7.20) of the form

$$y = x^r \sum_{0}^{\infty} a_n x^n, \quad (7.22)$$

where r and the coefficients $a_0, a_1, \ldots$ are yet unknown. Since x^r may be undefined for negative x, we limit the equation to the domain $x > 0$, where this difficulty does not happen. For $x < 0$, we will replace x^r by $|x|^r$. We substitute the series (7.22) and its derivatives within the differential equation (7.20) and try to find such $r, a_0, a_1, \ldots$ that the equation holds. In our study, the process is done only formally, and we do not discuss whether and where the series (7.22) converges.

It is convenient to write (7.22) as

$$y(x) = \sum_{0}^{\infty} a_n x^{n+r} = a_0 x^r + a_1 x^{r+1} + a_2 x^{r+2} + \cdots. \quad (7.23)$$

Without loss of generality, suppose that $a_0 \neq 0$, that is, the first term that is written, really appears and is not 0. Otherwise, if $a_0 = 0$, the series is

$$a_1 x^{r+1} + a_2 x^{r+2} + \cdots,$$

which does not differ from the previous one except by replacing the symbol r (whose value is unknown) by $r+1$ (which is equally unknown) and renumbering the coefficients. The derivatives of (7.23) are

$$y' = \sum_0^\infty a_n(n+r)x^{n+r-1} = a_0 r x^{r-1} + a_1(r+1)x^r + \cdots,$$

$$y'' = \sum_0^\infty a_n(n+r)(n+r-1)x^{n+r-2} = a_0 r(r-1)x^{r-2} + a_1(r+1)r x^{r-1} + \cdots.$$

These are not Taylor series, since r is not necessarily a positive integer. Therefore, the first term of the series of y is not a constant, and it does not disappear in the derivatives y', y''. Substituting the series for y, y', y'', Eq. (7.20) is

$$L[y] = x^2 \sum_0^\infty a_n(n+r)(n+r-1)x^{n+r-2}$$

$$+ x \left(\sum_0^\infty p_k x^k \right) \left(\sum_0^\infty a_n(n+r)x^{n+r-1} \right) + \left(\sum_0^\infty q_k x^k \right) \left(\sum_0^\infty a_n x^{n+r} \right) = 0$$

and after the extraction of the common factor x^r,

$$L[y] = x^r \left[\sum_0^\infty a_n(n+r)(n+r-1)x^n + \left(\sum_0^\infty p_k x^k \right) \left(\sum_0^\infty a_n(n+r)x^n \right) \right.$$

$$\left. + \left(\sum_0^\infty q_k x^k \right) \left(\sum_0^\infty a_n x^n \right) \right] = 0.$$

(7.24)

First, we calculate the coefficient of x^m (without x^r). For $m = 0$, the sum of the constant terms from all series is $a_0 [r(r-1) + p_0 r + q_0]$. Let's denote

$$F(r) \equiv r(r-1) + p_0 r + q_0.$$

(7.25)

This quadratic polynomial will appear again and again later on.

7.3 Solutions Around a Regular-Singular Point

For every integer $m \geq 1$, the power x^m in (7.24) (x^r excluded) results from a combination of all products $x^k x^n$ where $k + n = m$. The coefficient of x^m, $m \geq 1$, from all these combinations, is

$$x^m : \quad a_m(m+r)(m+r-1)$$
$$+ p_0 a_m(m+r) + p_1 a_{m-1}(m-1+r) + \cdots + p_m a_0 r$$
$$+ q_0 a_m + q_1 a_{m-1} + \cdots + q_m a_0 .$$

After this sum is arranged according to $a_m, a_{m-1}, \ldots, a_0$, we get

$$a_m \Big[(m+r)(m+r-1) + p_0(m+r) + q_0\Big] + \sum_{l=0}^{m-1} a_l \Big[(l+r)p_{m-l} + q_{m-l}\Big] .$$

Here, a_m is multiplied exactly by $F(m+r)$, where $F(r)$ is the quadratic polynomial (7.25). Hence, the coefficient of the power x^m in (7.24) is

$$a_m F(r+m) + \sum_{l=0}^{m-1} a_l \Big[(l+r)p_{m-l} + q_{m-l}\Big] .$$

To this end, $L[y]$ in (7.24) is written as

$$L[y] = x^r \Bigg[a_0 F(r) + \sum_{m=1}^{\infty} \Bigg(a_m F(r+m) + \sum_{l=0}^{m-1} a_l \Big[(l+r)p_{m-l} + q_{m-l}\Big] \Bigg) x^m \Bigg] . \tag{7.26}$$

$L[y] = 0$ if the coefficient of each power is zero:

$$a_0 F(r) = 0 , \tag{7.27}$$

$$a_m F(r+m) + \sum_{l=0}^{m-1} a_l \Big[(l+r)p_{m-l} + q_{m-l}\Big] = 0, \quad m = 1, 2, \ldots . \tag{7.28}$$

Since it was assumed that $a_0 \neq 0$, (7.27) compels

$$F(r) = r(r-1) + p_0 r + q_0 = 0 . \tag{7.29}$$

This quadratic equation is called the *indicial equation*, and its two roots r_1, r_2 are called the *indices* of the regular-singular point. The indicial equation is the same as the characteristic equation of Euler's equation (7.21), which we met as an approximation to the original differential equation.

Let's take one of the two indices, say r_1, and try to determine the corresponding coefficients $a_1, a_2, \ldots$ from (7.28). By the recursion formula

$$a_m = -\frac{\sum_{l=0}^{m-1} a_l\left[(l+r_1)p_{m-l} + q_{m-l}\right]}{F(m+r_1)}, \qquad m = 1, 2, \ldots \qquad (7.30)$$

each coefficient a_m is determined by his predecessors $a_0, a_1, \ldots, a_{m-1}$. To emphasize that these are the coefficients of the power series that correspond to the index r_1, we will denote them occasionally by $a_m(r_1)$. This process can be performed when no denominator is zero,

$$F(r_1 + m) \neq 0, \qquad m = 1, 2, \ldots,$$

that is, the indicial equation has no zero at any of the points $r_1+1, r_1+2, r_1+3, \ldots$.

We consider first the case when the indicial equation (7.29) has two different real roots r_2, r_1, and suppose that $r_2 < r_1$. Since r_1 is the rightmost zero, $F(r)$ cannot be zero at any of the points $r_1 + 1, r_1 + 2, \ldots$. Therefore, $F(r_1 + m) \neq 0$, and our algorithm is applicable for the index r_1, and it formally determines a solution

$$y_1(x) = \sum_{0}^{\infty} a_n(r_1) x^{n+r_1}.$$

The mere determination of the coefficients of the series does not imply its convergence. This must be proved separately, but the proof will not be given here.

When we try to repeat the same process for the smaller (left) index r_2 with

$$a_m(r_2) = -\frac{\sum_{l=0}^{m-1} a_l\left[(l+r_2)p_{m-l} + q_{m-l}\right]}{F(m+r_2)}, \qquad m = 1, 2, \ldots, \qquad (7.31)$$

$F(r)$ should not be zero at any point of the form $r_2 + m$ for $m = 1, 2, \ldots$. But $F(r)$ is zero again only at r_1 that is located to the right of r_2. Hence, the recursion may be executed if no point $r_2 + m$ coincides with r_1, i.e., the distance $r_1 - r_2$ is not a positive integer. Thus, we showed:

Theorem 7.2 *Let r_1, r_2 be two different real indices, the roots of the indicial equation (7.29) that correspond to the regular-singular point $x = 0$ and suppose that $r_2 < r_1$. To the bigger index, r_1, there corresponds a solution*

$$y_1(x) = x^{r_1} \sum_{0}^{\infty} a_n(r_1) x^n,$$

7.3 Solutions Around a Regular-Singular Point

where $a_n(r_1)$ is given by (7.30). If the difference $r_1 - r_2$ is not an integer, then also to the smaller index r_2, there corresponds a solution

$$y_2(x) = x^{r_2} \sum_0^\infty a_n(r_2) x^n,$$

and $a_n(r_2)$ is given by (7.31).

As mentioned above, we do not prove here the convergence of the series.

Exercise 7.2 What can be said about the form of the solutions if the indices r_1, r_2 are two conjugate complex numbers?

Example 7.5

$$3xy'' + y' + xy = 0.$$

After normalizing this equation, we will get the coefficients

$$P(x) = \frac{1/3}{x}, \qquad Q(x) = \frac{x}{3x} = \frac{x^2/3}{x^2},$$

whose numerators $q(x) = x^2/3$, $p(x) = 1/3$ are, of course, trivial power series around $x = 0$. So $x = 0$ is a regular-singular point for the equation. In the differential equation, we replace $y(x)$ by a series $\sum_0^\infty a_n x^{n+r}$, $a_0 \neq 0$, and so for its derivatives y', y'':

$$L[y] = 3x \sum_0^\infty a_n(n+r)(n+r-1)x^{n+r-2} + \sum_0^\infty a_n(n+r)x^{n+r-1} + x \sum_0^\infty a_n x^{n+r}$$

$$= x^r \left[\sum_0^\infty 3a_n(n+r)(n+r-1)x^{n-1} + \sum_0^\infty a_n(n+r)x^{n-1} + \sum_0^\infty a_n x^{n+1} \right].$$

In the last of the three summations, we shift the dummy index n by 2 to receive

$$= x^r \left[\sum_0^\infty 3a_n(n+r)(n+r-1)x^{n-1} + \sum_0^\infty a_n(n+r)x^{n-1} + \sum_2^\infty a_{n-2} x^{n-1} \right].$$

Now $n = 0, 1$ appears in only two of the three series, while $n \geq 2$ appears in all three. Therefore, we separate the first two terms that contain the powers x^{-1}, x^0. There is no reason to worry about the appearance of a negative power x^{-1}, because all the series are multiplied by x^r for some unknown r. After a reorganization

$$L[y] = x^r \left[\left(3a_0 r(r-1) + a_0 r\right)x^{-1} + \left(3a_1(1+r)r + a_1(1+r)\right)x^0 \right.$$
$$\left. + \sum_{n=2}^{\infty} \left(3a_n(n+r)(n+r-1) + a_n(n+r) + a_{n-2}\right)x^{n-1} \right] = 0.$$

Each coefficient in the power series must be zero:

$$a_0 [3r(r-1) + r] = 0,$$
$$a_1 [3(1+r)r + (1+r)] = 0,$$
$$a_n [3(n+r)(n+r-1) + (n+r)] + a_{n-2} = 0, \qquad n = 2, 3, \ldots.$$

After some simplification,

$$a_0 r(3r - 2) = 0,$$
$$a_1 (r+1)(3r+1) = 0, \qquad (7.32)$$
$$a_n (n+r)(3(n+r) - 2) + a_{n-2} = 0, \qquad n = 2, 3, \ldots.$$

Since $a_0 \neq 0$, the first of the equations is the indicial equation, which determines the corresponding indices $r_1 = 2/3$, $r_2 = 0$. Each index will be treated separately.

For the larger index $r_1 = 2/3$, Eqs. (7.32) ascertain $a_1 = 0$ and the recursion formula

$$a_n = a_n(2/3) = -\frac{a_{n-2}}{(n+\frac{2}{3})(3n)} = -\frac{a_{n-2}}{n(3n+2)}, \qquad n = 2, 3, \ldots.$$

$a_1 = 0$ and the recursion formula give rise to $a_3 = a_5 = \cdots = 0$. For the even indices, we get one by one that

$$a_2 = \frac{-a_0}{2 \cdot 8}, \quad a_4 = \frac{a_2}{4 \cdot 14} = \frac{+a_0}{(2 \cdot 4)(8 \cdot 14)}, \quad a_6 = \frac{-a_4}{6 \cdot 20} = \frac{-a_0}{(2 \cdot 4 \cdot 6)(8 \cdot 14 \cdot 20)},$$

and generally,

$$a_{2k} = \frac{(-1)^k a_0}{2^{2k} k! 4 \cdot 7 \cdot 10 \cdots (3k+1)}.$$

After omission of the arbitrary a_0, the first solution is

$$y_1(x) = x^{2/3} \left[1 + \sum_{k=1}^{\infty} \frac{(-1)^k x^{2k}}{2^{2k} k! 4 \cdot 7 \cdot 10 \cdots (3k+1)}\right].$$

7.3 Solutions Around a Regular-Singular Point

By the ratio test, it is clear that this series converges to every value of x.

For the left index $r_2 = 0$, (7.32) implies $a_1 = 0$ and the recursion formula

$$a_n = -\frac{a_{n-2}}{n(3n-2)}, \quad n = 2, 3, \ldots.$$

Here, again, $a_1 = a_3 = a_5 = \cdots = 0$. For the even indices, we get

$$a_2 = \frac{-a_0}{2 \cdot 4}, \quad a_4 = \frac{-a_2}{4 \cdot 10} = \frac{+a_0}{(2 \cdot 4) \cdot (4 \cdot 10)}, \quad a_6 = \frac{-a_4}{6 \cdot 16} = \frac{-a_0}{(2 \cdot 4 \cdot 6)(4 \cdot 10 \cdot 16)},$$

and in general,

$$a_{2k} = \frac{(-1)^k a_0}{2^{2k} k! \, 2 \cdot 5 \cdot 8 \cdots (3k-1)}.$$

A similar calculation leads us to a second solution

$$y_2(x) = 1 + \sum_{k=1}^{\infty} \frac{(-1)^k x^{2k}}{2^{2k} k! \, 2 \cdot 5 \cdot 8 \cdots (3k-1)}, \quad -\infty < x < \infty.$$

While $y_2(x)$ has derivatives of any order at the regular-singular point $x = 0$ (and at every other point), the solution $y_1(x)$ is continuous but not differentiable at this point. Thus, $y_1(x)$ solves the differential equation only in the interval $0 < x < \infty$ (or in the interval $-\infty < x < 0$).

The Frobenius method enables always to find one solution around a regular-singular point, the one that corresponds to the larger index. The calculation of a second solution encounters difficulties in two cases:

(i) If the difference $r_1 - r_2$ is a positive integer N, then the formula for a_N in (7.31) involves a zero denominator. That's why this algorithm is not applicable to calculate a second solution.
(ii) If $r_1 = r_2$, the Frobenius method allows the calculation of only one solution of the form $y_1 = \sum_0^\infty a_n x^{n+r_1}$. Using the same method for r_2, which equals r_1, does not lead to another, different solution.

Once we know one solution, it is theoretically possible to find a second solution by reduction of the order method. But this way is not practical since the first solution is given by an infinite series, and calculation of another solution using Eq. (4.11) is too difficult.

In the following two sections, we will try to describe another way to calculate a second solution for the problematic cases.

7.4 Equal Indices

We have seen in the previous section that if the indicial equation $F(r) = 0$ has two equal roots $r_1 = r_2$, we get the first solution $y_1(x) = y(x, r_1)$, but the recursion formula (7.31) does not provide a second solution. A second solution will be found by a method similar to the one that was used in Sect. 4.6 for equations with constant coefficients and characteristic polynomials with equal roots.

We consider r as a variable parameter, and analogously with (7.31), we define

$$a_n(r) = -\frac{\sum_{l=0}^{n-1} a_l\left[(l+r)p_{n-l} + q_{n-l}\right]}{F(n+r)}, \quad n = 1, 2, \ldots,$$

where a_0 is some constant. So we get recursively $a_1(r), a_2(r), \ldots$, all functions of r. For a constant a_0 and these $a_1(r), a_2(r), \ldots$, we define

$$y(x, r) = \sum_{n=0}^{\infty} a_n(r) x^{n+r}.$$

According to the definition of $a_1(r), a_2(r), \ldots$, all terms in the series (7.26), except the first one, are zero, and the result of the assignment in the differential operator will amount to

$$L[y(x, r)] = x^r \cdot a_0 F(r).$$

In the present case where $F(r)$ has one root of multiplicity two, we have $F(r) = (r - r_1)^2$. So

$$L[y(x, r)] = a_0 x^r (r - r_1)^2 \tag{7.33}$$

for all r. We differentiate both sides of (7.33) according to r. The derivative of the right-hand side according to the parameter r is

$$\frac{\partial}{\partial r}\left[a_0 x^r (r - r_1)^2\right] = a_0 \left[x^r \ln x \cdot (r - r_1)^2 + x^r \cdot 2(r - r_1)\right]$$

7.4 Equal Indices

The derivative of the left-hand side is

$$\frac{\partial}{\partial r} L[y(x,r)] = L\left[\frac{\partial}{\partial r} y(x,r)\right] = L\left[\sum_{n=0}^{\infty} \frac{\partial}{\partial r}\left(a_n(r)x^{n+r}\right)\right]$$

$$= L\left[\sum_{n=0}^{\infty} a_n(r)x^{n+r} \ln x + \sum_{n=0}^{\infty} \frac{\partial a_n(r)}{\partial r} x^{n+r}\right] \tag{7.34}$$

$$= L\left[y(x,r) \ln x + \sum_{n=0}^{\infty} \frac{\partial a_n(r)}{\partial r} x^{n+r}\right].$$

Finally, we place on both sides $r = r_1$. The right-hand side is zero due to the presence of the factors $(r - r_1)^2$, $(r - r_1)$, and on the left-hand side, there remains

$$L\left[y(x,r_1) \ln x + \sum_{n=0}^{\infty} \frac{\partial a_n(r_1)}{\partial r} x^{n+r_1}\right] = 0. \tag{7.35}$$

Here $y(x, r_1)$ is the already known first solution $y_1(x)$. The meaning of (7.35) is that we have found a second solution

$$y_2(x) = y_1(x) \ln x + \sum_{n=0}^{\infty} \frac{\partial a_n(r_1)}{\partial r} x^{n+r_1}.$$

Calculation of the coefficients of the power series $\left.\frac{\partial a_n(r_1)}{\partial r}\right|_{r=r_1}$ involves many difficulties. Not only do we have to recursively calculate every $a_1(r), a_2(r), \ldots$ as an explicit function of the variable r, but we also have to calculate their derivatives and only finally place in them the value $r = r_1$. Despite the great technical difficulty in performing these calculations, this method is useful for determining in principle the form of the second solution. Let's summarize it:

Theorem 7.3 *If for the singular-regular point $x = 0$ we have $r_1 = r_2$, then in addition to the first solution $y_1(x) = x^{r_1} \sum_{0}^{\infty} a_n(r_1)x^n$ that corresponds to the index r_1, there exists another solution of the form*

$$y_2(x) = y_1(x) \ln x + x^{r_1} \sum_{n=0}^{\infty} b_n x^n. \tag{7.36}$$

The solution (7.36) reminds us of the solution of the Euler equation with two equal indices. See Sect. 4.9. For $x < 0$, $\ln x$, x^r must be replaced by $\ln |x|$, $|x|^r$, respectively.

In practice, we try to determine the coefficients b_n by placing the assumed solution (7.36) into the differential equation and comparing the coefficients.

Example 7.6 $x = 0$ is a singular-regular point of the equation

$$x^2 y'' - 3xy' + (4 - 4x)y = 0$$

since the normalized coefficients are $P(x) = -3/x$, $Q(x) = (4-4x)/x^2$. When we substitute $y = \sum_0^\infty a_n x^{n+r}$, $a_0 \neq 0$, in the original equation, the result is

$$L[y] = x^2 \sum_{n=0}^{\infty} a_n (n+r)(n+r-1) x^{n+r-2} - 3x \sum_{n=0}^{\infty} a_n (n+r) x^{n+r-1}$$

$$+ (4 - 4x) \sum_{n=0}^{\infty} a_n x^{n+r}$$

$$= \sum_{n=0}^{\infty} a_n \Big[(n+r)(n+r-1) - 3(n+r) + 4\Big] x^{n+r} - 4 \sum_{n=0}^{\infty} a_n x^{n+r+1}$$

$$= \sum_{n=0}^{\infty} a_n (n+r-2)^2 x^{n+r} - 4 \sum_{n=1}^{\infty} a_{n-1} x^{n+r}$$

$$= a_0 (r-2)^2 x^r + \sum_{n=1}^{\infty} \Big[a_n (n+r-2)^2 - 4 a_{n-1} \Big] x^{n+r} = 0 \, .$$

The equation is satisfied for all x if the coefficient of each power is zero.

$$F(r) = (r - 2)^2 = 0 \, ,$$

$$a_n (n+r-2)^2 - 4 a_{n-1} = 0 \, , \qquad n = 1, 2, \ldots \, .$$

The indicial equation has two equal roots, $r_1 = r_2 = 2$. One solution that corresponds to $r_1 = 2$ is defined by the recursion formulas

$$a_n = \frac{4}{n^2} a_{n-1} = \frac{4}{n^2} \times \frac{4}{(n-1)^2} a_{n-2} = \cdots = \frac{4^n}{(n!)^2} a_0 \, ,$$

7.4 Equal Indices

and the corresponding solution is $y_1(x) = \sum_{n=0}^{\infty} \dfrac{4^n}{(n!)^2} x^{n+2}$. A second solution is of the form $y_2(x) = y_1(x) \ln x + \sum_{n=0}^{\infty} b_n x^{n+2}$, and its derivatives are

$$y_2'(x) = y_1'(x) \ln x + y_1(x) \dfrac{1}{x} + \sum_{n=0}^{\infty} b_n (n+2) x^{n+1},$$

$$y_2''(x) = y_1''(x) \ln x + 2y_1'(x) \dfrac{1}{x} - y_1(x) \dfrac{1}{x^2} + \sum_{n=0}^{\infty} b_n (n+2)(n+1) x^n.$$

We substitute the derivatives into the differential equation:

$$L[y] = x^2 \left[y_1''(x) \ln x + 2y_1'(x) \dfrac{1}{x} - y_1(x) \dfrac{1}{x^2} + \sum_{n=0}^{\infty} b_n (n+2)(n+1) x^n \right]$$

$$- 3x \left[y_1'(x) \ln x + y_1(x) \dfrac{1}{x} + \sum_{n=0}^{\infty} b_n (n+2) x^{n+1} \right]$$

$$+ (4 - 4x) \left[y_1(x) \ln x + \sum_{n=0}^{\infty} b_n x^{n+2} \right].$$

After some rearrangements

$$= \left[x^2 y_1'' - 3x y_1' + (4 - 4x) y_1 \right] \ln x + 2x y_1'(x) - 4 y_1(x)$$

$$+ \sum_{0}^{\infty} b_n (n+2)(n+1) x^{n+2} - 3 \sum_{0}^{\infty} b_n (n+2) x^{n+2}$$

$$+ 4 \sum_{0}^{\infty} b_n x^{n+2} - 4 \sum_{0}^{\infty} b_n x^{n+3} = 0.$$

Since $y_1(x)$ is a solution of the differential equation, the coefficient of $\ln x$ is zero, and the function $\ln x$ does not appear anymore in the equation. We are left with

$$2xy_1'(x) - 4y_1(x) + \sum_0^\infty b_n \left[(n+2)(n+1) - 3(n+2) + 4\right] x^{n+2} - 4 \sum_1^\infty b_{n-1} x^{n+2}$$

$$= 2x \sum_{n=0}^\infty \frac{4^n}{(n!)^2} (n+2) x^{n+1} - 4 \sum_{n=0}^\infty \frac{4^n}{(n!)^2} x^{n+2} + \sum_{n=0}^\infty b_n n^2 x^{n+2}$$

$$- 4 \sum_1^\infty b_{n-1} x^{n+2} = 0 .$$

After a reorganization of the series, all terms that correspond to $n = 0$ disappear, and there remains only

$$\sum_{n=1}^\infty \frac{4^n \cdot 2n}{(n!)^2} x^{n+2} + \sum_{n=1}^\infty \left[b_n n^2 - 4b_{n-1}\right] x^{n+2} = 0 .$$

This leads to the recursion formula

$$b_n = \frac{1}{n^2} \left[4b_{n-1} - \frac{2n \cdot 4^n}{(n!)^2} \right], \qquad n = 1, 2, \ldots,$$

which allows the calculation of the coefficients $b_1, b_2, \ldots$ one by one, as functions of b_0:

$$b_1 = 4b_0 - 8, \ b_2 = \frac{4b_1 - 16}{4} = 4b_0 - 12, \ b_3 = \frac{1}{9}\left[4b_2 - \frac{64}{6}\right] = \frac{16}{9} b_0 - \frac{176}{27}, \ \ldots .$$

Each coefficient $b_1, b_2, \ldots$ depends on b_0, which can be chosen arbitrarily. So the second solution consists of the sum of two functions, one of which is multiplied by the parameter b_0. Since a second-order differential equation has exactly two linearly independent solutions, it follows that the series multiplied by b_0 must be a solution by itself. Being a power series around 0, it is not a multiple of a solution containing the $\ln x$, so it must be a multiple of the first solution $y_1(x)$. To save this unnecessary calculation, we chose $b_0 = 0$ and get

$$b_1 = -8, \ b_2 = -12, \ b_3 = -\frac{154}{27}, \ \ldots .$$

and a second solution $y_2(x) = y_1(x) \ln x - 8x - 12x^2 - \frac{154}{27} x^3 + \cdots$.

7.5 Indices That Differ by a Positive Integer

If the difference $r_1 - r_2$ is a positive integer, say $r_1 - r_2 = N$, the calculation of the coefficients of a second solution by the recursion formula (7.31) fails. Indeed,

$$a_m(r_2) = -\frac{\sum_{l=0}^{m-1} a_l[(l+r_2)p_{m-l} + q_{m-l}]}{F(m+r_2)}$$

works for $m = 1, 2, \ldots, N - 1$; however, for $m = N$, the denominator is $F(N + r_2) = F(r_1) = 0$, which is unacceptable. To examine this situation, we go one step backward to Eq. (7.28),

$$a_m F(r_2 + m) + \sum_{l=0}^{m-1} a_l\left[(l+r_2)p_{m-l} + q_{m-l}\right] = 0 .$$

For $m = N$, this is

$$a_N \times 0 + \sum_{l=0}^{N-1} a_l\left[(l+r_2)p_{m-l} + q_{m-l}\right] = 0 . \qquad (7.37)$$

where $a_1, a_2, \ldots, a_{N-1}$ had been already calculated. Two situations may occur:
Case (i). It may happen fortunately that

$$\sum_{l=0}^{N-1} a_l\left[(l+r_2)p_{m-l} + q_{m-l}\right] = 0 .$$

If so, Eq. (7.37) is the identity $a_N \times 0 + 0 = 0$, and no contradiction occurs. On the contrary, equality (7.37) holds for any a_N. For the next indices, $m = N + 1, N + 2, \ldots$; we encounter no more zero denominators. Therefore, we choose an arbitrary a_N and get a second solution $\sum_0^\infty a_n(r_2)x^{n+r_2}$ by the recursion formula (7.28).

Exercise 7.3 Equation (7.17) may have only two linearly independent solutions. Explain how it is possible that the second solution depends on an arbitrary parameter.

Case (ii). If, on the other hand,

$$\sum_{l=0}^{N-1} a_l\left[(l+r_2)p_{m-l} + q_{m-l}\right] \neq 0 , \qquad (7.38)$$

equality (7.37) cannot hold for any choice of a_N, and it is impossible to find a second solution of the form $\sum_0^\infty a_n(r_2)x^{n+r_2}$. (This is the case for the Bessel equation of integer orders, which will be studied in Sect. 7.6). What is then the form of a second solution?

We will calculate a second solution by a method that is similar to the one in the case $r_1 = r_2$. We refer to r as a variable parameter and choose the first coefficient not as a constant number but rather as a simple function of r,

$$a_0 = a_0(r) = r - r_2 .$$

The consecutive coefficients $a_1, a_2, \ldots$ are determined by the recursion formula

$$a_m = a_m(r) = -\frac{\sum_{l=0}^{m-1} a_l \left[(l+r)p_{m-l} + q_{m-l}\right]}{F(m+r)}, \quad m = 1, 2, \ldots . \quad (7.39)$$

So all the coefficients $a_0(r), a_1(r), a_2(r), \ldots$ are rational functions (quotients of polynomials) of the variable r. Moreover, by our construction, each of the expressions $a_0(r), a_1(r), a_2(r), \ldots$ contains in its numerator the factor $r - r_2$. Since $F(r) = (r - r_1)(r - r_2)$ and $N = r_1 - r_2$, the denominator at the N-th step of the recursion is

$$F(N + r) = (N + r - r_1)(N + r - r_2) = (r - r_2)(r + N - r_2) .$$

That is, the denominator also contains the factor $r - r_2$. These two identical factors in the numerator and denominator cancel out; therefore, the fraction $a_N(r)$ is well-defined even for $r = r_2$,

$$a_N(r_2) = \lim_{r \to r_2} a_N(r) \neq 0 .$$

All other coefficients $a_1(r), a_2(r), \ldots, a_{N-1}(r), a_{N+1}(r), a_{N+2}(r), \ldots$ do not pose any difficulty of division by zero even for $r = r_2$, and the series $a_1(r), a_2(r), \ldots$ is defined properly for all r. For these $a_n(r)$ and the variable r, we define the function

$$y(x, r) = \sum_{n=0}^{\infty} a_n(r) x^{n+r}$$

and substitute it into the original differential equation. The coefficients $a_1(r), a_2(r), a_3(r)$, and ... were defined by (7.39) so that all terms in the series (7.26), except for the first one, become zero, and there remains only

$$L[y(x, r)] = a_0 F(r) x^r .$$

But $a_0 = r - r_2$, $F(r) = (r - r_1)(r - r_2)$, so

$$L[y(x, r)] = (r - r_1)(r - r_2)^2 x^r . \quad (7.40)$$

7.5 Indices That Differ by a Positive Integer

From here on, we follow the method that was used in the case of $r_1 = r_2$. The derivative of the left-hand side of (7.40) according to r is, as in (7.34),

$$\frac{\partial}{\partial r} L[y(x,r)] = L\left[y(x,r) \ln x + \sum_{n=0}^{\infty} \frac{\partial a_n(r)}{\partial r} x^{n+r} \right]$$

while the derivative of the right-hand side is

$$\frac{\partial}{\partial r}\left[(r-r_1)(r-r_2)^2 x^r\right] = (r-r_2)^2 x^r + 2(r-r_1)(r-r_2) x^r + (r-r_1)(r-r_2)^2 x^r \ln x .$$

These two derivatives are identically equal, and for $r = r_2$, the result is

$$L\left[\ln x \sum_{n=0}^{\infty} a_n(r_2) x^{n+r_2} + \sum_{n=0}^{\infty} \frac{\partial a_n(r_2)}{\partial r} x^{n+r_2} \right] = 0 .$$

So we obtained another solution of the differential equation.

It remains to identify the series $\sum_{n=0}^{\infty} a_n(r_2) x^{n+r_2}$, which is multiplied by $\ln x$. We already commented that the expressions $a_0(r), a_1(r), a_2(r), \ldots, a_{N-1}(r)$ contain in their numerators the factor $r - r_2$; hence, for $r = r_2$, the first N terms of the series are all zero, and there remains only $\sum_{n=N}^{\infty} a_n(r_2) x^{n+r_2}$, $a_N(r_2) \neq 0$. After shifting the dummy index from n to $n + N$, the series becomes

$$\sum_{n=0}^{\infty} a_{n+N}(r_2) x^{n+N+r_2}, \qquad a_N(r_2) \neq 0 .$$

Since $r_2 + N = r_1$, the replacement of n by $N + n$ in the recursion formula (7.39) for $a_n(r_2)$ changes the denominator $F(n + r_2)$ to $F(n + N + r_2) = F(n + r_1)$. But this becomes exactly the recursion formula for $r = r_1$ and $x^{n+N+r_2} = x^{n+r_1}$, so we identify the series in question with the first solution $y_1(x)$, possibly multiplied by some constant factor A, arising from the first element $\lim_{r \to r_2} a_N(r) \neq 0$. This proves:

Theorem 7.4 *If $r_1 - r_2$ is a positive integer, then around the singular-regular point $x = 0$, there exists, in addition to the solution $y_1(x)$, a second solution of the form*

$$y_2(x) = A y_1(x) \ln x + x^{r_2} \sum_{n=0}^{\infty} b_n x^n ,$$

where *A* is a constant number. If the equality (7.37) does not comprise a contradiction, the second solution is of the form

$$y_2(x) = x^{r_2} \sum_{n=0}^{\infty} b_n x^n ,$$

namely, A = 0.

7.6 The Bessel Equation

The differential equation

$$x^2 y'' + xy' + (x^2 - \alpha^2) y = 0 \qquad (7.41)$$

is called the Bessel[5] equation of order α, $\alpha \geq 0$. Bessel equations of different orders appear in many applications. We will apply them to demonstrate the various possibilities for the indices r_1, r_2.

The point $x = 0$ is a singular-regular point for the Bessel equation because the coefficients of the normalized equation are

$$P(x) = \frac{1}{x} , \qquad Q(x) = \frac{x^2 - \alpha^2}{x^2} ,$$

and the numerators

$$p(x) \equiv 1 , \qquad q(x) = -\alpha^2 + 0 \cdot x + 1 \cdot x^2 + 0 \cdot x^3 + \cdots ,$$

are, in fact, finite power series around $x = 0$. Hence, we look for a solution of the form $y = \sum_{n=0}^{\infty} a_n x^{n+r}$, $a_0 \neq 0$. After substituting y and its derivatives in Eq. (7.41), we get

$$L[y] = x^2 \sum_{n=0}^{\infty} a_n (n+r)(n+r-1) x^{n+r-2} + x \sum_{n=0}^{\infty} a_n (n+r) x^{n+r-1}$$

$$+ (x^2 - \alpha^2) \sum_{0}^{\infty} a_n x^{n+r}$$

---
[5] Friedrich Wilhelm Bessel, 1784–1846.

7.6 The Bessel Equation

$$= \sum_0^\infty a_n\left[(n+r)(n+r-1)+(n+r)-\alpha^2\right]x^{n+r} + \sum_{n=0}^\infty a_n x^{n+r+2}$$

$$= \sum_0^\infty a_n\left[(n+r)^2-\alpha^2\right]x^{n+r} + \sum_{n=2}^\infty a_{n-2} x^{n+r} = 0.$$

The coefficients are all zero if

$$a_0\left[r^2-\alpha^2\right]=0, \qquad n=0, \qquad (7.42)$$

$$a_1\left[(1+r)^2-\alpha^2\right]=0, \qquad n=1, \qquad (7.43)$$

$$a_n\left[(n+r)^2-\alpha^2\right]+a_{n-2}=0, \qquad n=2,3,\ldots. \qquad (7.44)$$

The indicial equation (7.42) determines the indices

$$r_1 = \alpha \geq 0, \qquad r_2 = -\alpha \leq 0.$$

First, we consider the solution that corresponds to $r_1 = \alpha \geq 0$. For it (7.43) is

$$a_1[2\alpha + 1] = 0,$$

and since $2\alpha + 1 \neq 0$, it follows that $a_1 = 0$. (7.44) becomes

$$a_n = -\frac{a_{n-2}}{n(n+2\alpha)}, \qquad n = 2, 3, \ldots$$

and $a_1 = 0$ implies $a_3 = a_5 = \ldots = 0$. The even coefficients are, recursively,

$$a_2 = -\frac{a_0}{2(2+2\alpha)}, \qquad a_4 = -\frac{a_2}{4(4+2\alpha)} = +\frac{a_0}{2\cdot 4(2+2\alpha)(4+2\alpha)},$$

and in general

$$a_{2k} = \frac{(-1)^k a_0}{2\cdot 4\cdots(2k)(2+2\alpha)(4+2\alpha)\cdots(2k+2\alpha)}$$

$$= \frac{(-1)^k a_0}{2^{2k} k!(1+\alpha)(2+\alpha)\ldots(k+\alpha)}, \qquad k = 1, 2, 3, \ldots. \qquad (7.45)$$

To sum up, a solution is given by the series

$$\sum_{k=0}^\infty a_{2k} x^{2k+\alpha} = a_0 \sum_{k=0}^\infty \frac{(-1)^k}{2^{2k} k!(1+\alpha)(2+\alpha)\cdots(k+\alpha)} x^{2k+\alpha}.$$

It is customary to divide this series by 2^α and to omit the arbitrary constant a_0. So we get a solution of Eq. (7.41)

$$y_1(x) = \sum_{k=0}^{\infty} \frac{(-1)^k}{k!(1+\alpha)(2+\alpha)\cdots(k+\alpha)} \left(\frac{x}{2}\right)^{2k+\alpha}. \tag{7.46}$$

According to the ratio test, this series converges for every value of x. Its differentiability at $x = 0$ depends on the power x^α.

In the literature, it is agreed to define the *Bessel function of order* α as a solution that is a multiple of $y_1(x)$ by a certain constant. The precise definition will be given in Chap. 8, Eq. (8.18), as soon as the *Gamma function* is defined.

Now we turn to the second solution of the Bessel equation Eq. (7.41). As long as the difference of the indices $r_1 - r_2 = 2\alpha$ is not an integer, no difficulty arises to find a second solution. To the index $r_2 = -\alpha < 0$, there corresponds a second solution

$$y_2(x) = \sum_{k=0}^{\infty} \frac{(-1)^k}{k!(1-\alpha)(2-\alpha)\cdots(k-\alpha)} \left(\frac{x}{2}\right)^{2k-\alpha}. \tag{7.47}$$

When $\alpha = 0$, the indices are equal, $r_1 = r_2 = 0$. The first solution of Eq. (7.41) is

$$J_0(x) = \sum_{k=0}^{\infty} \frac{(-1)^k}{(k!)^2} \left(\frac{x}{2}\right)^{2k} \tag{7.48}$$

for all x, and it satisfies the initial value conditions $J_0(0) = 1$, $J_0'(0) = 0$. It is called the *Bessel function of order 0*. The second solution is of the form (7.36), and due to the presence of $\ln|x|$, this solution is defined only for $x \neq 0$.

When $r_1 - r_2 = 2\alpha$ is a positive integer N, that is, $\alpha = \frac{1}{2}, 1, \frac{3}{2}, 2, \ldots$, the general theory predicts difficulties to find a second solution. The recursion formula (7.44) becomes for $r = r_2 = -\alpha$

$$a_n n(n - 2\alpha) + a_{n-2} = 0, \qquad n = 2, 3, \ldots, \tag{7.49}$$

and for the integer $N = 2\alpha$, it is

$$a_N \times 0 + a_{N-2} = 0. \tag{7.50}$$

But for $\alpha = \frac{1}{2}, \frac{3}{2}, \frac{5}{2}, \ldots$, no real problem arises since $N = 2\alpha$ is an odd integer and it is already known that $a_1 = a_3 = a_5 = \ldots = a_{N-2} = a_N = \ldots = 0$. Thus (7.50) holds trivially. On the other hand, for the integer values $\alpha = 1, 2, 3, \ldots$, $N = 2\alpha$ is an even integer, and $a_N \times 0 + a_{N-2} = 0$ cannot determine a_N. See (7.38). In this case, a second solution must contain $\ln|x|$ as described in Theorem 7.4.

7.6 The Bessel Equation

For example, the first solution of Bessel equation of order $\alpha = 1$ is

$$J_1(x) = \sum_{k=0}^{\infty} \frac{(-1)^k}{k!(k+1)!} \left(\frac{x}{2}\right)^{2k+1}$$

but any other independent solution contains $\ln|x|$.

We demonstrate the behavior of the recursion formula (7.45) for a Bessel equation of order $\alpha = 1/2$. Equation (7.45) becomes for $\alpha = 1/2$

$$a_{2k}(1/2) = \frac{(-1)^k a_0}{2 \cdot 4 \cdots (2k)(2+1)(4+1) \cdots (2k+1)} = \frac{(-1)^k}{(2k+1)!} a_0,$$

If a_0 is omitted, we get a solution

$$y_1(x) = x^{1/2} \sum_{k=0}^{\infty} \frac{(-1)^k x^{2k}}{(2k+1)!} = x^{-1/2} \sum_{k=0}^{\infty} \frac{(-1)^k x^{2k+1}}{(2k+1)!} = \frac{\sin x}{\sqrt{x}}.$$

Similarly, for the second index $\alpha = -1/2$,

$$a_{2k}(-1/2) = \frac{(-1)^k a_0}{2 \cdot 4 \cdots (2k)(2-1)(4-1) \cdots (2k-1)} = \frac{(-1)^k}{(2k)!} a_0,$$

and the corresponding solution of the differential equation is

$$y_2(x) = x^{-1/2} \sum_{k=0}^{\infty} \frac{(-1)^k x^{2k}}{(2k)!} = \frac{\cos x}{\sqrt{x}}.$$

Note that the solution $y_1(x) = \dfrac{\sin x}{\sqrt{x}}$ is continuous at $x = 0$ but has no derivative there. In contrast, the solution $y_2(x) = \dfrac{\cos x}{\sqrt{x}}$ is not even continuous at $x = 0$.

Do solutions of Bessel equations of other orders α also have a relation to trigonometric functions, or is there at least some external similarity among them? Here, one of the weaknesses of power series solutions shows up. Even though we can write a solution explicitly as an infinite power series, it is impossible to guess from the series the qualitative behavior of the function that it describes. (Can you sketch the solutions we found in Example 7.5?) In the following example, we show an alternative approach to the investigation of a Bessel equation by a change of variables.

Fig. 7.1 Bessel functions $J_0(x)$, $J_1(x)$

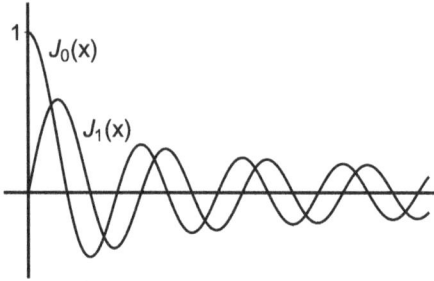

Example 7.7 Let us choose a new variable $v(x) = \sqrt{x}\, y(x)$ in the Bessel equation $x^2 y'' + x y' + (x^2 - \alpha^2) y = 0$. The derivatives of $y = v x^{-1/2}$ are

$$y' = v' x^{-1/2} + v \left(-\frac{1}{2}\right) x^{-3/2},$$

$$y'' = v'' x^{-1/2} + 2 v' \left(-\frac{1}{2}\right) x^{-3/2} + v \left(-\frac{1}{2}\right)\left(-\frac{3}{2}\right) x^{-5/2}.$$

With the new variable v, the Bessel equation becomes

$$x^2 \left(v'' x^{-1/2} - v' x^{-3/2} + \frac{3}{4} v x^{-5/2}\right) + x \left(v' x^{-1/2} - \frac{1}{2} v x^{-3/2}\right) + (x^2 - \alpha^2) v x^{-1/2} = 0.$$

After multiplying by $x^{3/2}$ and rearrangement, there remains the normalized equation

$$v'' + \left(1 + \frac{1/4 - \alpha^2}{x^2}\right) v = 0. \tag{7.51}$$

For $\alpha = 1/2$, this is the differential equation $v'' + v = 0$ whose solutions are $\cos x$, $\sin x$. This explains the form of the solutions of the Bessel equation of order $\alpha = 1/2$. However, it is possible to conclude from this also about the behavior of Bessel equations of other orders. When $x \to \infty$, Eq. (7.51) is "close" to the equation $v'' + v = 0$, and it can be expected that solutions of (7.51) will be "similar" in some sense to the trigonometric functions. This hypothesis must of course be given precise content. It is not done here, but Fig. 7.1 shows solutions of Bessel equations of orders 0 and 1, which hint about the behavior of solutions of equations of other orders.

Problems

7.1 Find two linearly independent solutions of the differential equation $y'' + \sin x \, y' + \cos x \, y = 0$ using power series around $x = 0$. Calculate the terms of the series up to the power x^4.

7.2

(a) Find two linearly independent solutions of the differential equation
$y'' + t^2 y' + t y = 0$ as power series around $t = 0$.
(b) Find the first five terms of the power series solution of the initial value problem

$$y'' + t^2 y' + t y = e^t, \quad y(0) = 0, \ y'(0) = 0.$$

in the neighborhood of $t = 0$

7.3 Solve the Chebyshev equation $(1 - x^2) y'' - x y' + \alpha^2 y = 0$ by power series around $x = 0$. Show that when the constant α is a non-negative integer m, one of the solutions is a polynomial of order m.

7.4 Solve the Hermite equation $y'' - 2x y' + \alpha y = 0$ as power series around $x = 0$. Show that when the constant α is an even non-negative number $2m$, one of the solutions is a polynomial of order m.

7.5 Solve the Legendre equation of order 3, $(1 - x^2) y'' - 2x y' + 12 y = 0$ as a power series around $x = 0$, and find its polynomial solution.

7.6 Show that all solutions of the equation $(x^2 + 1) y'' - 4x y' + 6 y = 0$ are polynomials, and find them explicitly.

7.7 Solve the Airy equation $y'' + x y = 0$ as a power series around $x = 0$.

7.8

(a) Find all the solutions of the equation $y'' + 2x y' + 2 y = 0$ by expanding them to power series around 0.
(b) Identify the series that is an even function as a familiar elementary function.
(c) Find a formula for the second solution by reducing the order of the equation.

7.9 The Euler's equation $x^2 y'' + 2x y' - 6 y = 0$ has the solutions x^2, x^{-3}. Solve this equation also by the expansion of its solution to power series around the regular point $x = 1$.
Hint: Let $t = x - 1$; then, $x^2 = (t + 1)^2 = t^2 + 2t + 1$.

7.10 In the introduction to Chap. 7, we mentioned the difficulty to understand the shape of the function that a power series represents, for example, the series $y(x) = \sum_{0}^{\infty} (-1)^n \frac{x^{3n}}{(3n)!}$. Show that $y(x)$ is the solution of the initial value problem $y^{(3)} + y = 0$, $y(0) = 1$, $y'(0) = y''(0) = 0$, and find it explicitly.

7.11 Find the power series solutions near $x = 0$ to the four initial value problems in Problem 4.13, (b), (c), of Chap. 4. How are the initial conditions expressed in the power series corresponding to each of the solutions?

7.12 Solve the Legendre equation of order 2, $(1 - x^2)y'' - 2xy' + 6y = 0$ around the regular-singular point $x = 1$. It is convenient to change variables $t = x - 1$.

7.13 Solve the following equations by power series around an appropriate regular-singular point. Some of the equations were solved in Chap. 4, Problems 4.19 and 4.20 by reduction of order.

(a) $(x - 1)y'' - xy' + y = 0$.
(b) $x(x + 1)y'' - 2y' - 2y = 0$.
(c) $xy'' + (x - 2)y' - 3y = 0$.
(d) $xy'' + y' - 4y = 0$.
(e) $ty'' - (1 + 3t)y' + 3y = 0$.
(f) $xy'' + (x - 1)y' - y = 0$.
(g) $x(x - 1)y'' + (3x - 1)y' + y = 0$.
(h) $xy'' - (x + N)y' + Ny = 0$, N is a positive integer.

7.14 Solve the equation $2x^2y'' - xy + (1 + x)y = 0$ around the regular-singular point $x = 0$. Verify that the solutions are $\sqrt{x} \cos \sqrt{2x}$, $\sqrt{x} \sin \sqrt{2x}$.

7.15 Solve the Laguerre equation $xy'' + (1 - x)y' + \alpha y = 0$ by series around its regular-singular point $x = 0$. Show that when α is a non-negative integer m, one of the solutions is a polynomial of order m.

7.16 Find two linearly independent solutions of the equation

$$x^2 y'' + \sin x \ y' + \cos x \ y = 0$$

as power series around the regular-singular point $x = 0$. Calculate explicitly the first four terms in each series.

7.17 The equation $xy'' + y' - 4y = 0$ has two equal indices at the regular-singular point $x = 0$. Find two linearly independent solutions of the equation around this point.

7.18 The equation $(x^2 - x)y'' + (3x - 1)y' + y = 0$ has a regular-singular point $x = 0$ with the indices $r_1 = r_2 = 0$. Find one solution by the Frobenius method, and identify it explicitly as a known elementary function. Find a second solution by reduction of the order of the equation.

7.19 What happens if around a *regular* point, where the coefficients of a differential equation can be expanded to power series, we look for solutions of the form $\sum_{n=0}^{\infty} a_n x^{n+r}$, which corresponds to a *regular-singular* point?

7.20 For the equation

$$x^2 y'' + (1 + 3x)y + y = 0$$

7.6 The Bessel Equation

the point $x = 0$ is neither regular nor regular-singular. Check that if despite this we look for a solution from the form $\sum_{n=0}^{\infty} a_n x^{n+r}$, we accept $r = 0$, a recursion formula $a_{n+1} = -(n+1)a_n$, and the series

$$y(x) = a_0 \sum_{n=0}^{\infty} (-1)^n n! \, x^n .$$

For which values x does this series converge? What is the conclusion about a solution around a point that is not regular-singular?

Chapter 8
The Laplace Transform

8.1 Definition of the Laplace Transform

In Chap. 4, we introduced two methods of solving non-homogeneous linear equations. The method of variation of parameters is applicable for any non-homogeneous linear equation. The second method, of the undetermined coefficients, is used only for differential equations with constant coefficients and with exponential, trigonometric, and polynomial functions on their right-hand side. In this chapter, we will discuss another method that is used mostly to solve equations with constant coefficients. This is the *Laplace*[1] *transform* method.

Why is another method needed to solve a problem that we already know how to deal with? The answer lies in several of its particularly convenient features:

1. The Laplace transform method turns the solution of a differential equation into an algebraic problem and uses preprepared tables.
2. The Laplace transform naturally handles initial value problems.
3. The Laplace transform treats effectively equations where the right-hand side contains piecewise continuous functions, such as step functions.

Definition 8.1 Given a function $f(t)$ defined for $0 \leq t < \infty$. The *Laplace transform* of $f(t)$, which will henceforth be denoted as $F(s)$ or $\mathscr{L}\{f(t)\}(s)$, is

$$F(s) = \mathscr{L}\{f(t)\}(s) = \int_0^\infty e^{-st} f(t)\, dt , \qquad (8.1)$$

provided that the integral exists.

---
[1] Pierre-Simon Laplace, 1749–1827.

For ease of writing, we will write the Laplace transform as $\mathscr{L}\{f(t)\}$, but it must be remembered that $F(s)$, the transform of $f(t)$, is a function of another variable s. Thus, the Laplace transform maps a function $f(t)$ into another function $F(s)$.

Laplace used this transform for his research in probability. The Laplace transform is one of numerous integral transforms. We have already encountered in Eq. (4.37) the integral $y(x) = \int_a^x K(x,t) f(t) \, dt$, where $K(x,t)$ is the Cauchy kernel. The most important among the integral transforms is the Fourier transform

$$F(s) = \mathscr{F}\{f(t)\} = \int_{-\infty}^{\infty} e^{-ist} f(t) \, dt \ .$$

But there are many others, such as the Weierstrass transform

$$W(s) = \int_{-\infty}^{\infty} e^{(s-t)^2/4} f(t) \, dt$$

related to the heat conduction equation, the Poisson integral

$$u(r, \theta) = \int_0^{2\pi} \frac{1 - r^2}{1 - 2r \cos(\theta - t) + r^2} f(t) \, dt$$

related to analytic functions and harmonic functions in the unit circle, and more.

The integral (8.1) is a generalized integral due to the unbounded domain of integration, so it must be understood as

$$\int_0^{\infty} e^{-st} f(t) \, dt = \lim_{T \to \infty} \int_0^T e^{-st} f(t) \, dt \ .$$

This limit does not always exist, so we need some assumptions that guarantee its existence.

To ensure that the integral $\int_0^T e^{-st} f(t) \, dt$ is defined for every T, we assume that $f(t)$ is a *piecewise continuous* function, that is, $f(t)$ is continuous in every finite interval $[0, T]$ except for at most a finite number of jump of discontinuities. For such function, the Riemann integral exists.

To guarantee that the integral converges when $T \to \infty$, $f(t)$ should not grow "too quickly." For that, we assume that there are constants $M > 0$, c so that

$$|f(t)| \leq M e^{ct}, \qquad 0 \leq t < \infty \ .$$

Such a function is said to be of *exponential order of magnitude*.

8.1 Definition of the Laplace Transform

Theorem 8.1 *If the function $f(t)$ is piecewise continuous and $|f(t)| \leq Me^{ct}$ for $0 \leq t < \infty$, then its Laplace transform $F(s) = \mathscr{L}\{f(t)\}$ exists for all $s > c$. Moreover,*

$$\lim_{s \to \infty} F(s) = 0.$$

Proof Recall the comparison criterion for generalized integrals: if $|p(t)| \leq q(t)$ and the generalized integral $\int_0^\infty q(t)\,dt$ exists, so does the integral $\int_0^\infty p(t)\,dt$. Moreover, $p(t)$ is even absolutely integrable. In our case, $|e^{-st} f(t)| \leq Me^{ct}e^{-st} = Me^{-(s-c)t}$, and when $s > c$,

$$\int_0^T |e^{-st} f(t)|\,dt \leq \int_0^T Me^{-(s-c)t}\,dt = \frac{M}{s-c}\left[1 - e^{-(s-c)T}\right] < \frac{M}{s-c}.$$

So $\int_0^\infty |e^{-st} f(t)|\,dt$ converges, and also the Laplace transform $\int_0^\infty e^{-st} f(t)\,dt$ exists for all $s > c$. The last inequality also implies that $F(s) \to 0$ when $s \to \infty$.

Unless we restrict the size of $f(t)$, it does not necessarily have a Laplace transform. For example, for the function $f(t) = e^{t^2}$, the integral

$$\int_0^\infty e^{-st} e^{t^2}\,dt = \int_0^\infty e^{t(t-s)}\,dt$$

diverges for every s.

A fundamental property of the Laplace transform is its linearity:

Theorem 8.2 *If $\mathscr{L}\{f_1(t)\} = F_1(s)$ exists for $s > c_1$ and $\mathscr{L}\{f_2(t)\} = F_2(s)$ exists for $s > c_2$, then every linear combination $k_1 f_1(t) + k_2 f_2(t)$ has its Laplace transform*

$$\mathscr{L}\{k_1 f_1(t) + k_2 f_2(t)\} = k_1 F_1(s) + k_2 F_2(s)$$

for $s > \max\{c_1, c_2\}$.

This feature follows directly from the linearity of integrals.

Example 8.1 The Laplace transform of the function $f(t) \equiv 1$ is

$$\mathscr{L}\{1\} = \int_0^\infty e^{-st} \times 1\,dt = \left.\frac{e^{-st}}{-s}\right|_{t=0}^\infty = \frac{1}{s}.$$

Example 8.2 The Laplace transform of the function $f(t) = e^{at}$ is

$$\mathscr{L}\{e^{at}\} = \int_0^\infty e^{-st} e^{at}\,dt = \left.\frac{e^{(a-s)t}}{a-s}\right|_{t=0}^\infty = \frac{1}{s-a}, \quad s > a.$$

Example 8.3 For the Laplace transform of $\sin(\omega t)$, we need the integral

$$\int e^{\alpha t} \sin(\omega t)\,dt = \frac{e^{\alpha t}}{\alpha^2 + \omega^2}\Big[\alpha \sin(\omega t) - \omega \cos(\omega t)\Big]$$

that can be calculated by two integrations by parts (but we may take its value from a table of integrals). In our case, we choose $\alpha = -s$ and get for all $s > 0$

$$\mathscr{L}\{\sin(\omega t)\} = \int_0^\infty e^{-st} \sin(\omega t)\,dt$$

$$= \frac{e^{-st}}{s^2 + \omega^2}\Big[-s\sin(\omega t) - \omega \cos(\omega t)\Big]_{t=0}^\infty = \frac{\omega}{s^2 + \omega^2}\,. \tag{8.2}$$

Similarly, we get

$$\mathscr{L}\{\cos(\omega t)\} = \frac{s}{s^2 + \omega^2}\,, \quad s > 0\,. \tag{8.3}$$

The last two transformations can also be calculated using complex numbers. Due to the linearity,

$$\mathscr{L}\{\cos(\omega t)\} + i\mathscr{L}\{\sin(\omega t)\} = \mathscr{L}\{\cos(\omega t) + i\sin(\omega t)\} = \mathscr{L}\{e^{i\omega t}\}$$

$$= \int_0^\infty e^{-st} e^{i\omega t}\,dt = \int_0^\infty e^{(-s+i\omega)t}\,dt$$

$$= \frac{1}{-s+i\omega} e^{(-s+i\omega)t}\Big|_{t=0}^\infty = 0 - \frac{1}{-s+i\omega}$$

$$= \frac{s}{s^2 + \omega^2} + i\frac{\omega}{s^2 + \omega^2}\,.$$

Equations (8.2) and (8.3) follow by comparing the real and imaginary parts.

A useful property of the Laplace transform discusses the effect of multiplication by an exponential function.

Theorem 8.3 *If* $\mathscr{L}\{f(t)\} = F(s)$ *for* $s > c$, *then*

$$\mathscr{L}\{e^{at} f(t)\} = F(s-a)\,, \quad s > c+a. \tag{8.4}$$

Indeed,

$$\mathscr{L}\{e^{at} f(t)\} = \int_0^\infty e^{-st} e^{at} f(t)\,dt = \int_0^\infty e^{-(s-a)t} f(t)\,dt = F(s-a)\,.$$

8.2 Initial Value Problems

Applying Theorem 8.3 to the formulas (8.2) and (8.3), we conclude that

$$\mathscr{L}\{e^{at}\sin(\omega t)\} = \frac{\omega}{(s-a)^2 + \omega^2}, \quad s > a,$$

$$\mathscr{L}\{e^{at}\cos(\omega t)\} = \frac{s-a}{(s-a)^2 + \omega^2}, \quad s > a. \tag{8.5}$$

Later, we will consider transforms of additional functions and summarize them in Table 8.1 on page 296.

After we defined the Laplace transform $F(s) = \mathscr{L}\{f(t)\}$ of a given function $f(t)$, we will also define the inverse operation. This definition confronts us with a fundamental problem. As we know, functions that differ from each other only at a finite number of points (say $f(t) = \tilde{f}(t)$ for all $t \neq c$ but $f(c) \neq \tilde{f}(c)$) have the same definite integral, and they also have the same Laplace transform $F(s)$. Therefore, the inverse Laplace transform of $F(s)$ is defined only up to "very few points" (a concept that requires a careful definition). To avoid this difficulty, the next definition is restricted to continuous functions.

Definition 8.2 If $f(t)$ is a continuous function and $F(s) = \mathscr{L}\{f(t)\}$ for $s > c$, it is said that $f(t)$ is the *inverse Laplace transform* of $F(s)$,

$$f(t) = \mathscr{L}^{-1}\{F(s)\}.$$

The proof of the existence of the inverse Laplace transform and its integral formula is a difficult problem. In this framework, we content ourselves to identify the inverse transform with aid of tables, such as Table 8.1.

8.2 Initial Value Problems

To use the Laplace transform for initial value problems, we first show some of its properties. They are summarized later in Table 8.2. What is, for example, the relation between the Laplace transform of a function and its derivative?

Theorem 8.4

(i) *If $f(t)$ is a continuous function such that $|f(t)| \leq Me^{ct}$ for all $0 \leq t < \infty$ and $f'(t)$ is piecewise continuous, then $\mathscr{L}\{f'(t)\}$, $\mathscr{L}\{f(t)\}$ exist for all $s > c$ and*

$$\mathscr{L}\{f'(t)\} = s\mathscr{L}\{f(t)\} - f(0), \quad s > c. \tag{8.6}$$

(ii) *If $f'(t)$, $f(t)$ are continuous functions so that $|f'(t)| \leq M_1 e^{ct}$, $|f(t)| \leq M_0 e^{ct}$ for all $0 \leq t < \infty$ and $f''(t)$ is piecewise continuous, then*

$$\mathscr{L}\{f''(t)\} = s^2 \mathscr{L}\{f(t)\} - sf(0) - f'(0), \quad s > c. \tag{8.7}$$

Proof For the sake of simplicity, suppose that $f'(t)$ has one jump discontinuity at $t=a$. We split the interval $[0, T]$ to $[0, a]$ and $[a, T]$, and integrate by parts on each subinterval.

$$\int_0^T e^{-st} f'(t)\, dt = \int_0^a e^{-st} f'(t)\, dt + \int_a^T e^{-st} f'(t)\, dt$$

$$= \left(e^{-st} f(t) \Big|_{t=0}^{a^-} + s \int_0^a e^{-st} f(t)\, dt \right) + \left(e^{-st} f(t) \Big|_{t=a^+}^{T} + s \int_a^T e^{-st} f(t)\, dt \right)$$

$$= \left(e^{-sc} f(a^-) - f(0) \right) + s \int_0^a e^{-st} f(t)\, dt + \left(e^{-sT} f(T) - e^{-sa} f(a^+) \right) + s \int_0^a e^{-st} f(t)\, dt \ .$$

As $f(t)$ is continuous, $f(a^-) = f(a^+)$ and the corresponding terms cancel out, the two integrals add to a single integral, and there remains

$$\int_0^T e^{-st} f'(t)\, dt = e^{-sT} f(T) - f(0) + s \int_0^T e^{-st} f(t)\, dt \ .$$

If $f'(t)$ has more discontinuity points, we split $[0, T]$ into more subintervals and obtain the same result. Note that without the continuity of $f(t)$, this argument fails. In view of $|f(T)| \le M_0 e^{cT}$, we have $\lim_{T \to \infty} |e^{-sT} f(T)| \le \lim_{T \to \infty} |M_0 e^{(c-s)T}| = 0$ when $s > c$. This completes the proof of (8.6).

For the second derivative, we repeat the same argument twice and get

$$\mathscr{L}\{f''(t)\} = s\mathscr{L}\{f'(t)\} - f'(0) = s\left[s\mathscr{L}\{f(t)\} - f(0)\right] - f'(0) \ ,$$

which is the desired identity (8.7). Similarly, we get that if $f, f', \ldots, f^{(n-1)}$ are continuous, and $f^{(n)}$ is piecewise continuous, then

$$\mathscr{L}\{f^{(n)}(t)\} = s^n \mathscr{L}\{f(t)\} - s^{n-1} f(0) - s^{n-2} f'(0) - \cdots - f^{(n-1)}(0) \ .$$

Now we can use the Laplace transform to solve simple initial value problems.

Example 8.4 Let us solve, for example, the initial value problem

$$y'' - 5y' + 6y = 0 \ ,$$
$$y(0) = 3, \quad y'(0) = -1 \ .$$

We apply the Laplace transform to both sides of the differential equation and use its linearity,

$$\mathscr{L}\{y''\} - 5\mathscr{L}\{y'\} + 6\mathscr{L}\{y\} = 0 \ .$$

8.2 Initial Value Problems

Using (8.7), (8.6), the equation becomes

$$\left[s^2\mathscr{L}\{y(t)\} - sy(0) - y'(0)\right] - 5\left[s\mathscr{L}\{y(t)\} - y(0)\right] + 6\mathscr{L}\{y(t)\} = 0.$$

With the notation $Y(s) = \mathscr{L}\{y(t)\}$, it is

$$(s^2 - 5s + 6)Y(s) + (5 - s)y(0) - y'(0) = 0,$$

and with the initial value conditions $y(0) = 3$, $y'(0) = -1$, we finally receive

$$Y(s) = \frac{3s - 16}{s^2 - 5s + 6} = \frac{3s - 16}{(s - 2)(s - 3)}. \tag{8.8}$$

Now we need to invert the Laplace transform, namely, to find a function $y(t)$ whose Laplace transform is (8.8). This is written symbolically as $y(t) = \mathscr{L}^{-1}\{Y(s)\}$. For this purpose, we simplify (8.8) by expanding $Y(s)$ into partial fractions

$$\frac{3s - 16}{s^2 - 5s + 6} = \frac{a}{s - 2} + \frac{b}{s - 3} = \frac{a(s - 3) + b(s - 2)}{(s - 2)(s - 3)},$$

compare the coefficients in the numerator

$$s: \quad 3 = a + b,$$
$$1: \quad -16 = -3a - 2b,$$

and get $a = 10$, $b = -7$. So

$$Y(s) = \frac{3s - 16}{s^2 - 5s + 6} = \frac{10}{s - 2} - \frac{7}{s - 3}. \tag{8.9}$$

According to Example 8.2, $\mathscr{L}\{e^{at}\} = \dfrac{1}{s - a}$ and $\mathscr{L}^{-1}\left\{\dfrac{1}{s - a}\right\} = e^{at}$. By the linearity of the Laplace transform, it follows that the solution of the initial value problem is

$$y(t) = \mathscr{L}^{-1}\{Y(s)\} = 10e^{2t} - 7e^{3t}.$$

Example 8.5 Now we solve a problem with a non-homogeneous equation

$$y'' - 5y' + 6y = 8e^t,$$
$$y(0) = 3, \quad y'(0) = -1.$$

Applying the Laplace transform brings us this time to

$$(s^2 - 5s + 6)Y(s) + (5 - s)y(0) - y'(0) = \frac{8}{s-1},$$

namely,

$$Y(s) = \frac{3s - 16}{s^2 - 5s + 6} + \frac{8}{(s-1)(s^2 - 5s + 6)}.$$

We expand again $Y(s)$ into partial fractions, and after some calculations, we get

$$Y(s) = \frac{10}{s-2} - \frac{7}{s-3} + 8\left(\frac{1/2}{s-1} - \frac{1}{s-2} + \frac{1/2}{s-3}\right)$$

$$= \frac{2}{s-2} - \frac{3}{s-3} + \frac{4}{s-1}.$$

By the inverse transform, we get the solution $y(t) = 2e^{2t} - 3e^{3t} + 4e^{t}$.

These simple initial value problems can, of course, be solved by other methods, and the use of the Laplace transform does not have currently any unique advantage.

A similar solution method is generally suitable for linear equations with constant coefficients

$$ay'' + by' + cy = q(t).$$

We activate the Laplace transform and get

$$a\left[s^2 \mathscr{L}\{y(t)\} - sy(0) - y'(0)\right] + b\left[s\mathscr{L}\{y(t)\} - y(0)\right] + c\mathscr{L}\{y(t)\} = Q(s),$$

where $Q(s) = \mathscr{L}\{q(t)\}$. So

$$Y(s) = \frac{(as+b)y(0) + ay'(0)}{as^2 + bs + c} + \frac{Q(s)}{as^2 + bs + c}. \tag{8.10}$$

The denominator $as^2 + bs + c$ is exactly the characteristic polynomial that corresponds to the operator on the left-hand side of the differential equation, and the first term of (8.10) is a rational function. If the roots r_1, r_2 of the characteristic polynomial are real, we expand the first term of (8.10) into partial fractions of the form

$$\frac{c_1}{s-r_1} + \frac{c_2}{s-r_2}, \tag{8.11}$$

8.2 Initial Value Problems

and their inverse transformation is $c_1 e^{r_1 t} + c_2 e^{r_2 t}$. If the roots of the characteristic polynomial are conjugate complex numbers $\alpha \pm i\beta$, then

$$as^2 + bs + c = a[(s-\alpha)^2 + \beta^2].$$

In this case, we will write the first term of (8.10) in the form of

$$\frac{c_1(s-\alpha) + c_2 \beta}{(s-\alpha)^2 + \beta^2}. \tag{8.12}$$

Its inverse transformation will be according to (8.5), $c_1 e^{\alpha t} \cos \beta t + c_2 e^{\alpha t} \sin \beta t$. For example, the two functions

$$F_1(s) = \frac{1}{s^2 - 4s - 12} = \frac{1/8}{s-6} - \frac{1/8}{s+2}, \quad F_2(s) = \frac{1}{s^2 - 4s + 13} = \frac{1}{(s-2)^2 + 9}$$

have inverse Laplace transformations

$$\mathscr{L}^{-1}\left\{\frac{1}{s^2 - 4s - 12}\right\} = \frac{1}{8}e^{6t} - \frac{1}{8}e^{-2t}, \quad \mathscr{L}^{-1}\left\{\frac{1}{s^2 - 4s + 13}\right\} = \frac{1}{3}e^{2t} \sin(3t),$$

respectively.

The second expression in Eq. (8.10) may be considered as a product of two Laplace transforms,

$$Q(s) \frac{1}{as^2 + bs + c} = Q(s)G(s).$$

To calculate an inverse transform of such product, we need additional tools that will be developed in Sect. 8.4.

The calculations of the inverse Laplace transforms in the previous examples are particular cases of a general result.

Theorem 8.5 *Let the polynomials $Q(s)$, $P(s)$ be of degrees $q < p$, respectively. If $P(s)$ has p simple roots $s_1, \ldots, s_p$, then the inverse Laplace transform of the quotient $Q(s)/P(s)$ is*

$$\mathscr{L}^{-1}\left\{\frac{Q(s)}{P(s)}\right\} = \sum_{i=1}^{p} \frac{Q(s_i)}{P'(s_i)} e^{s_i t}.$$

Proof Since the zeros of $P(s)$ are simple, it may be factorized as $P(s) = a(s - s_1) \cdots (s - s_p)$, and the quotient can be expanded into partial fractions

$$\frac{Q(s)}{P(s)} = \frac{a_1}{s - s_1} + \cdots + \frac{a_k}{s - s_k} + \cdots + \frac{a_p}{s - s_p}.$$

Multiplying both sides by $s - s_k$ and letting $s \to s_k$, we use that $P(s_k) = 0$, $P'(s_k) \neq 0$, and find

$$a_k = \lim_{s \to s_k} \frac{Q(s)}{P(s)}(s - s_k) = Q(s_k) \lim_{s \to s_k} \frac{s - s_k}{P(s) - P(s_k)} = Q(s_k) \frac{1}{P'(s_k)}.$$

The theorem follows by the linearity of the transform and $\mathscr{L}^{-1}\left\{\dfrac{1}{s - s_k}\right\} = e^{s_k t}$.

8.3 Additional Features of the Laplace Transform

We prove several more properties of the Laplace transform, which will be summarized in Table 8.2.

Theorem 8.6 *If* $\mathscr{L}\{f(t)\} = F(s)$, *then* $\mathscr{L}\{f(ct)\} = \dfrac{1}{c} F\left(\dfrac{s}{c}\right)$ *for all* $c > 0$.

This follows by the change of variables $t = u/c$, $dt = du/c$, in the integrals

$$\mathscr{L}\{f(ct)\} = \int_0^\infty e^{-st} f(ct) \, dt = \int_0^\infty e^{-s(u/c)} f(u) \, \frac{du}{c}$$

$$= \frac{1}{c} \int_0^\infty e^{-(s/c)u} f(u) \, du = \frac{1}{c} F\left(\frac{s}{c}\right).$$

Theorem 8.7

$$\mathscr{L}\{tf(t)\} = -\frac{d}{ds} \mathscr{L}\{f(t)\}. \tag{8.13}$$

Proof By direct differentiation of $\mathscr{L}\{f(t)\}$,

$$\frac{d}{ds} \mathscr{L}\{f(t)\} = \frac{d}{ds} \int_0^\infty e^{-st} f(t) \, dt = \int_0^\infty \frac{\partial}{\partial s}\left[e^{-st} f(t)\right] dt$$

$$= \int_0^\infty (-t) e^{-st} f(t) \, dt = -\mathscr{L}\{tf(t)\}.$$

Repeating this argument n times, we get

$$\mathscr{L}\{t^n f(t)\} = (-1)^n F^{(n)}(s). \tag{8.14}$$

Exercise 8.1 Calculate $\mathscr{L}\{t \sin(\omega t)\}$, $\mathscr{L}\{t^2 \sin(\omega t)\}$.

8.3 Additional Features of the Laplace Transform

Theorem 8.8 *If $\mathscr{L}\{f(t)\} = F(s)$ and $\lim_{t \to 0^+} f(t)/t$ exists, then*

$$\mathscr{L}\left\{\frac{f(t)}{t}\right\} = \int_s^\infty F(u)\,du.$$

Proof Let $F(u) = \int_0^\infty e^{-ut} f(t)\,dt$. The required right-hand side becomes, after changing the order of integration,

$$\int_s^\infty F(u)\,du = \int_{u=s}^\infty \left(\int_{t=0}^\infty e^{-ut} f(t)\,dt\right) du = \int_{t=0}^\infty \left(\int_{u=s}^\infty e^{-ut}\,du\right) f(t)\,dt.$$

The internal integral is $\displaystyle\int_{u=s}^\infty e^{-ut}\,du = \left.\frac{e^{-ut}}{-t}\right|_{u=s}^\infty = 0 + \frac{e^{-st}}{t}$, ; therefore, the previous double integral equals

$$= \int_0^\infty \frac{e^{-st}}{t} f(t)\,dt = \mathscr{L}\left\{\frac{f(t)}{t}\right\}.$$

Exercise 8.2 Show by Theorem 8.8 that $\mathscr{L}\left\{\dfrac{\sin t}{t}\right\} = \dfrac{\pi}{2} - \arctan(s)$. By the definition of the Laplace transform, conclude that $\displaystyle\int_0^\infty \frac{\sin t}{t}\,dt = \frac{\pi}{2}$.

Theorem 8.9 *If $\mathscr{L}\{f(t)\} = F(s)$, then* $\mathscr{L}\left\{\displaystyle\int_0^t f(\tau)\,d\tau\right\} = \dfrac{1}{s} F(s)$.

Proof Let $g(t) = \int_0^t f(\tau)\,d\tau$. If $f(t)$ is of exponential order of magnitude, that is, $|f(\tau)| \le Me^{c\tau}$, then $g(t)$ is such as well since

$$|g(t)| \le \int_0^t Me^{c\tau}\,d\tau = \frac{M}{c}(e^{ct} - 1) < \frac{M}{c} e^{ct}.$$

Therefore, the Laplace transform of $g(t)$ exists. $g(t)$ and $g'(t) = f(t)$ satisfy $\mathscr{L}\{g'(t)\} = s\mathscr{L}\{g(t)\} - g(0)$. But $g(0) = 0$, so

$$\mathscr{L}\left\{\int_0^t f(\tau)\,d\tau\right\} = \mathscr{L}\{g(t)\} = \frac{1}{s}\mathscr{L}\{g'(t)\} = \frac{1}{s}\mathscr{L}\{f(t)\} = \frac{F(s)}{s}.$$

Next we show several applications of Laplace transforms and their properties.

Theorem 8.10 *For non-negative integer n,*

$$\mathscr{L}\{t^n\} = \frac{n!}{s^{n+1}}, \qquad s > 0. \tag{8.15}$$

Proof This is a simple application of Theorem 8.7. We start with $\mathscr{L}\{1\} = \dfrac{1}{s}$, then by $\mathscr{L}\{tf(t)\} = -\dfrac{d}{ds}F(s)$, we have

$$\mathscr{L}\{t\} = \mathscr{L}\{t \times 1\} = -\frac{d}{ds}\left(\frac{1}{s}\right) = \frac{1}{s^2}, \qquad s > 0,$$

next

$$\mathscr{L}\{t^2\} = \mathscr{L}\{t \times t\} = -\frac{d}{ds}\left(\frac{1}{s^2}\right) = \frac{2}{s^3},$$

and so on. Equation (8.15) follows after n steps.

This method is suitable for calculating the Laplace transform of positive, integer powers. Due to the importance of the topic, we show another method, which is also suitable for non-integer powers. According to the definition

$$\mathscr{L}\{t^p\} = \int_0^\infty e^{-st} t^p \, dt \, .$$

In integral calculus, it is shown with the help of a comparison criterion that this integral converges at $t = \infty$ for all $s > 0$ and converges at $t = 0$ for all $p > -1$. Therefore, it is assumed from here on that $p > -1$. We replace the variable of integration t by $u = st$, $du = s\,dt$, and get

$$\mathscr{L}\{t^p\} = \int_0^\infty e^{-st} t^p \, dt = \int_0^\infty e^{-u} \left(\frac{u}{s}\right)^p \frac{du}{s} = \frac{1}{s^{p+1}} \int_0^\infty e^{-u} u^p \, du \, .$$

Here we meet one of the most important functions of mathematical analysis, the *Gamma function*,

$$\Gamma(p) = \int_0^\infty e^{-u} u^{p-1} \, du \, , \quad p > 0 \, .$$

With this notation, the Laplace transform of t^p is

$$\mathscr{L}\{t^p\} = \frac{\Gamma(p+1)}{s^{p+1}}, \quad s > 0, \tag{8.16}$$

for all $p > -1$.

We show several features of the Gamma function. The most important of the features is the recurrence formula

$$\Gamma(p+1) = p\,\Gamma(p), \quad p > 0, \tag{8.17}$$

8.3 Additional Features of the Laplace Transform

which is verified by the following integration by parts:

$$\Gamma(p+1) = \int_0^\infty e^{-u} u^p \, du = -e^{-u} u^p \Big|_{u=0}^\infty - \int_0^\infty -e^{-u} p u^{p-1} \, du$$

$$= 0 + p \int_0^\infty e^{-u} u^{p-1} \, du = p\,\Gamma(p).$$

In addition, $\Gamma(1) = \int_0^\infty e^{-u} u^0 \, du = -e^{-u}\Big|_{u=0}^\infty = 1$. Therefore, for an integer $p = n$, we get recursively

$$\Gamma(n+1) = n\Gamma(n) = n(n-1)\Gamma(n-1) = \ldots = n!\,\Gamma(1) = n!\,.$$

In this case, the formulas (8.16) and (8.15) are identical.

For non-integer values of p, it is not simple to calculate $\Gamma(p)$. We will do it for $p = 1/2$. With $u = v^2$, $du = 2v\,dv$,

$$\Gamma\left(\frac{1}{2}\right) = \int_0^\infty e^{-u} u^{-1/2} \, du = \int_0^\infty e^{-v^2} v^{-1} \cdot 2v \, dv = 2 \int_0^\infty e^{-v^2} \, dv = \int_{-\infty}^\infty e^{-v^2} \, dv,$$

and we know from integral calculus that $\int_{-\infty}^\infty e^{-v^2} \, dv = \sqrt{\pi}$. Therefore,

$$\mathscr{L}\{t^{-1/2}\} = \frac{\Gamma(1/2)}{s^{1/2}} = \sqrt{\frac{\pi}{s}}, \qquad s > 0.$$

Exercise 8.3 Calculate the values of $\Gamma(3/2)$, $\Gamma(5/2)$, ..., and $\Gamma(n+1/2)$ by the recurrence formula (8.17).

Remark (Bessel Functions) Recall that in Sect. 7.6, Eq. (7.46), we wrote a solution of the Bessel equation that corresponds to the index $r_1 = \alpha$, as

$$y_1(x) = \sum_{k=0}^\infty \frac{(-1)^k}{k!\,(1+\alpha)(2+\alpha)\ldots(k+\alpha)} \left(\frac{x}{2}\right)^{2k+\alpha}.$$

It is customary to define the *Bessel function of order* α as

$$J_\alpha(x) = \sum_{k=0}^\infty \frac{(-1)^k}{k!\,\Gamma(k+\alpha+1)} \left(\frac{x}{2}\right)^{2k+\alpha}. \tag{8.18}$$

Since

$$\Gamma(k+\alpha+1) = (k+\alpha)\Gamma(k+\alpha) = \ldots$$
$$= (k+\alpha)(k+\alpha-1)\cdots(\alpha+1)\Gamma(\alpha+1),$$

$J_\alpha(x)$ differs from the solution $y_1(x)$ by the constant factor $\Gamma(\alpha+1)$. For instance, for $\alpha = 1/2$,

$$J_{1/2}(x) = \frac{1}{\Gamma(3/2)} \frac{\sin x}{\sqrt{x}} = \sqrt{\frac{2}{\pi x}} \sin(x) \,.$$

Example 8.6 We return to the resonance problem $y'' + \omega_0^2 y = \cos \omega t$, which was discussed in Example 4.27, for $\omega \neq \omega_0$ and for $\omega = \omega_0$. The Laplace transform of the equation is

$$\left[s^2 Y(s) - s y(0) - y'(0)\right] + \omega_0^2 Y(s) = \mathscr{L}\{\cos \omega t\} = \frac{s}{s^2 + \omega^2} \,,$$

that is,

$$Y(s) = y(0) \frac{s}{s^2 + \omega_0^2} + y'(0) \frac{1}{s^2 + \omega_0^2} + \frac{s}{(s^2 + \omega_0^2)(s^2 + \omega^2)} \,. \tag{8.19}$$

(i) If $\omega \neq \omega_0$ we break down the last term into partial fractions,

$$\frac{s}{(s^2 + \omega_0^2)(s^2 + \omega^2)} = \frac{1}{\omega_0^2 - \omega^2} \left[\frac{s}{s^2 + \omega^2} - \frac{s}{s^2 + \omega_0^2}\right]. \tag{8.20}$$

With the help of Table 8.1, we get the inverse transform

$$y(t) = y(0) \cos \omega_0 t + \frac{y'(0)}{\omega_0} \sin \omega_0 t + \frac{\cos \omega t - \cos \omega_0 t}{\omega_0^2 - \omega^2} \,.$$

The first two terms are solutions of the homogeneous equation and take care of the initial value conditions, and the last term is a particular solution of the non-homogeneous equation.

(ii) If $\omega = \omega_0$, then (8.19) is

$$Y(s) = y(0) \frac{s}{s^2 + \omega_0^2} + y'(0) \frac{1}{s^2 + \omega_0^2} + \frac{s}{(s^2 + \omega_0^2)^2},$$

and we need the inverse transform of $s/(s^2 + \omega_0^2)^2$. Fortunately, we have

$$\frac{s}{(s^2 + \omega_0^2)^2} = -\frac{d}{ds}\left(\frac{1/2}{s^2 + \omega_0^2}\right). \tag{8.21}$$

Together with (8.13) (Theorem 8.7) and by (8.2), this yields

$$\frac{s}{(s^2+\omega_0^2)^2} = -\frac{d}{ds}\left(\frac{1/2}{s^2+\omega_0^2}\right) = -\frac{d}{ds}\left(\mathscr{L}\left\{\frac{\sin\omega_0 t}{2\omega_0}\right\}\right) = \mathscr{L}\left\{t\,\frac{\sin\omega_0 t}{2\omega_0}\right\}.$$

Consequently, the solution of the differential equation is

$$y(t) = y(0)\cos\omega_0 t + y'(0)\frac{\sin\omega_0 t}{\omega_0} + \frac{t\sin\omega_0 t}{2\omega_0},$$

as was found in Example 4.27.

8.4 Convolution

Definition 8.3 Given two functions $g(t)$, $f(t)$, piecewise continuous on $[0,\infty)$. Their *convolution* is the function $f * g$ that is defined by

$$(f * g)(t) = \int_{r=0}^{t} f(t-r)g(r)\,dr, \qquad 0 \le t < \infty.$$

Convolution of functions is commutative. By change of variables $p = t - r$, $dp = -dr$,

$$(f * g)(t) = \int_{r=0}^{t} f(t-r)g(r)\,dr = \int_{p=t}^{0} f(p)g(t-p)\,(-dp)$$

$$= \int_{p=0}^{t} g(t-p)f(p)\,dp = (g * f)(t).$$

The importance of convolution for the Laplace transform and its inverse transform lies in the following feature:

Theorem 8.11 If $\mathscr{L}\{f(t)\} = F(s)$, $\mathscr{L}\{g(t)\} = G(s)$, then $\mathscr{L}\{(f*g)(t)\} = F(s)G(s)$. In other words, the inverse transform of the product $F(s)G(s)$ is the convolution

$$\mathscr{L}^{-1}\{F(s)G(s)\} = (f * g)(t).$$

Proof

$$\mathscr{L}\{(f*g)(t)\} = \int_{t=0}^{\infty} e^{-st}\left[\int_{r=0}^{t} f(t-r)g(r)\,dr\right]dt.$$

The domain of integration in the (t,r)-plane is the sector $0 \le t < \infty$, $0 \le r \le t$, which is trapped between the positive horizontal t axis and the bisector $r = t$. After a change of the order of integration, we write this domain as $0 \leq r < \infty$, $r \leq t < \infty$, and the integral will be

$$= \int_{r=0}^{\infty} \left[\int_{t=r}^{\infty} e^{-st} f(t-r) g(r) \, dt \right] dr = \int_{r=0}^{\infty} g(r) \left[\int_{t=r}^{\infty} e^{-st} f(t-r) \, dt \right] dr \, .$$

In the inner integral, we replace the variable t by $p = t - r$, $dp = dt$, and the domain $r \leq t < \infty$ becomes $0 \leq p < \infty$:

$$= \int_{r=0}^{\infty} g(r) \left[\int_{p=0}^{\infty} e^{-s(p+r)} f(p) \, dp \right] dr$$

$$= \left[\int_{r=0}^{\infty} e^{-sr} g(r) \, dr \right] \times \left[\int_{p=0}^{\infty} e^{-sp} f(p) \, dp \right] = G(s) F(s) \, .$$

This identity allows us to calculate the inverse transformation of functions of multiplicative form. For example, in (8.10), the term $\dfrac{Q(s)}{as^2 + bs + c}$ is of the form $Q(s)H(S)$, where $Q(s) = \mathscr{L}\{q(t)\}$ and

$$H(s) = \frac{1}{as^2 + bs + c} \, .$$

Of course, $\mathscr{L}^{-1}\{Q(s)\} = q(t)$, and the inverse transform $\mathscr{L}^{-1}\{H(s)\} = h(t)$ was already calculated by factorizing the denominator or by writing it as a complete square as in (8.11), (8.12). Hence, by Theorem 8.11,

$$\mathscr{L}^{-1}\left\{ \frac{Q(s)}{as^2 + bs + c} \right\} = (h * q)(t) = \int_0^t h(t-r) q(r) \, dr \, .$$

Example 8.7 The equation $y'' + y = q(t)$ was solved in Example 4.21 by variation of the parameter method. Now, we apply the Laplace transform to get $\left(s^2 Y(s) - s y(0) - y'(0) \right) + Y(s) = Q(s)$, with $Q(s) = \mathscr{L}\{q(t)\}$, that is,

$$Y(s) = y(0) \frac{s}{s^2 + 1} + y'(0) \frac{1}{s^2 + 1} + \frac{Q(s)}{s^2 + 1} \, .$$

Because of $\mathscr{L}^{-1}\left\{ \dfrac{1}{s^2+1} \right\} = \sin t$, it follows that $\mathscr{L}^{-1}\left\{ \dfrac{Q(s)}{s^2+1} \right\} = (q * \sin)(t)$. Finally,

$$y(t) = y(0) \cos t + y'(0) \sin t + \int_0^t \sin(t-r) q(r) \, dr \, .$$

Compare this with the solution of Example 4.21.

8.5 Step Functions

Example 8.8 Convolution is not a magic instrument, and it is not necessarily the best choice in every situation. In Example 8.6, Part (i), we calculated $\mathscr{L}^{-1}\{s/((s^2+\omega_0^2)(s^2+\omega^2))\}$, $\omega \neq \omega_0$, by the expansion to partial fractions (8.20). The same result can be achieved by a systematic use of convolution

$$\mathscr{L}^{-1}\left\{\frac{s}{(s^2+\omega_0^2)(s^2+\omega^2)}\right\} = \frac{1}{\omega_0}\mathscr{L}^{-1}\left\{\frac{s}{s^2+\omega_0^2}\right\} * \mathscr{L}^{-1}\left\{\frac{\omega_0}{s^2+\omega^2}\right\}$$

$$= \frac{1}{\omega_0}\cos(\omega_0 s) * \sin(\omega s) = \frac{1}{\omega_0}\int_0^t \cos\omega_0(t-r)\sin\omega r\, dr$$

$$= \frac{1}{2\omega_0}\int_0^t \left[\sin(\omega_0 t - \omega_0 r + \omega r) + \sin(\omega_0 t - \omega_0 r - \omega r)\right] dr$$

$$= \frac{1}{2\omega_0}\left[-\frac{\cos(\omega_0 t - \omega_0 r + \omega r)}{\omega - \omega_0} + \frac{\cos(\omega_0 t - \omega_0 r - \omega r)}{\omega + \omega_0}\right]_{r=0}^t$$

$$= \frac{1}{2\omega_0}\left[\frac{\cos(\omega_0 t) - \cos(\omega t)}{\omega - \omega_0} - \frac{\cos(\omega_0 t) - \cos(\omega t)}{\omega + \omega_0}\right] = \frac{\cos(\omega_0 t) - \cos(\omega t)}{\omega^2 - \omega_0^2},$$

which is a considerably longer calculation.

In Part (ii) of Example 8.6, $\mathscr{L}^{-1}\{s/(s^2+\omega_0^2)^2\}$ was calculated by an ad hoc trick (8.21). The same result can be achieved by use of a convolution

$$\mathscr{L}^{-1}\left\{\frac{s}{(s^2+\omega_0^2)^2}\right\} = \frac{1}{\omega_0}\mathscr{L}^{-1}\left\{\frac{s}{s^2+\omega_0^2}\right\} * \mathscr{L}^{-1}\left\{\frac{\omega_0}{s^2+\omega_0^2}\right\}$$

$$= \frac{1}{\omega_0}\int_0^t \cos\omega_0(t-r)\sin\omega_0 r\, dr = \frac{1}{\omega_0}\int_0^t \frac{1}{2}\left[\sin\omega_0 t - \sin\omega_0(2r-t)\right] dr$$

$$= \frac{1}{2\omega_0}\left[r\sin\omega_0 t - \frac{\cos\omega_0(2r-t)}{2\omega_0}\right]_{r=0}^t = \frac{t\sin\omega_0 t}{2\omega_0}.$$

The calculation of the convolution is longer than the solution in Example 8.6, but at least it is obtained systematically.

8.5 Step Functions

All the differential equations that we have solved so far by the Laplace transform can be solved also by other long-known methods. The Laplace transform proves its effectiveness and is very useful for non-homogeneous linear differential equations, such as

$$ay'' + by' + cy = q(t), \tag{8.22}$$

in which the right-hand side $q(t)$ is discontinuous and the points of discontinuity of $q(t)$ are of the type of a "jump." As a motivation for this situation, think about an electric network that is switched from "off" to "on."

It is clear that when $q(t)$ is discontinuous, solution $y(t)$ of Eq. (8.22) cannot have two continuous derivatives. When $q(t)$ is piecewise continuous, we will look for a solution such that $y(t)$ and $y'(t)$ are continuous but $y''(t)$ is only piecewise continuous. The points of discontinuity of $y''(t)$ will be at the points of discontinuity of $q(t)$.

A very simple function with a jump is the *unit step function*

$$u_c(t) = \begin{cases} 0, & 0 \leq t < c, \\ 1, & c \leq t, \end{cases} \tag{8.23}$$

which is also called the *Heaviside* function.[2] Its Laplace transform is

$$\mathscr{L}\{u_c(t)\} = \int_0^\infty e^{-st} u_c(t)\, dt = \int_c^\infty e^{-st}\, dt$$

$$= \frac{e^{-st}}{-s}\bigg|_{t=c}^\infty = \frac{e^{-cs}}{s}, \quad s > 0.$$

$u_c(t)$ is used as a building block to write other discontinuous functions. For example, the pulse function, whose value is 1 in the interval $[a, b]$ and 0 elsewhere may be written as $g(t) = u_a(t) - u_b(t)$, and its Laplace transform is

$$\mathscr{L}\{g(t)\} = \frac{e^{-as} - e^{-bs}}{s}, \quad s > 0.$$

To write, for example, the step function

$$g(t) = \begin{cases} a, & 0 \leq t < 1, \\ b, & 1 \leq t < 3, \\ c, & 3 \leq t < 5, \\ d, & 5 \leq t, \end{cases}$$

we add for each point of discontinuity a step function $u_c(t)$ multiplied by the appropriate jump and get that

$$g(t) = a + (b-a)u_1(t) + (c-b)u_3(t) + (d-c)u_5(t),$$

$$\mathscr{L}\{g(t)\} = \frac{a}{s} + (b-a)\frac{e^{-s}}{s} + (c-b)\frac{e^{-3s}}{s} + (d-c)\frac{e^{-5s}}{s}.$$

---

[2] Oliver Heaviside, 1850–1925.

8.5 Step Functions

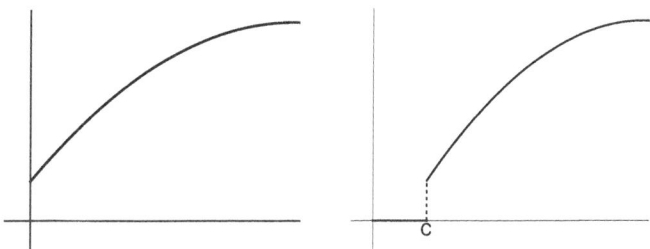

Fig. 8.1 A function $f(t)$ and its shift to the right, $u_c(t)f(t-c)$

Given a function $f(t)$ that is defined for $0 \leq t < \infty$, let's shift its graph to the right by c and complete its value in the interval $0 \leq t < c$ by 0. Thus, we obtained a new function

$$g(t) = \begin{cases} 0, & 0 \leq t < c, \\ f(t-c), & c \leq t. \end{cases}$$

It is convenient to write this function as $g(t) = u_c(t)f(t-c)$. Note that also $u_c(t)$ itself is a shift $u_0(t-c)$. See Fig. 8.1.

Theorem 8.12 *If $\mathscr{L}\{f(t)\} = F(s)$, then the transform of its right shift by c is*

$$\mathscr{L}\{u_c(t)f(t-c)\} = e^{-cs}F(s).$$

Proof The Laplace transform of the shifted function is

$$\mathscr{L}\{u_c(t)f(t-c)\} = \int_0^\infty e^{-st}u_c(t)f(t-c)\,dt = \int_c^\infty e^{-st}f(t-c)\,dt.$$

By a change of the variable to $p = t - c$, $dp = dt$, we get

$$= \int_{p=0}^\infty e^{-s(p+c)}f(p)\,dp = e^{-cs}\int_0^\infty e^{-sp}f(p)\,dp = e^{-cs}F(s).$$

This result is often formulated in the opposite direction: if $\mathscr{L}^{-1}\{F(s)\} = f(t)$, then

$$\mathscr{L}^{-1}\{e^{-cs}F(s)\} = u_c(t)f(t-c). \tag{8.24}$$

Example 8.9 We will solve the differential equation

$$y'' - 4y' + 3y = \begin{cases} \sin t, & 0 \leq t < \pi, \\ 0, & \pi \leq t. \end{cases}$$

The right-hand side $q(t)$ is continuous, but its definition changes at $t = \pi$ from $\sin t$ to 0. Hence, it is best to write it as

$$q(t) = \sin t + u_\pi(t)[0 - \sin t] = \sin t + u_\pi(t) \sin(t - \pi) .$$

The Laplace transform of the equation is

$$\left(s^2 Y(s) - s y(0) - y'(0)\right) - 4\left(s Y(s) - y(0)\right) + 3 Y(s)$$

$$= \mathscr{L}\{\sin t + u_\pi(t) \sin(t - \pi)\} = \frac{1}{s^2 + 1} + \frac{e^{-\pi s}}{s^2 + 1} .$$

To simplify the calculation, we choose the initial value conditions $y'(0) = 0$, $y(0) = 0$, which leads to

$$Y(s) = \frac{1 + e^{-\pi s}}{(s^2 - 4s + 3)(s^2 + 1)} .$$

We try a partial fractions expansion

$$\frac{1}{(s^2 - 4s + 3)(s^2 + 1)} = \frac{A}{s - 3} + \frac{B}{s - 1} + \frac{Cs + D}{s^2 + 1} ,$$

and compare the coefficients in the numerators

$$1 \equiv A(s^2 + 1)(s - 1) + B(s^2 + 1)(s - 3) + (Cs + D)(s^2 - 4s + 3) .$$

$$\begin{aligned}
s^3 : \quad & A + B + C && = 0 , \\
s^2 : \quad & -A - 3B - 4C + D && = 0 , \\
s : \quad & A + B + 3C - 4D && = 0 , \\
1 : \quad & -A - 3B + 3D && = 1 ,
\end{aligned}$$

hence $A = 1/20$, $B = -5/20$, $C = 4/20$, $D = 2/20$ and

$$Y(s) = \frac{1}{20} \left[\frac{1}{s - 3} - \frac{5}{s - 1} + \frac{4s + 2}{s^2 + 1} \right] \left(1 + e^{-\pi s}\right) .$$

By already known inverse transforms, we have

$$\mathscr{L}^{-1} \left\{ \frac{1}{s - 3} - \frac{5}{s - 1} + \frac{4s + 2}{s^2 + 1} \right\} = e^{3t} - 5e^t + 4 \cos t + 2 \sin t ,$$

8.5 Step Functions

therefore

$$y(t) = \mathscr{L}^{-1}\{Y(s)\} = \frac{1}{20}\left[e^{3t} - 5e^t + 4\cos t + 2\sin t\right]$$
$$+ \frac{u_\pi(t)}{20}\left[e^{3(t-\pi)} - 5e^{t-\pi} - 4\cos t - 2\sin t\right].$$

It can also be written as

$$y(t) = \begin{cases} \left(e^{3t} - 5e^t + 4\cos t + 2\sin t\right)/20, & 0 \le t < \pi, \\ \left(e^{3t} + e^{3(t-\pi)} - 5e^t - 5e^{t-\pi}\right)/20, & \pi \le t. \end{cases}$$

Check that the solution $y(t)$, and its first- and second-order derivatives are continuous at $x = \pi$.

The principles of Laplace transform are applicable also for systems of differential equations.

Example 8.10 Consider the system of equations

$$\frac{dx}{dt} = 3x - 4y,$$
$$\frac{dy}{dt} = x - y, \quad x(0) = 3, \ y(0) = 5.$$

Let $\mathscr{L}\{x(t)\} = X(s)$, $\mathscr{L}\{y(t)\} = Y(s)$. Applying the transform to the equations of the system, we have

$$sX(s) - 3 = 3X(s) - 4Y(s),$$
$$sY(s) - 5 = X(s) - Y(s),$$

that is,

$$(s-3)X(s) + 4Y(s) = 3,$$
$$-X(s) + (s+1)Y(s) = 5.$$

By Cramer's law, the solution of this algebraic system is

$$X(s) = \frac{\begin{vmatrix} 3 & 4 \\ 5 & s+1 \end{vmatrix}}{\begin{vmatrix} s-3 & 4 \\ -1 & s+1 \end{vmatrix}} = \frac{3s - 17}{s^2 - 2s + 1} = \frac{3}{s-1} - \frac{14}{(s-1)^2},$$

$$Y(s) = \frac{\begin{vmatrix} s-3 & 3 \\ -1 & 5 \end{vmatrix}}{\begin{vmatrix} s-3 & 4 \\ -1 & s+1 \end{vmatrix}} = \frac{5s-12}{s^2-2s+1} = \frac{5}{s-1} - \frac{7}{(s-1)^2}.$$

The inverse transform yields

$$x(t) = 3e^t - 14te^t, \quad y(t) = 5e^t - 7te^t.$$

8.6 The Dirac Function

As we mentioned in Sect. 4.9, a homogeneous linear differential equation usually describes the self-motion of a model without applying an external force, while that of the corresponding non-homogeneous equation describes the behavior of that model under the influence of an external force. In this chapter, we discuss equations of the form

$$ay'' + by' + cy = q(t)$$

when $q(t)$ has a unique behavior: the values of $q(t)$ are very large over a very short time interval, and outside this interval of time, its value is zero. As an example, you can think of the force that exerts a hammer hitting the head of a nail or a projectile penetrating a heavy block: in both cases, it is difficult to estimate the duration of the action (very short) and the force applied at the same time (very large), but its effect is clear. For such a force $q(t)$, we integrate Newton's second law, $my''(t) = q(t)$, as

$$m \frac{dy}{dt} \bigg|_0^t = \int_0^t q(t)\, dt.$$

That is the change of the momentum

$$mv - mv_0 = \int_0^t q(t)\, dt.$$

It turns out that the important concept is not necessarily the height of the peak of the graph or the width of its operation range but its integral, i.e., the area that is bounded beneath it. In a mechanical system, this size is called the *impulse* of the force $q(t)$.

An idealized model for $q(t)$ can be any of the functions in Fig. 8.2. For example,

$$q_h(t) = \begin{cases} 1/2h, & -h \le t \le h, \\ 0, & \text{otherwise}. \end{cases}$$

8.6 The Dirac Function

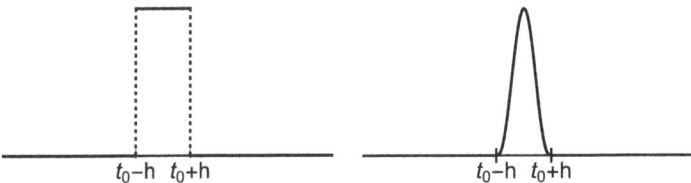

Fig. 8.2 Impulse functions

At any fixed point $t \neq 0$ and a sufficiently small h, we have $q_h(t) = 0$, and for all $h > 0$,

$$\int_{-\infty}^{\infty} q_h(t)\, dt = 1 .$$

As an idealization of $q_h(t)$, we would like to define the *delta function of Dirac*,[3] $\delta(t)$, which maintains the features

$$\delta(t) = 0 \quad \text{for all } t \neq 0 ,$$
$$\int_{-\infty}^{\infty} \delta(t)\, dt = 1 . \qquad (8.25)$$

The problematic nature of this definition is obvious. It is clear that no function in the usual sense of calculus has these properties! The value of an integral does not depend on the value of the function at a single point, and if a function is zero at every point except one, its integral (Riemann integral) is zero. Despite these difficulties the delta "function" of Dirac, $\delta(t)$, and its shifts $\delta(t - t_0)$ are useful tools. The reason for this is that in the context of the Laplace transform, we are not interested in the pointwise values of these "functions" but rather in the integrals in which they appear. For every $t_0 > 0$ and $h < t_0$, we have, due to the concentration of $q_h(t - t_0)$ in the interval $[t_0 - h, t_0 + h]$, that for any continuous function $f(t)$

$$\int_0^{\infty} f(t) q_h(t - t_0)\, dt = \int_{t_0-h}^{t_0+h} f(t) q_h(t - t_0)\, dt .$$

According to the integral intermediate value theorem, there exists a point $\tilde{t}$ in the interval $[t_0 - h, t_0 + h]$ so that

$$= f(\tilde{t}) \int_{t_0-h}^{t_0+h} q_h(t - t_0)\, dt = f(\tilde{t}) .$$

---
[3] Paul Dirac, 1902–1986.

When $h \to 0$, then $\tilde{t} \to t_0$, so

$$\lim_{h \to 0} \int_0^\infty f(t) q_h(t - t_0)\,dt = f(t_0)\,.$$

Since we consider the $\delta(t-t_0)$ as an idealization of the impulse $q_h(t-t_0)$, we define the action of the Dirac function on a function $f(t)$ by

$$\int_{t=0}^\infty f(t)\delta(t - t_0)\,dt = f(t_0)\,. \tag{8.26}$$

This intuitive definition may be given another meaning. Equation (8.26) matches to the function $f(t)$ the number $f(t_0)$. As such, $\delta(t - t_0)$ can be considered as an operator from the space of continuous functions to the real numbers, which is defined by $\delta(t - t_0) : f(t) \to f(t_0)$. A precise mathematical justification is given in the *theory of distributions*. Here we will be content with (8.26) as a definition.

If we take in Eq. (8.26), $f(t) = e^{-st}$, we get the Laplace transform of $\delta(t - t_0)$:

$$\mathscr{L}\{\delta(t - t_0)\} = \int_0^\infty e^{-st}\delta(t - t_0)\,dt = e^{-st_0}\,, \qquad s > 0\,.$$

The convolution of $\delta(t)$ and of $f(t)$ is, again by (8.26),

$$(f * \delta)(t) = \int_0^\infty f(t - r)\delta(r)\,dr = \int_0^\infty f(s)\delta(t - s)\,ds = f(t)\,.$$

The Dirac delta function and the unit step function are closely related. Since the effect of the impulse $\delta(u - c)$ is concentrated at c,

$$\int_{-\infty}^t \delta(u - c)\,du = \begin{cases} 0, & 0 \le t < c\,, \\ 1, & c \le t\,, \end{cases}$$

However, the right-hand side is the definition of the unit step function $u_c(t)$. So

$$\int_{-\infty}^t \delta(u - c)\,du = u_c(t)\,.$$

It is tempting to formulate the inverse relation as *"the derivative of the unit step function is the Dirac function."* This statement looks reasonable, but it has no meaning in classical calculus, since the step function is not differentiable at c and $\delta(t - c)$ is not a function.

The Dirac function is useful for differential equations with impulses.

8.6 The Dirac Function

Example 8.11

$$y'' - 5y' + 6y = 8\delta(t-1) + 5\delta(t-4),$$
$$y(0) = 0, \quad y'(0) = 0.$$

By applying the Laplace transform, $(s^2 - 5s + 6)Y(s) = 8e^{-s} + 5e^{-4s}$,

$$Y(s) = \left(\frac{1}{s-3} - \frac{1}{s-2}\right)(8e^{-s} + 5e^{-4s}).$$

Since $\mathscr{L}^{-1}\left\{\frac{1}{s-3} - \frac{1}{s-2}\right\} = e^{3t} - e^{2t}$, we get according to (8.24) the solution

$$y(t) = \mathscr{L}^{-1}\{Y(s)\}$$
$$= 8u_1(t)\left(e^{3(t-1)} - e^{2(t-1)}\right) + 5u_4(t)\left(e^{3(t-4)} - e^{2(t-4)}\right).$$

Note that the solution is continuous at $t = 1, 4$, but it is not differentiable there.

When an impulse acts at the initial point $t = 0$, say

$$ay'' + by' + cy = \delta(t), \quad y(0) = 0, \quad y'(0) = 0, \quad (8.27)$$

we mean that the system is at rest until $t = 0$, and then it is activated by a unit pulse.

Exercise 8.4

(i) A harmonic oscillator $y'' + y = 0$ starting at rest is excited by a unit impulse at $t = 0$, that is,

$$y'' + y = \delta(t), \quad y(0) = 0, \quad y'(0) = 0.$$

Solve this initial value problem.

(ii) Let another unit impulse act on the same system after a half period π,

$$y'' + y = \delta(t) + \delta(t - \pi), \quad y(0) = 0, \quad y'(0) = 0.$$

Find the solution and explain its behavior for $0 \leq t < \infty$.

(iii) Finally suppose that the second unit impulse acts on the system after a complete period 2π,

$$y'' + y = \delta(t) + \delta(t - 2\pi), \quad y(0) = 0, \quad y'(0) = 0.$$

Solve and explain the differences between the three initial value problems and their physical interpretations.

Table 8.1 Some Laplace transforms

$f(t)$	$F(s) = \mathcal{L}\{f(t)\}$	
1	$\dfrac{1}{s}$,	$s > 0$
e^{at}	$\dfrac{1}{s-a}$,	$s > a$
t^n, integer n,	$\dfrac{n!}{s^{n+1}}$,	$s > 0$
t^p, $p > -1$	$\dfrac{\Gamma(p+1)}{s^{p+1}}$,	$s > 0$
$e^{at}t^n$	$\dfrac{n!}{(s-a)^{n+1}}$,	$s > a$
$\sin at$	$\dfrac{a}{s^2+a^2}$,	$s > 0$
$\cos at$	$\dfrac{s}{s^2+a^2}$,	$s > 0$
$e^{at}\sin bt$	$\dfrac{b}{(s-a)^2+b^2}$,	$s > a$
$e^{at}\cos bt$	$\dfrac{s-a}{(s-a)^2+b^2}$,	$s > a$
$t\sin at$	$\dfrac{2as}{(s^2+a^2)^2}$,	$s > 0$
$t\cos at$	$\dfrac{s^2-a^2}{(s^2+a^2)^2}$,	$s > 0$
$\sinh at$	$\dfrac{a}{s^2-a^2}$,	$\lvert s\rvert > \lvert a\rvert$
$\cosh at$	$\dfrac{s}{s^2-a^2}$,	$\lvert s\rvert > \lvert a\rvert$
$u_c(t)$, $c > 0$	$\dfrac{e^{-cs}}{s}$,	$s > 0$
$\delta(t-c)$	e^{-cs},	$s > 0$
$J_0(t)$	$1/\sqrt{s^2+1}$,	$s > 0$

Example 8.12 Let $z(t)$ denote the solution of the impulse equation

$$az'' + bz' + cz = \delta(t), \quad z(0) = 0, \ z'(0) = 0$$

and $y(t)$ the solution of

$$ay'' + by' + cy = q(t), \quad y(0) = 0, \ y'(0) = 0.$$

The transform of the first equation is $(as^2 + bs + c)\mathcal{L}\{z\} = \mathcal{L}\{\delta(t)\} = 1$, so

$$\mathcal{L}\{z\} = \frac{1}{as^2 + bs + c}.$$

8.7 Some Nice Examples

Table 8.2 Some properties of the Laplace transform

$f(t)$	$F(s) = \mathscr{L}\{f(t)\}$
$f(ct)$	$\frac{1}{c}F\left(\frac{s}{c}\right)$, $c > 0$
$e^{ct}f(t)$	$F(s-c)$
$tf(t)$	$-F'(s)$
$t^n f(t)$	$(-1)^n F^{(n)}(s)$
$\frac{f(t)}{t}$	$\int_s^\infty F(r)\,dr$
$f'(t)$	$sF(s) - f(0)$ c
$f''(t)$	$s^2 F(s) - sf(0) - f'(0)$ c
$f^{(n)}(t)$	$s^n F(s) - s^{n-1} f(0) - \cdots - f^{(n-1)}(0)$
$\int_0^t f(\tau)\,d\tau$	$\frac{1}{s}F(s)$
$u_c(t)f(t-c)$	$e^{-cs}F(S)$
$(f*g)(t) = \int_0^t f(t-r)g(r)\,dr$	$F(s)G(s)$

Similarly, the transform of the second equation is $(as^2 + bs + c)\mathscr{L}\{y\} = \mathscr{L}\{q\}(s)$, and

$$\mathscr{L}\{y\} = \frac{\mathscr{L}\{q\}(s)}{as^2 + bs + c} = \mathscr{L}\{z\}\mathscr{L}\{q\}.$$

Consequently, $y(t) = (z*q)(t) = \int_0^\infty z(t-r)q(r)\,dr$. This formula reminds the solution of nonhomogeneous equations by the Cauchy kernel (Eq. 4.37). Compare with Example 4.21

8.7 Some Nice Examples

Laplace transforms has useful applications that are not directly related to differential equations with constant coefficients.

Example 8.13 Let us calculate the Laplace transform of $J_0(t)$, the Bessel function of order 0. According to (7.48), the power series expansion of $J_0(t)$ is

$$J_0(t) = \sum_{k=0}^{\infty} \frac{(-1)^k t^{2k}}{(k!)^2 2^{2k}}.$$

By the linearity of the Laplace transform and by $\mathscr{L}\{t^{2k}\} = \dfrac{(2k)!}{s^{2k+1}}$,

$$\mathscr{L}\{J_0(t)\} = \sum_{k=0}^{\infty} \frac{(-1)^k}{(k!)^2 \, 2^{2k}} \frac{(2k)!}{s^{2k+1}} = \frac{1}{s} \sum_{k=0}^{\infty} \frac{(-1)^k}{k!} \frac{1 \cdot 3 \cdot 5 \cdots (2k-1)}{2^k} \frac{1}{s^{2k}}$$

$$= \frac{1}{s} \sum_{k=0}^{\infty} \frac{1}{k!} \left(-\frac{1}{2}\right)\left(-\frac{3}{2}\right) \cdots \left(-k + \frac{1}{2}\right) \left(\frac{1}{s^2}\right)^k$$

and by the binomial series, this is

$$= \frac{1}{s}\left(1 + \frac{1}{s^2}\right)^{-1/2} = (s^2 + 1)^{-1/2}.$$

Another method to calculate $\mathscr{L}\{J_0(t)\}$ is based on the fact that $J_0(t)$ is also the solution of the differential equation $t^2 y'' + t y' + (t^2 - 0^2) y = 0$ and the initial value conditions $y(0) = 1$, $y'(0) = 0$. Thus, we start to calculate $\mathscr{L}\{ty'' + y' + ty\}$, a differential operator whose coefficients are not constants. Let $\mathscr{L}\{y\} = Y(s)$. Then,

$$\mathscr{L}\{ty\} = -\frac{d}{ds}\mathscr{L}\{y\} = -Y'(s), \quad \mathscr{L}\{y'\} = sY(s) - y(0),$$

$$\mathscr{L}\{ty''\} = -\frac{d}{ds}\mathscr{L}\{y''\} = -\frac{d}{ds}\left(s^2 Y(s) - sy(0) - y'(0)\right)$$

$$= -s^2 Y'(s) - 2sY(s) + y(0).$$

Taking into account the initial values, the transform of the equation becomes

$$\mathscr{L}\{ty'' + y' + ty\} = \left(-s^2 Y'(s) - 2sY(s) + 1\right) + (sY(s) - 1) + (-Y'(s)) = 0,$$

and after reorganization,

$$\frac{Y'(s)}{Y(s)} = -\frac{s}{s^2 + 1}.$$

By integration of this separated equation, $\ln|Y(s)| = -\dfrac{1}{2}\ln(s^2+1) + \ln C$, i.e., $Y(s) = C(s^2+1)^{-1/2}$. $y(0) = 1$ fixes $C = 1$, so

$$\mathscr{L}\{J_0(t)\} = (s^2 + 1)^{-1/2}.$$

Theorem 8.13

$$\int_0^{\infty} \mathscr{L}\{f\}(u) g(u)\, du = \int_0^{\infty} f(u) \mathscr{L}\{g\}(u)\, du \tag{8.28}$$

provided that the integrals converge.

8.7 Some Nice Examples

Proof We plug the definition $\mathscr{L}\{f\}(u) = \int_0^\infty e^{-ut} f(t)\,dt$ into the left-hand side of (8.28):

$$\int_0^\infty \mathscr{L}\{f\}(u)g(u)\,du = \int_0^\infty \left[\int_0^\infty e^{-ut} f(t)\,dt\right] g(u)\,du = \int_0^\infty \int_0^\infty f(t)g(u) e^{-ut}\,dt\,du\,.$$

We do the same with the right-hand side of (8.28), reverse the order of integration (assuming that it is allowed), and conclude that it is also equal to the same double integral. Equality (8.28) reminds of integration by parts.

Example 8.14 Identity (8.28) may be used to compute some definite integrals, for example, $\int_0^\infty \dfrac{\sin u}{u}\,du$, a well-known example of a convergent but not absolutely convergent integral. The convergence is usually proved by Dirichlet test[4] but the proof of the convergence gives no hint about the value of the integral. To evaluate the integral, we identify in (8.28) $\dfrac{1}{u}$ as $\mathscr{L}\{1\}(u)$ and $\sin(u)$ as $g(u)$ and get

$$\int_0^\infty \frac{\sin u}{u}\,du = \int_0^\infty \mathscr{L}\{1\}(u)\cdot\sin u\,du = \int_0^\infty 1\cdot\mathscr{L}\{\sin\}(u)\,du = \int_0^\infty \frac{du}{u^2+1}\,du = \frac{\pi}{2}\,.$$

This integral was evaluated in Exercise 8.2 by another method.

Exercise 8.5 Calculate $\displaystyle\int_0^\infty \frac{e^{-at} - e^{-bt}}{t}\,dt$ and $\mathscr{L}\left\{\dfrac{e^{-at} - e^{-bt}}{t}\right\}$.

Example 8.15 We show another method to calculate definite integrals with the aid of known Laplace transforms. Suppose we want to evaluate the integral $\int_0^\infty t e^{-t} \sin t\,dt$. By (8.13),

$$\int_0^\infty t e^{-st} \sin t\,dt = \mathscr{L}\{t \sin t\} = -\frac{d}{ds}\mathscr{L}\{\sin t\} = -\frac{d}{ds}\left\{\frac{1}{s^2+1}\right\} = \frac{s}{(s^2+1)^2}$$

and for $s = 1$, it follows that $\int_0^\infty t e^{-t} \sin t\,dt = 1/4$.

Example 8.16 Calculate the integral $\displaystyle\int_0^1 x^{p-1}(1-x)^{q-1}\,dx$.

Let $f(t) = t^{q-1}$, $g(t) = t^{p-1}$. Their convolution is

$$h(t) = (f * g)(t) = \int_{r=0}^t r^{p-1}(t-r)^{q-1}\,dr,$$

---

[4] $\int_0^\infty f(x)g(x)\,dx$ converges, provided that $|\int_0^x f(t)\,dt|$ is bounded and $g(x)$ decreases monotonically to 0.

and our integral will be $h(1)$. According to Eq. (8.16),

$$\mathscr{L}\{t^{p-1}\} = \frac{\Gamma(p)}{s^p}, \quad \mathscr{L}\{t^{q-1}\} = \frac{\Gamma(q)}{s^q}, \quad s > 0$$

and

$$\mathscr{L}\{h(t)\} = \mathscr{L}\{f(t)\} \times \mathscr{L}\{g(t)\} = \frac{\Gamma(p)\Gamma(q)}{s^{p+q}}.$$

Therefore,

$$h(t) = \mathscr{L}^{-1}\left\{\frac{\Gamma(p)\Gamma(q)}{s^{p+q}}\right\} = \frac{\Gamma(p)\Gamma(q)}{\Gamma(p+q)} t^{p+q-1},$$

and our integral is

$$\int_0^1 r^{p-1}(1-r)^{q-1}\, dr = h(1) = \frac{\Gamma(p)\Gamma(q)}{\Gamma(p+q)}.$$

Problems

8.1 Solve the following initial value problems:
(a) $y'' - 5y' + 4y = e^{3t}, \quad y(0) = 2, \ y'(0) = 3$.
(b) $2y'' + y' - y = e^{3t}, \quad y(0) = 2, \ y'(0) = 3$.
(c) $y'' + 2y' + y = e^{3t}, \quad y(0) = 2, \ y'(0) = 3$.
(d) $y'' + 2y' + y = e^{-t}, \quad y(0) = 2, \ y'(0) = 3$.
(e) $y'' - y = t, \quad y(0) = 2, \ y'(0) = 3$.
(f) $y'' + y' - 6y = e^{2t}, \quad y(0) = 0, \ y'(0) = 0$.
(g) $y'' - 2y' + 7y = \sin t, \quad y(0) = 0, \ y'(0) = 0$.

8.2 According to Theorem 8.7, if $F(s) = \mathscr{L}\{f(t)\}$, then $F'(s) = -\mathscr{L}\{tf(t)\}$. This can be stated as "If $\mathscr{L}^{-1}\{F'(s)\} = -tf(t)$ then $\mathscr{L}^{-1}\{F(s)\} = f(t)$." Use this feature to calculate $\mathscr{L}^{-1}\{\arctan s\}$ and $\mathscr{L}^{-1}\left\{\ln\left(\frac{s+1}{s-1}\right)\right\}$.

8.3 (a) Show by expansion of the product to partial fractions that

$$\mathscr{L}^{-1}\left\{\frac{1}{(s+a)(s+b)}\right\} = \frac{e^{-at} - e^{-bt}}{b-a}.$$

8.7 Some Nice Examples

(b) Show the same formula by the convolution

$$\mathscr{L}^{-1}\left\{\frac{1}{(s+a)(s+b)}\right\} = \mathscr{L}^{-1}\left\{\frac{1}{s+a}\right\} * \mathscr{L}^{-1}\left\{\frac{1}{s+a}\right\}.$$

8.4 Solve the following initial value problems:

(a) $y'' + y = \begin{cases} \sin t, & 0 \leq t < \pi, \\ \cos t, & \pi \leq t, \end{cases}$ $y(0) = 0, \; y'(0) = 0$.

(b) $y'' - y = \begin{cases} 0, & 0 \leq t < 2, \\ 1, & 2 \leq t, \end{cases}$ $y(0) = 0, \; y'(0) = 0$.

(c) $y'' - y' - 2y = \begin{cases} e^{7t}, & 0 \leq t < 1, \\ e^{7t} + 1, & 1 \leq t, \end{cases}$ $y(0) = 0, \; y'(0) = 0$.

8.5 Given the initial value problem

$$y'' + 4y' + 7y = u_\pi(t) - u_{2\pi}(t) + u_{3\pi}(t) - + \ldots + u_{17\pi}(t),$$
$$y(0) = 0, \; y'(0) = 0.$$

What is the value of $y(1)$? Hint: Think before you calculate.

8.6 Solve the two initial value problems

$$y' = \delta(t) + \delta(t-1) + \delta(t-2) + \ldots, \quad y(0) = 0,$$
$$y' = \delta(t) - \delta(t-1) + \delta(t-2) - + \ldots, \quad y(0) = 0,$$

and draw their solutions.

8.7 Calculate the inverse Laplace transforms of the functions

$$\frac{1}{s(s^2+1)}, \quad \frac{1}{s^2(s^2+1)}, \quad \frac{1}{s^3(s^2+1)}.$$

8.8 Calculate $\mathscr{L}^{-1}\{F(s)\}$ of the following functions using two different methods, expansion to partial fractions and convolution:

$$\frac{1}{s^2(s^2+1)^2}, \quad \frac{1}{s(s^2+2s+2)}.$$

8.9 Calculate $\mathscr{L}^{-1}\{F(s)\}$ of the functions

$$\frac{1}{(s^2+9)^2}, \quad \frac{s}{(s^2+9)^2}, \quad \frac{s^2}{(s^2+9)^2}, \quad \frac{1}{s^4-1}.$$

8.10 Solve the following initial value problems:
(a) $4y'' + 4y' + 9y = 5\delta(t-1) + 6\delta(t-4)$, $y(0) = 0$, $y'(0) = 0$.
(b) $y'' + 3y' + 2y = 2\delta(t) + 3\delta(t-1)$, $y(0) = 0$, $y'(0) = 0$.

8.11 Use Example 8.13 to show that $\int_0^\infty J_0(t)\,dt = 1$.

8.12 Calculate $\displaystyle\int_0^\infty \frac{\sin^2 t}{t^2}\,dt$.

8.13 Calculate $\mathscr{L}\{\sin t\}$ by the Taylor series $\sin t = \displaystyle\sum_{n=0}^\infty (-1)^n t^{2n+1}/(2n+1)!$.
Remember that Laplace transforms are defined for some $s > c$.

8.14 If $f(t)$ is a periodic function of period T, that is, $f(t+T) = f(t)$, then
$$\mathscr{L}\{f(t)\} = \frac{1}{1 - e^{-sT}} \int_0^T e^{-st} f(t)\,dt.$$

8.15 Consider the series $f(t) = \displaystyle\sum_{n=0}^\infty (-1)^n u_n(t)$ of step functions. Draw its graph, explain why it converges, and show that it is periodic. Find its Laplace transform.

8.16 Solve the initial value problem
$$\frac{dx}{dt} = x + y,$$
$$\frac{dy}{dt} = 4x + y, \quad x(0) = 5,\ y(0) = 6,$$
and compare the result with Example 5.3.

Appendix A
The Orbits of the Planets

In the years 1609–1619, Johannes Kepler[1] published his three laws of planetary motion:

- Each planet moves in an elliptical orbit with the sun at one of its foci.
- The line joining the sun to a planet sweeps out during its motion equal areas in equal intervals of time.
- The second power of the orbital period of a planet is proportional to the third power of the length of the major axis of its elliptical orbit.

Kepler's conclusions stemmed from the years-long astronomic measurements of the astronomer Tycho Brahe[2] about the orbit of Mars. Kepler's laws were a work of fitting a curve to numerical data, but Kepler did not try to explain the cause of the motion of the planets.

During the years 1660–1684, Isaac Newton[3] invented the basics of mechanics and differential calculus. One of his greatest successes was to prove Kepler's laws on the basis of the law of gravitation. In this chapter, we utilize what we had learned about differential equations to demonstrate this achievement. This is not the way that Newton did it.

By our notation, if the vector $\mathbf{r}(t)$ indicates the position of a body at time t, its acceleration obeys Newton's second law, which states that

$$m\,\mathbf{r}''(t) = \mathbf{F}\,. \tag{A.1}$$

---

[1] Johannes Kepler, 1571–1630.
[2] Tycho Brahe, 1546–1601.
[3] Isaac Newton, 1643–1727.

Newton's law of gravitation states that the gravitational force exerted by a point-like body with mass M on another point-like body with mass m, where $\mathbf{r}$ is the vector from M to m and r is its length, is

$$\mathbf{F} = -\frac{GMm}{r^2}\frac{\mathbf{r}}{r} . \tag{A.2}$$

These two principles form the vector differential equation

$$\mathbf{r}''(t) = -GM\frac{\mathbf{r}}{r^3} , \tag{A.3}$$

which determines the movement of one body relative to the other. Our goal is to solve this equation. This enables us to highlight various methods that we met in the previous chapters.

Let's start with some physical considerations. Let the body m be at a certain moment at the point $\mathbf{r}_0$, and its speed at that moment is $\mathbf{v}_0$. These two vectors span a plane in the three-dimensional space, the body has no velocity, and no force acts on it in a perpendicular direction to this plane. Therefore, its motion will take place in the same plane, and the problem of the movement of two point-bodies is a two-dimensional problem. The plane in which the earth moves is called the *ecliptic plane* because it is the plane where the solar eclipses occur. Let us fix this plane as our (x, y) plane, place the origin of the axes at the body M, and mark the location of the other body by $\mathbf{r} = (x, y)$. In this notation, the components of the vector equation (A.3) are written as a system of second-order autonomous differential equations,

$$x''(t) = -GM\frac{x}{r^3} , \tag{A.4}$$

$$y''(t) = -GM\frac{y}{r^3} , \quad r = \sqrt{x^2 + y^2}. \tag{A.5}$$

We switch to polar coordinates by

$$x = r\cos\theta, \quad y = r\sin\theta , \tag{A.6}$$

or equivalently,

$$r^2 = x^2 + y^2, \quad \theta = \arctan(y/x) .$$

The derivative of θ with respect to t is

$$\frac{d\theta}{dt} = \frac{(y'x - yx')/x^2}{1 + (y/x)^2} = \frac{y'x - yx'}{x^2 + y^2} ,$$

A The Orbits of the Planets   305

that is, $r^2\theta' = y'x - yx'$. After another differentiation, we substitute x'' and y'' by (A.4), (A.5) and have

$$(r^2\theta')' = (y'x - yx')' = y''x - yx'' = \left(-k\frac{y}{r^3}\right)x - \left(k\frac{x}{r^3}\right)y \equiv 0.$$

So $r^2\theta'$ is constant, and we denote it by

$$r^2\theta' \equiv L. \tag{A.7}$$

Physically Eq. (A.7) is the conservation of angular momentum. This equality also confirms Kepler's second law, because the area swept by the motion of the vector $\mathbf{r}(t)$ is given in polar coordinates by the integral

$$\frac{1}{2}\int_{\theta_1}^{\theta_2} r^2 \, d\theta = \frac{1}{2}\int_{t_1}^{t_2} r^2(t)\frac{d\theta}{dt}\, dt = \frac{1}{2}\int_{t_1}^{t_2} L \, dt = \frac{L}{2}(t_2 - t_1) \tag{A.8}$$

which depends only on the length $t_2 - t_1$ of the time interval.

Our goal is to construct from (A.4), (A.5) a second-order differential equation for $r(t)$. From the derivative of $r^2 = x^2 + y^2$, we get $rr' = xx' + yy'$, and after a second differentiation,

$$\begin{aligned}(rr')' = (xx' + yy')' &= (xx'' + x'^2) + (yy'' + y'^2) \\ &= (x'^2 + y'^2) + (xx'' + yy''),\end{aligned} \tag{A.9}$$

which we want to translate to r, θ coordinates. The derivatives of (A.6) according to t are

$$x' = r'\cos\theta - r\sin\theta \cdot \theta', \qquad y' = r'\sin\theta + r\cos\theta \cdot \theta',$$

which lead to $x'^2 + y'^2 = r'^2 + r^2\theta'^2$. (This equation multiplied by $\frac{1}{2}m$ is the kinetic energy in Cartesian and polar coordinates, respectively). The last term of (A.9) is

$$xx'' + yy'' = x\left(-GM\frac{x}{r^3}\right) + y\left(-GM\frac{y}{r^3}\right) = -GM\frac{x^2+y^2}{r^3} = -\frac{GM}{r},$$

(which expresses potential energy). Substitution of the last two equalities and $(rr')' = rr'' + r'^2$ in Eq. (A.9) gives

$$rr'' + r'^2 = \left(r'^2 + r^2\theta'^2\right) - \frac{GM}{r}.$$

Finally, according to (A.7), we have $\theta'(t) = L/r^2$, so after reorganization, we receive the second-order differential equation

$$r''(t) = \frac{L^2}{r^3} - \frac{GM}{r^2}. \tag{A.10}$$

This equation is of the form discussed in the Sect. 2.6

So far we have treated all unknowns as functions of t. But to draw the required orbit in polar coordinates, it is necessary to consider r as a function of θ. For reasons that will become clear soon, it is also useful to replace r with a new variable $u = 1/r$. We perform the two changes of variables simultaneously, that is, $r(t) = 1/u(\theta)$, and differentiate $r(t)$ according to the chain rule. To prevent any confusion, the variables t and θ are stated explicitly at each differentiation.

$$\frac{dr}{dt} = \frac{d}{d\theta}\left(\frac{1}{u(\theta)}\right)\frac{d\theta}{dt} = -\frac{1}{u^2(\theta)}\frac{du}{d\theta}\frac{d\theta}{dt}.$$

By (A.7), $d\theta/dt = L/r^2 = Lu^2$, so

$$\frac{dr}{dt} = -\frac{1}{u^2}\frac{du}{d\theta}Lu^2 = -L\frac{du}{d\theta}.$$

By an additional differentiation

$$\frac{d^2r}{dt^2} = \frac{d}{dt}\left(-L\frac{du}{d\theta}\right) = \frac{d}{d\theta}\left(-L\frac{du}{d\theta}\right)\frac{d\theta}{dt}$$

$$= -L\frac{d^2u}{d\theta^2}\frac{L}{r^2} = -L^2u^2\frac{d^2u}{d\theta^2}.$$

Thus, in terms of $u = 1/r$ and θ, Eq. (A.10) becomes

$$-L^2u^2\frac{d^2u}{d\theta^2} = L^2u^3 - GMu^2$$

which is equivalent to

$$\frac{d^2u}{d\theta^2} + u = \frac{GM}{L^2}.$$

This is a non-homogeneous linear equation with constant coefficients, and its general solution is

$$u(\theta) = \frac{GM}{L^2} + c_1\cos\theta + c_2\sin\theta = \frac{GM}{L^2} + c\cos(\theta - \theta_0)$$

A The Orbits of the Planets

Fig. A.1 An ellipse with eccentricity $\varepsilon = 0.6$

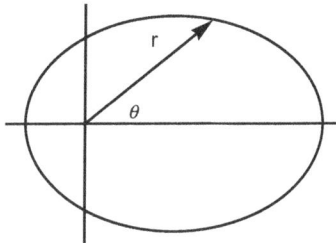

which depends on two arbitrary parameters c, θ_0 (see Problem 6.14 at the end of Chap. 6). By rotation of the polar system, it can be assumed that $\theta_0 = 0$, and so $r = 1/u$ is

$$r = \frac{R}{1 - \varepsilon \cos \theta} \tag{A.11}$$

where $R = L^2/GM$, $\varepsilon = -cL^2/GM$.

Equation (A.11) is the polar equation of the conic sections (ellipse, parabola, hyperbola), which have one of the focal points at the origin. The parameter ε is called the *eccentricity* of the orbit: $\varepsilon = 0$ corresponds to a circle; for $0 < \varepsilon < 1$, the orbit is an ellipse (see Fig. A.1); for $\varepsilon = 1$, it is a parabola; and $\varepsilon > 1$ indicates a hyperbola. Earth's orbit, which is close to a circle, has an eccentricity $\varepsilon = 0.017$, while Haley's comet[4] with a very elongated orbit corresponds to $\varepsilon = 0.967$.

Let's summarize the properties of Eq. (A.11) in the following theorem:

Theorem A.1 *For $\varepsilon \neq 1$, the polar equation (A.11) is equivalent to the Cartesian equation*

$$(1 - \varepsilon^2)^2 \left(x - \frac{R\varepsilon}{1 - \varepsilon^2} \right)^2 + (1 - \varepsilon^2) y^2 = R^2 . \tag{A.12}$$

For $0 < \varepsilon < 1$, it is an ellipse, and for $1 < \varepsilon$, it is a hyperbola. For $\varepsilon = 1$, the polar equation (A.11) describes a parabola and is equivalent to the Cartesian equation

$$2Rx = y^2 - R^2 .$$

The proof is done by direct calculation. We substitute $x = \dfrac{R \cos \theta}{1 - \varepsilon \cos \theta}$, $y = \dfrac{R \sin \theta}{1 - \varepsilon \cos \theta}$ in (A.12) and verify that indeed

$$(1 - \varepsilon^2)^2 \left(\frac{R \cos \theta}{1 - \varepsilon \cos \theta} - \frac{R\varepsilon}{1 - \varepsilon^2} \right)^2 + (1 - \varepsilon^2) \left(\frac{R \sin \theta}{1 - \varepsilon \cos \theta} \right)^2 \equiv R^2 .$$

---

[4] Edmond Halley, 1656–1742.

For $\varepsilon = 1$, the identity $2R \dfrac{R\cos\theta}{1-\varepsilon\cos\theta} = \left(\dfrac{R\sin\theta}{1-\varepsilon\cos\theta}\right)^2 - R^2$ must be verified.

We derive now Kepler's third law. Let T denote the time of a complete period around the elliptic orbit. As in Eq. (A.8), the area of the ellipse is

$$A = \frac{1}{2}\int_0^T r^2\,d\theta = \frac{1}{2}\int_0^T L\,dt = LT/2\,.$$

On the other hand, the area of a canonical ellipse $x^2/a^2 + y^2/b^2 = 1$ with semi-axes a, b is πab. The long axis of the shifted ellipse (A.12) is determined by its intersection points with the x-axis, and its length is

$$2a = \left(\frac{R\varepsilon}{1-\varepsilon^2} + \frac{R}{1-\varepsilon^2}\right) - \left(\frac{R\varepsilon}{1-\varepsilon^2} - \frac{R}{1-\varepsilon^2}\right) = \frac{2R}{1-\varepsilon^2} = \frac{2L^2/GM}{1-\varepsilon^2}\,.$$

That is, $a(1 - \varepsilon^2) = L^2/GM$. The short semi-axis is

$$b = \max\{y\} = \frac{R}{\sqrt{1-\varepsilon^2}} = a\sqrt{1-\varepsilon^2}\,.$$

Thus, the area of the ellipse is $A = \pi ab = \pi a^2\sqrt{1-\varepsilon^2}$, and the period time is

$$T = 2A/L = 2\pi a^2\sqrt{1-\varepsilon^2}/L\,.$$

From the formulas for T and a, we have that the ratio T^2/a^3 equals

$$\frac{T^2}{a^3} = \frac{4\pi^2 a^4(1-\varepsilon^2)/L^2}{a^3} = \frac{4\pi^2 a(1-\varepsilon^2)}{L^2} = \frac{4\pi^2}{GM}\,,$$

a constant that depends neither on the mass m of a planet that surrounds M nor on the shape of its orbit. That is Kepler's third law.

In this chapter, we discussed the motion of two bodies only. What happens when n bodies act on each other according to Newton's laws and gravity? The system of differential equations that corresponds to this problem is called the *n-body problem*, and it continues to engage the scientific community. Poincaré's[5] studies on this topic led to the discovery of the phenomenon of chaos.

---

[5] Henri Poincaré, 1854–1912.

Appendix B
Historical Notes

The theory of differential equations was created at the same time as the birth of differential and integral calculus. In the years 1660–1680, Isaac Newton and Gottfried Wilhelm Leibniz[1] laid the cornerstones of calculus. Their simultaneous activity caused endless disputes between themselves and between the camps of their fans about the question of the priority of the discovery. This struggle keeps historians of science busy throughout centuries.

Newton and Leibniz realized that many geometric questions, such as calculating tangents and curvature of curves and minimum problems on the one hand and calculating areas, volumes, and arcs on the other, are reflections of two fundamental ideas which are called in our language "differentiation" and "integration." Both discovered that these two operations are opposite to each other ("The fundamental theorem of the differential and integral calculus"). They developed algorithms to solve these problems, and each of them used his own notation. In Newton's terminology, any quantity that changes with time is called a flow and is marked by $x, y, \ldots$. Its velocity was called fluxion and marked by $\dot{x}$. "Infinitely small" quantities were marked by "o." Leibniz contributed to the theory the idea of differentials, the notation dx, dy, and the sign of integral $\int$ formed from the letter S, (for Summa, the Latin word for sum).

Newton classified first-order differential equations into three fundamental forms, which in today's notation will be written as $\frac{dy}{dx} = f(y)$, $\frac{dy}{dx} = f(x)$ and $\frac{dy}{dx} = f(x, y)$, where $f(x, y)$ is a polynomial of x and y. Equations of the third kind were solved by Newton with the help of power series.

Leibniz discovered the techniques of separation of variables, solution of first-order homogeneous linear equations, and solution by a change of variables.

---

[1] Gottfried Wilhelm Leibniz, 1646–1716.

Among the first problems that had been studied by differential equations were geometrical and physical problems, such as:

- The inverse problem of tangents: knowing the tangents of a curve, find its equation.
- The isochrone (tautochrone) problem (Huygens[2]). A body slides without friction along a curve under the influence of gravitation. For which curve the time needed to slide to its lowest point is independent of its starting point?
- The brachistochrone problem (Johann Bernoulli[3]). A body slides without friction along a curve under the influence of gravitation. For which curve the sliding time from the start point to the endpoint is as short as possible?
- The catenary problem (Huygens). Which the curve describes the shape of a flexible chain hanging from its two endpoints?

The members of the Bernoulli family left their mark on the mathematics of the seventeenth century. The most important of them were the two brothers Jacob (James) Bernoulli[4] and Johann (Jean) Bernoulli, and the son of the latter, Daniel Bernoulli.[5] Their family relationship could easily compete with any modern-day melodrama series.

Jacob Bernoulli developed the method of separation of variables and is known for his contribution to probability theory. It was in his writings that the name "differential equation" (aequationes differentiales) appeared for the first time. His younger brother, Johann, developed many of the integration methods. He considered the separation of variables as the fundamental way to solve differential equations and discovered the existence of integrating factors. In the next generation, his son Daniel contributed mainly to the field of hydrodynamics.

Brook Taylor wrote in 1715 about initial value and boundary value problems. He solved equations by power series (which are called today Taylor series) and noticed the existence of singular solutions.

The outstanding mathematician at the beginning of the eighteenth century was Leonhard Euler who dealt with every possible mathematical area. In the field of differential equations, he found in 1743 the solutions of the linear, homogeneous equations with constant coefficients. Euler attached great importance to integrating factors and considered them the key to solving differential equations.

Joseph Louis Lagrange generalized the concept of the integrating factor to linear equations of any order. For a given equation

$$p_0(x)y^{(n)} + p_1(x)y^{(n-1)} + \ldots + p_0(x)y = 0$$

---

[2] Christiaan Huygens, 1629–1695.
[3] Johann Bernoulli, 1667–1748.
[4] Jacob (James, Jacques) Bernoulli, 1655–1705.
[5] Daniel Bernoulli, 1700–1782.

B Historical Notes

he suggested to find a function $\mu(x)$ such that multiplying the equation by $\mu(x)$ will make it an exact derivative of a certain expression, which enables immediate integration and lowering the order of the derivatives by one. Lagrange found that the condition for this is that $\mu(x)$ will be the solution of the equation

$$(-1)^n(p_0(x)\mu)^{(n)} + (-1)^{n-1}(p_1(x)y)^{(n-1)} + \ldots - (p_1(x)\mu)' + p_0(x)y = 0.$$

This equation is called today the *adjoint equation*. In 1774, Lagrange discovered the variation of the parameter method. Lagrange's main contribution to science was in the field of analytical mechanics.

Until the end of the eighteenth century, all activity about differential and integral calculus was carried out formally, without feeling any need to justify any step. All the concepts that are known today to a student in his first course, such as "limit" and "continuous function," have not yet been born. People of the seventeenth and eighteenth centuries used infinite series without limitation, and the question of the convergence of series was not asked at all. Derivatives of all known functions were calculated (because non-smooth functions were not used), and no one doubted that each function has an integral. Until that time, researchers of differential equations focused on finding solution methods and explicit solutions to specific equations.

All this changed with the work of Augustin Louis Cauchy. At the beginning of the nineteenth century, Cauchy defined limits with the aid of ε, δ and gave a precise foundation for the basic concepts of differential calculus. However, he did not distinguish still between pointwise convergence and uniform convergence of a series of functions. This issue awaited its solution for another twenty years, till the days of Karl Weierstrass.

In the field of differential equations, Cauchy was the first to ask whether an initial value problem

$$y' = f(x, y), \qquad y(x_0) = y_0,$$

must have a solution or not, regardless of whether we know how to solve it explicitly. In his lectures in 1821, he proved the first existence theorem (without uniqueness) for initial value problems. In 1850, Cauchy published another existence theorem for ordinary differential equations and partial equations (the majorant method). In 1891, Emile Picard proved an existence and uniqueness theorem by the method of successive iterations. This method has since been adopted for various functional equations.

In the years 1829–1837, Jacques C. F. Sturm and Joseph Liouville developed the theory bearing their names about boundary value problems for differential equations and expansion of functions to a series of eigenfunctions.

A big step forward was made at the beginning of the twentieth century by Henri Poincaré. Poincaré discussed the qualitative and geometric properties of trajectories and the critical points of nonlinear systems. The inspiration for his work came from investigating the laws of motion of objects in the solar system ("the n-body problem"). Poincaré's achievements were a first step in the direction of dynamical systems theory and chaos theory.

Index

A
Abel, N.H., 108
Abel's formula, 108, 169
Airy equation, 267
Analytic function, 235
Asymptotic stability, 80, 126, 161, 204, 217
Autonomous equation, 70
Autonomous system, 199
    linear, 205
    non-linear, 220

B
Bernoulli, J., 41, 310
Bernoulli's equation, 41
Bessel equation, 246, 262
Bessel function, 264, 283
Bessel, F.W., 262
Boundary value problem, 84
Brahe, T., 303

C
Canonical form, 179
Cauchy, A.L., 51, 140, 311
Cauchy kernel, 140, 191
Characteristic polynomial, 118, 249
Comparison of coefficients, 237
Conic sections, 307
Convergence radius, 234
Convolution, 141, 285
Critical point, 203

D
Diagonalizable matrix, 175, 177
Dirac function, 293
Dirac, P., 293
Direction field, 33, 71, 204
Domain of definition, 51
Domain of definition of a solution, 50

E
Eigenvalue, 173
    algebraic multiplicity, 175
    geometric multiplicity, 175
Eigenvector, 173
Equation of a homogeneous type, 20
Equation with constant coefficients, 117
    general solution, 118
Equilibrium point, 203, 220
    center, 213
    improper node, 211
    node, 207
    saddle, 208
    spiral, 214
    star, 210
Euler, L., 120, 310
Euler's equation, 129, 247, 249
Euler's formula, 120
Exact equation, 23, 25
Existence and uniqueness theorem, 50, 83, 97, 159
Exponential order of magnitude, 272

F
First-order linear equation
    general solution, 9
    homogeneous, 8
    non-homogeneous, 11

Frobenius, F.G., 246
Frobenius method, 246
Fundamental matrix, 171, 190
Funnel, 73

G
Gamma function, 282
General solution, 171
Global solution, 77, 98, 162
Gronwall's lemma, 62
Growth of population, 2

H
Heaviside function, 288
Heaviside, O., 288
Hermite equation, 267
High order linear equation, 99
 general solution, 133
 homogeneous, 99
 non-homogeneous, 132
 particular solution, 133
Hypergeometric equation, 246

I
Indices, 249
 differ by integer, 259
 equal, 254
Indicial equation, 249
Initial value problem, 10, 50, 97, 159, 276
Integrating factor, 12, 29, 310
Isocline, 35, 71, 73

J
Jacobian, 221
Jordan, C., 181
Jordan chain, 185

K
Kepler, J., 303

L
Laplace, P.S., 271
Laplace transform, 271
Laurent series, 245
Legendre equation, 241, 246, 268
Legendre polynomial, 243
Leibniz, G.W., 309
Lindelöf, E., 50

Linear combination, 100, 164
Linear dependence, 96, 100, 165
Linearization, 221
Linearly dependent solutions, 100
Linearly independent solutions, 118, 125
Linear operator, 99
Linear space, 100, 164
Local solution, 49, 51, 68
Logistic equation, 3, 19, 34, 63, 80–82

N
Newton, I., 303, 309
Normalized equation, 1, 7, 12, 83, 111, 115, 137, 236, 241

O
Orthogonal curves, 37

P
Partial fractions, 277
Pendulum, 5, 33, 223
Phase plane, 200
Phase portrait, 200
Picard, E., 50, 311
Piecewise continuous function, 272
Planets, 303
Poincaré, H., 308, 311
Population growth, 8
Potential function, 25
Power series, 234
Principle of superposition, 100

R
Radioactive decay, 3, 8
Reduction of order, 113
Regular point, 236
Regular-singular point, 244
Resonance, 148
Roots coalesce, 122

S
Separable equation, 16
Separation of variables, 9, 16, 68
Singular point, 244
Singular solution, 9, 17, 21
SIR model, 4
Small perturbations, 218
Spanning of the solution space, 102
Stability, 80, 126, 161, 188, 204, 217

Step function, 288
Sturm's separation theorem, 110
Successive iterations, 54, 57
System of differential equations, 159
    with constant coefficients, 173
    non-homogeneous, 189
    homogeneous, 164
    linear, 162

T
Taylor, B., 310
Taylor formula, 220
Taylor series, 233
Tchebycheff equation, 246, 267
Trace of matrix, 169, 176

Trajectory, 200
Trivial solution, 100

U
Undetermined coefficients, 141

V
Vandermonde determinant, 118
Variation of parameters, 13, 134, 189

W
Weierstrass, K., 311
Wronskian, 104, 166

The manufacturer's authorised representative in the EU is Springer Nature Customer Service Centre GmbH, Europaplatz 3, 69115 Heidelberg, Germany. If you have any concerns regarding our products, please contact ProductSafety@springernature.com

Printed and bound by CPI Group (UK) Ltd, Croydon, CR0 4YY

26/03/2026

02078965-0003